T0181012

Springer Geophysics

The Springer Geophysics series seeks to publish a broad portfolio of scientific books, aiming at researchers, students, and everyone interested in geophysics. The series includes peer-reviewed monographs, edited volumes, textbooks, and conference proceedings. It covers the entire research area including, but not limited to, geodesy, planetology, geodynamics, geomagnetism, paleomagnetism, seismology, and tectonophysics.

More information about this series at http://www.springer.com/series/10173

Shenglai Yang

Fundamentals
of Petrophysics

Second Edition

Shenglai Yang
Department of Petroleum Engineering
China University of Petroleum
Beijing
China

Springer Geophysics
ISBN 978-3-662-57221-4   ISBN 978-3-662-55029-8   (eBook)
DOI 10.1007/978-3-662-55029-8

Jointly published with Petroleum Industry Press, Beijing, China.

The print edition is not for sale in China Mainland. Customers from China Mainland please order the print book from Petroleum Industry Press.

Printed on acid-free paper

This Springer imprint is published by Springer Nature
The registered company is Springer-Verlag GmbH Germany
The registered company address is: Heidelberger Platz 3, 14197 Berlin, Germany

Preface

This book presents fundamental physical and physicochemical knowledge involved in oil and gas development engineering, such as physical and chemical phenomena, physical process. It is arranged to provide the knowledge of porous rock properties of reservoir rocks, and properties of fluids, i.e., gases, hydrocarbon liquids, and aqueous solutions, and the mechanism of multiphase fluid flow in porous media. The book also brings together the application of the above theories and knowledge.

This textbook is written to serve the undergraduate teaching of the petroleum engineering major. It focuses on the introduction of basic conceptions, terms, definitions, and theories, placing more emphasis on the width of knowledge rather than the depth. What should be learned through the course includes the definitions of some essential physical parameters, the physical process, important physical phenomena, influencing factors, and engineering applications. In addition, the measuring methods and experimental procedures for the key parameters are also included.

Through the study of this book, good foundation of knowledge frame for petroleum engineering should be established for students, which is benefit for both their further studying and their work in the future when they are engaged in job.

The book is originally arranged to match a 48-class-hour course, but flexibility is permitted during teaching and studying process.

The editing work of this book began from year 2002. In 2004, It is translated to English from the textbook PetroPhysics (Yang 2004, in Chinese) published by the Petroleum Industry Press, Beijing, China, and was used in teaching in China University of Petroleum, Beijing. It is Mr. Dong Hua, Lü Wenfeng, Li Baoquan, Yang Sisong, Guo Xiudong, Sheng Zhichao, Sun Xiaojuan, Sun Rong, Hang Dazhen, and Wu Ming, who made contribution during the translation and compiling.

In 2008, Ms. Ma Kai helped to recheck and recompile the textbook again. In 2010, Ms. Li Min and Mr. Cao Jie helped and made some corrections.

I would like to express my sincere thanks and appreciations to all people who have helped in the preparation of this book and given any helps in anyways.

Due to limitations of my knowledge and ability, this book may contain mistakes, wrong sentences, and so on. Therefore, it is welcomed for the interested readers to make a correction and let me known, if any mistake is found.

Beijing, China   Shenglai Yang

Contents

Contents

Introduction

Why to Exploit Oil and Gas

Energy is the power to push the industrialization and modernization forwards. The development and utilization of energy is what a country's competitive strength and comprehensive national power hinge on.

Oil and Gas is an important kind of energy, high-quality material for chemical industry, and an important warfare material. Therefore, petroleum industry and oil and gas trade has been attached great attention by the super great power countries because it is important to both the energy safety and national defense of a country. Today, it is impossible to imagine what the society should be without the oil and gas. It can be say that petroleum is standing on an essential and indispensable point in the economic development of the world.

To meet increasing demands for oil and gas and maintain steadily growth in the national economy, the following policy should be taken:

(1) Explore and discover new oil and gas fields to enlarge the figure of petroleum reserves;
(2) Adopt advanced technology to develop the oil & gas reservoirs effectively;
(3) Apply EOR method to improve the oil & gas recovery and increase the production rate in the developing reservoirs;
(4) Develop energy-saving techniques and find alternative or new energy to reduce the amount of consumption for oil and gas;
(5) Cooperate internationally to engage in the development of world petroleum resources.

For every student who will work in the petroleum industry, related knowledge and techniques are required to study and hold reliably, in order to take on the above work successfully.

What is Petrophysics

Petrophysics is a science about the physical property of oil & gas, its formations, and physicochemical phenomena involved in an oil & gas reservoirs and during the exploiting of an oil & gas reservoirs. It primarily includes 3 parts:

(1) The physical properties of the reservoir fluids, including physical properties of oil, water, and gas at high pressure and high temperature, and the law that governs their phase-changing.
(2) The physical properties of the reservoir rocks, including porosity, permeability, saturation, and sensibility of the formation rocks.
(3) The physical properties of multiphase fluids in porous media and their seepage mechanisms.

A petroleum reservoir is geological structure which consists with formation rock, which is a porous medium, and the oil and gas is contained in the porous medium.

The reservoir, which is the objectives what we study, is buried thousands of meters below the earth surface, the reservoir fluids are under states of high pressure and high temperature. At this state, the crude oil dissolves large volume of gas; therefore, the physical properties of the underground reservoir fluids are vastly different from those of the subaerial ones.

This textbook is aimed to introduce the fundamental physical properties of the reservoir fluids and rocks, involving terms, concepts, and related knowledge, to help the students to build foundations for their professional studying in the future.

History of the Petrophysics

Petrophysics is a young science with history about 80 years. In the 1930s, some oil field engineers in America and Russia noticed the influence of the reservoir fluids on the production rate of oil wells and then gave some preliminary test and measurement on the fluid property. During the 1930s, a monograph about reservoir fluids was compiled in Russia already. In 1949, M. Mosket, an American, wrote a book named 'Physical Principles of Oil Production,' in which the related research and practical information present in the first half of the twentieth century were collected and gathered. It illustrated reservoir fluids information from a physics perspective and guided oil field development technically for various drive-typed reservoirs. In 1956, Ф.И. Котяхов, Russian, wrote a book named 'Fundamental of Petrophysics.' This was the first book especially for the petrophysics, and it found the base for this subject.

In China, professional team in researching and teaching on petrophysics was set up during 1950s. The first Petroleum University in China, Beijing Petroleum College, was founded in 1953. IN BPC, Russian expert Ш.К.Гиматудинов was invited to teach the course petrophysics, and postgraduates researching on petrophysics were firstly trained.

From 1950s and on, with the discovery and development of large oil fields, such as Daqing oil field, Liaohe oil field, Dagang oil field, the science on petrophysics in China was vigorously promoted. Research institutes and laboratories were established in every oil field, as well in the capital city, Beijing. A large number of researches on the petrophysical and EOR study were made and achievements had yielded.

During the 1960s, several books that recorded the research achievements in China were published in succession, and large numbers of articles and papers were published on magazines and journals.

Nowadays, the science of petrophysics has already become an indispensable part of the knowledge system in the development of oil fields, and it is sure that it will play more important roles in the coming days, to conquer more complicated problems we will face during the oil field development.

In the world, some monographs were published in the 1970s, and the physical properties of carbonate reservoirs were systematically illustrated in these books from an exploiting perspective. In those books, fractured reservoirs were also studied for its deposits, forming mechanisms, and recovery calculation, etc.

In 1977, two books wrote by Russian were published. The mechanism of the reservoir oil's seepage in the formation was probed in the books from a physicochemical standpoint. The books stressed that only by strengthening the study on the physicochemical mechanisms of the formations, the recovery ratio could be enhanced.

At the end of the twentieth century, the researches were focused on the following study, such as carbonate formations, clasolites, induced porosity, the non-Newtonianism of the reservoir fluids, and the application of the phase-state equations. At the same time, the modern experimental and measuring techniques and the application of computing technology had also made new progress, too.

Up to now, the petrophysics science has been developed preliminarily. A comparative theory-integrated system is framed. And the experimental and measuring techniques are also standardized, for example, standards for conventional core analysis, special core analysis, EOR experiments, flow test for the formation sensibility, and physical analysis of the reservoir fluids have already developed.

However, it should be noted that, due to the heterogeneity of the reservoir rocks and fluids and the complexity of some physical process in oil formations, lack of mature cognition for some phenomena still exists here and there. As research progressing, the science of petrophysics will certainly continue to develop.

Looking forward to the twenty-first century and further, the science of petrophysics will continue to develop and step onto a new stage in both breadth and depth. Its developments will have the following characteristics:

(1) Synthesis: By the infiltration and mutual coordination with other near disciplines, new borderline theories may be formed.

(2) Innovation: New theories may have to be applied to solve new problems.
(3) Practicality: Developing process will be simulated truly and really reservoir conditions by using new experiment technology.

The Aim and Learning Methods for Petrophysics

Petrophysics is a core course for students majoring in petroleum engineering, or oil field chemistry and geology. It is the aim of this course for students to have a firm grasp of the fundamentals and professional knowledge and skills about petroleum engineering.

As a science branch, petrophysics is based upon the experiments, so both theories and experiments are important. Therefore, students have to pay enough attention on the laboratory experiment and finish some exercise as homework involved in this book.

The SI unit system is adopted in this book, while some English engineering units are also briefly introduced. For example, when citing some particular equations like Darcy's law, the original Darcy units are reserved. It is necessary for the readers to master the unit system that are used in concerned equations.

Part I
Physico-Chemical Properties
of Reservoir Fluids

Introduction

The oil, natural gas, and water which reserved in oil or gas reservoirs are called reservoir fluids. Buried thousands of meters below the land surface, the reservoir fluids in primitive underground state are subjected to an environment of high pressure and high temperature. High pressure enables the crude oil to dissolve large volume of gas, so that the physical properties of the underground reservoir fluids are vastly different from those of the subaerial ones.

Along with the movement of the reservoir fluids from the formations to the wellbores and then to the oil-tank on surface, their pressure (P), temperature (T), and volume will be continuously changing, i.e., the pressure and temperature decrease. As P and T change, some other phenomena occur such as degasification, shrinkage of crude oil, expansion of gas, and wax precipitation. For these changes do have some effects on the oil and gas production rate. So, it is necessary to study the phenomena and laws to determine and optimize oil and gas production accurately.

At the stages of exploration of an oil field, the fluid physical properties are very useful to predict and determine the reservoir type, to ascertain the existence of gas-cap, to determine whether the gas will condensate in the formation or not, etc. Above work requires a great deal of deep and thorough knowledge about the fluid physiochemical properties as well as the laws of phase-changing that governing the process.

During reservoir development, the static and dynamic changes of the fluid parameters are indispensable to the accomplishment of the production management. The fluid parameters include the fluid volume factor, solubility coefficient, compression coefficient, viscosity of oil, and so on. These parameters are of great significance to gain efficient and economical development of the oil fields.

Chapter 1
Chemical Composition and Properties of Reservoir Fluids

Petroleum is a mixture of naturally occurring hydrocarbons, which may exist in solid, liquid, or gaseous states depending upon the conditions of reservoir pressure and temperature. Petroleum in liquid or gaseous states is called oil or nature gas separately.

Both the crude oil and natural gas are hydrocarbons in terms of chemical structure. It is already confirmed that the hydrocarbons in petroleum are made up mainly of three kinds of saturated hydrocarbons: alkanes, aromatics, and cycloalkanes. In fact, unsaturated hydrocarbons such as alkene and alkyne are not usually present in crude petroleum. Alkanes are also called paraffin series whose general chemical formula is C_nH_{2n+2}. Owing to the variation of molecular weight, alkanes exist in forms different from each other. For example, at room pressure and temperature, the C_1–C_4 are in a gaseous state, and they are the predominant contents of natural gas. The C_5–C_{15} are in a liquid state, and they are the predominant contents of crude oil; and the C_{16+} is in solid state, commonly known as paraffin.

At reservoir condition, the solid hydrocarbons exist in oil in the form of solute or crystal. Therefore, both the crude oil and the natural gas are hydrocarbons in terms of chemical structure.

1.1 Chemical Properties of Crude Oil

1.1.1 The Elemental Composition of Oil

By anglicizing oil's elemental composition, it comprises predominantly of carbon and hydrogen elements and also contains a small quantity of sulfur, oxygen, and nitrogen, as well as some others such as trace elements.

© Petroleum Industry Press and Springer-Verlag GmbH Germany 2017
S. Yang, *Fundamentals of Petrophysics*, Springer Geophysics,
DOI 10.1007/978-3-662-55029-8_1

The percentages of the carbon and hydrogen contents in crude oil are listed in Table 1.1. As shown in the table, crude oil consists of approximately 83–87 % carbon and 11–14 % hydrogen, and the sum of C + H up to 95–99 % by weight. For some oil from oil fields in China, the hydrogen-to-carbon molar ratio tends to be comparatively high.

The non-carbon and non-hydrogen elements make a total contribution less than 1–5 % and are the subordinate elements in crude oil. However, these elements all exist in oil as derivatives of the hydrocarbons. The derivative compounds bearing these elements take up a much greater percentage than 1–5 % and have strong effect on the properties of oil. For the reasons talked above, the existence of these elements should not be ignored.

Generally, compared with other countries, crude oil produced in China possesses lower sulfur content and higher nitrogen content as shown in Table 1.2. Taking Dagang and Daqing oil as examples, their sulfur percentages are both only 0.12 %. The oil from Gudao and Jianghan has the highest sulfur content in China and is still lower than that of other countries. For the nitrogen content, it is usually larger than 0.3 % for crude oil in China, and this high level is quite rare in foreign countries. An investigation once taken among 210 samples from some abroad oil fields shows that only 31 of them possess nitrogen content above 0.3 %.

Besides, crude oil also contains a quite tiny quantity of trace elements, including vanadium, nickel, iron, cobalt, magnesium, calcium, and aluminum, and each individual of them takes up a content less than 0.03 %.

Table 1.1 Hydrogen and carbon contents and their molar ratio in crude oil

Oilfield	Elemental content			
	Content of C (%wt)	Content of H (%wt)	Content of C + H (%wt)	H/C molar ratio
Daqing, China	85.7	13.3	99.0	1.86
Shengli, China	86.3	12.2	98.5	1.68
Karamay, China	86.1	13.3	98.4	1.85
Dagang, China	85.7	13.4	99.1	1.88
Canada	83.4	10.4	93.8	1.60
Mexico	84.2	11.4	95.6	1.62
Iran	85.4	12.8	98.2	1.80
Columbia	83.6	11.9	95.5	1.67
Romania	87.2	11.3	98.5	1.56
Russian	83.9	12.3	96.2	1.76

Table 1.2 Sulfur and nitrogen contents in Crude Oil

Oilfield	Elemental content	
	Sulfur (wt%)	Nitrogen (wt%)
Daqing, China	0.12	0.13
Shengli, China	0.80	0.41
Gudao, China	1.8–2.0	0.50
Dagang, China	0.12	0.23
Renqiu, China	0.30	0.38
Jianghan, China	1.83	0.30
Highest content in world	5.50	0.77
Lowest content in world	0.02	0.02

1.1.2   The Hydrocarbon Compounds in Crude Oil

It is reasonable to classify the compounds contained in crude oil into two general groups: hydrocarbons and non-hydrocarbons. The relative contents of hydrocarbons and non-hydrocarbons are not fixed but vary a lot in different reservoirs. For example, the hydrocarbon content makes a contribution up to 90 % in some types of light oil, while this value may also be lower than 50 %, and even as low as to 10–30 % in heavy oil.

1. The Alkane in Crude Oil

Alkane, sometimes referred to as chain hydrocarbon, is acyclic branched or unbranched hydrocarbons with the general formula C_nH_{2n+2}. Obviously, alkine is saturated hydrocarbons. Alkanes are chemically inactive and inert to react with other materials unless they are subjected to the conditions of heating, catalyst, or light environments, in which they can engage in kinds of reactions such as halogenation, sulfonation, oxidation, and cracking.

The hydrocarbon materials in crude oil include normal alkanes and isoparaffins, and the former group usually takes up a larger share. In particular, the paraffin-base oil often holds normal alkanes in an amount as rich as the total sum of the isoparaffins with the same carbon number. For instance, analyses of the gasoline fractions (60–140 °C) of the Daqing oil show that it possesses a 38 % share of normal alkanes, while only a 15 % share of its counterpart isoparaffins. However, in the asphalt-base oil, as shown in some foreign oil fields (e.g., the Borneo oil), the normal alkanes take a smaller share.

Among the various isoparaffins, the most abundant ones are the derivatives with two or three methyls, while the ones with four methyls or other big branches occur in a small amount.

2. The Cycloalkanes in Crude Oil

Cycloalkanes (also called naphthenes) are chemicals with one or more carbon rings to which hydrogen atoms are attached according to the formula C_nH_{2n}. As another major component in petroleum, the cycloalkane is chemically stable, too. Organic chemists have already synthesized cycloalkanes with varying number of carbon atoms, but only cyclopentane and cyclohexane, which, respectively, contain five and six carbons in a single ring, can abundantly exist in crude oil.

Apart from the single-ringed ones, in oil there also exist cycloalkanes with two or more rings. The two rings of a double-ring cycloalkane may be uniform, both being cyclopentanes or cyclohexanes, following a "one cyclopentane, one cyclo-hexane" pattern. Besides, the rings are dominantly connected in parallel. What should be noted is that cycloalkanes with three or more rings are also present in crude oil.

3. The Aromatic Hydrocarbons in Crude Oil

The existence of benzenoid aromatic hydrocarbons is quite common in all kinds of crude oil. According to related materials, it is said that more than forty kinds of benzenoid homologues, which hold a variety of alkyl side chains with different carbon numbers, were once separated from the oil produced in America. The side chains can be either alkyl or naphthenic. Through the analysis on the gasoline fractions (<160 °C) of the oil from China's Daqing, Shengli, Renqiu, and Xinjiang oil fields, among more than 100 kinds of non-numeric hydrocarbons that are quantitatively measured, there are about 10 benzenoid aromatic types.

1.1.3   The Non-hydrocarbon Compounds in Crude Oil

In crude oil, the elements such as sulfur, nitrogen, and oxygen exist in the form of hydrocarbon compounds or colloidal and bituminous substances. Due to the impurities, they are collectively called non-hydrocarbon compounds. Obviously, these compounds belong to polar substance.

1. Oxygen-containing chemicals: naphthenic acid, phenol, fatty acid, etc.
2. Sulfur-containing compounds: sulfureted hydrogen, thiol, thioether, thiofuran, etc. Besides, there is also elemental sulfur in crude oil.
3. Nitrogen-containing compounds: some heterocyclic compounds such as pyrrole, pyridine, quinoline, indole, and carbazole.
4. Colloid and asphaltene: colloid and asphaltene in crude oil belong to non-hydrocarbons, and they are mostly high molecular and heterocyclic compounds that comprise of oxygen, sulfur, and nitrogen elements. In addition, they possess high or moderate interfacial activity and have great influence on such oil

properties as color, density, viscosity, and interfacial tension. In a word, to know the characters of these compounds is of utmost importance in the development of oil reservoirs.

1.1.4 Molecular Weight, Wax Content, and Colloid and Asphaltene Content in Crude Oil

1. The Molecular Weight

The smallest molecule present in oil and gas is methane and its molecular weight is 16, while the greatest one is possessed by the asphaltene, which can be as large as several thousand. Because oil and gas is a mixture of many molecules, the molecular weight of crude oil varies in a wide range of several couples of hundred.

2. Wax Content

Wax content, expressed as a percentage, is the ratio of the wax, or called paraffin, to the ozokerite at normal temperature and pressure. Wax, a kind of white or thin yellow solid substance, composed of high alkanes, i.e., C_{16}–C_{35}, possesses a molecular weight of 300–500 and a melting point of 60–90 °C. Ozokerite, with atom number ranging from 36 to 55, is a naturally occurring odoriferous mineral wax (or paraffin) that predominantly comprised of complicated crystalline hydrocarbons with high boiling point. Its molecular weight range is 500–730, and its melting point range is 60–90 °C. Deep under the ground, the wax and ozokerite are usually dissolved in crude oil in a colloidal state, and they will crystallize from the oil when oil is produced through the wellbore with pressure and temperature drop.

3. Colloid Content

Colloid in crude oil is a kind of polycyclic aromatic hydrocarbon compounds bearing subordinate elements such as oxygen, nitrogen, and sulfur. Colloid usually has comparatively high molecular weight about 300–1000. It is usually dissolved in oil in a semisolid and dispersed state. It can be soluted in organic solvents such as sherwood oil, lubricating oil, gasoline, and chloroform.

The colloid content refers to the mass fraction of the colloid in crude oil, ranging from 5 % to 20 %.

4. Asphaltene Content

Asphaltene is non-pure-hydrocarbon compounds with high molecular weight (>1000) and is a kind of polycyclic aromatic black solid. It is insoluble in alcohol or sherwood oil but dissolves readily in benzene, chloroform, and carbon dioxide.

In crude oil, asphaltene exists in a very tiny amount usually less than 1 %. In fact, high asphaltene content leads to poor oil quality.

Colloid and asphaltene, with gradual gradation in molecular weight from one to the other, are distinguished from each other unobviously. The difference between them is as follows: (i) The colloid is a viscous liquid, while the asphaltene is a formless and fragile solid; (ii) the colloid is soluble in low molecular hydrocarbons, while the asphaltene is not.

5. Sulfur Content

Sulfur content is the mass fraction of the sulfur in crude oil. Although the sulfur content is usually less than 1 %, the presence of this little amount of sulfur has a quite big effect on the properties of crude oil.

The related data about the properties of crude oil is shown in Tables 1.3, 1.6, and 1.7.

1.2   Physical Properties and Classification of Crude Oil

1.2.1   Physical Properties of Crude Oil

The most commonly considered physical properties of crude oil are color, density, viscosity, freezing point, solubility, calorific value, fluorescence, optical activity, etc. In this lesson, the physical properties of the commercial stock tank-oil will be discussed.

1. Color

Owing to different components, crude oil has various colors. It can be in the colors of chocolate brown, dark brown, dark green, yellow, brownish yellow, light red, and so on. The color of oil depends on the relative contents of the light and the heavy hydrocarbons, and the colloid and the asphaltene contents in it. Generally speaking, the more the colloid and asphaltene, the darker the oil will be. So, to some extent, the color of the oil reflects the content of heavy constituents in it.

2. Density and Specific Gravity

The oil density is the mass per unit volume of oil under conditions of specified conditions of pressure and temperature. Its mathematical expression is:

$$\rho_o = \frac{m_o}{V_o} \tag{1.1}$$

where

ρ_o the density of crude oil, kg/m^3;
m_o the mass of crude oil, kg;
V_o the volume of crude oil, m^3.

Table 1.3 Properties of the oil samples from oil fields in China

Properties	Oilfield							
	Daqing	Shengli	Renqiu	Liaohe	HeiYoushan	Karamay No. 3 low freezing point oil	Gudao, Sheng li	Yangsanmu, Dagang
Specific gravity (d_4^{20})	0.8554	0.9005	0.8837	0.8662	0.9149	0.8839	0.9495	0.9437
Viscosity (50 °C), centipoises	17.3	75.1	50.5	7.8	316.2[a]	29.2	316.8	595.8
Freezing point (°C)	30	28	36	17	−22	−54	2	−4
Wax content (%)	26.2	14.6	22.8	13.5[b]	0.77	1.05	4.9	/
Asphaltene content (%)	0	5.1	2.5	0.17	1.36	0.53	2.9	0.4
Colloid content (%)	8.9[c]	23.2	23.2	14.4	21.2	13.3	24.8	21.8
Remanent charcoal (%)	2.9	6.4	6.7	3.59	5.3	3.8	7.4	6.0
Supplementary details	High- or medium-density oil				Low freezing point and high-density oil			

[a]Viscosity at 40 °C
[b]Distillation method
[c]Anglicized data from the alumina adsorption chromatograph

The specific gravity of tank-oil is defined as the ratio of the density of tank-oil (ρ_o) to the density of water (ρ_w), both under specified conditions of pressure and temperature. In China and former Soviet Union, the specific gravity is customarily referred to as the ratio of the density of the 1 atm and 20 °C crude oil to the density of the 1 atm and 4 °C water, and the expression for it is d_4^{20}. In petroleum industries in some Western countries such as UK and USA, the standards are temperature of 60 K (15.6 °C) and the atmospheric pressure, and the Greek letter γ_o is used to indicate the specific gravity. Since the temperature conditions are not the same, γ_o and d_4^{20} differ from each other both in physical significance and in numerical value. The conversion relationship between them is: $\gamma_o \approx 1.002 - 1.004 d_4^{20}$.

$$\gamma_o = \frac{\rho_o(P, T)}{\rho_w(15C, 1\,\text{atm})} \qquad \gamma_{oi} = \frac{\rho_o(P_i, T_i)}{\rho_w(15C, 1\,\text{atm})} = \frac{\rho_{oi}}{\rho_w(15C, 1\,\text{atm})}$$

In Western countries, the API (American Petroleum Institute) scale for crude oil is also used. The API gravity (API) can be expressed mathematically as follows:

$$\text{API} = \frac{141.5}{\gamma_o} - 131.5 \tag{1.2}$$

In the equation, γ_o is the specific gravity of the crude oil at 1 atm and 60 K (15.6 °C).

Based on Eq. (1.2), the API of water is 10. The API of crude oil increases with the decrease in its counterpart specific gravity γ_o. In fact, the API also increases as the solubility of gas in oil raises under certain conditions of pressure and temperature. In other words, the API is in direct proportion to the solubility of the gas, and this is the advantage of the API gravity to indicate the density of underground crude oil.

3. Freezing Point

The freezing point of crude oil is defined as the critical temperature at which the liquid crude oil solidifies and loses its liquidity at a specified pressure. It associates much with the contents of the wax, colloid and asphaltene, and light hydrocarbons in the crude oil, and the influencing factors are rather complicated. Generally speaking, the larger its light hydrocarbon content, the lower its freezing point; the larger its heavy-hydrocarbon content, especially the wax, the higher its freezing point is. The freezing point of crude oil is usually within the range of −56 to 50 °C, when the freezing point above 40 °C, which is called "high-freezing crude oil."

Most of the crude oil in China is wax-base one, and the high wax content inevitably results in high freezing point. For example, Daqing crude oil, having a wax content up to 25–30 %, possesses a freezing point about 25–30 °C.

4. Viscosity

The reservoir fluids—oil, gas, and water—are all viscous fluids. Viscosity, caused by molecular friction within oil, is a material property that measures a fluid's resistance to flow. It is defined as the ratio of the shear stress per unit area within the flowing liquid to the velocity gradient at any point. The viscosity signifies the level of difficulty that the liquid meets in flowing. High viscosity has stronger resistance, so it is difficult to flow.

As shown in Fig. 1.1, viscous fluid is filled in any two flat plates parallel to each other. Distance between the two plates is dy, the plate area is A, the velocity of the nether plate with its attaching fluid is v, the upper plate with its attaching fluid is $v + dv$, and the inner frictional resistance force between plates is F. According to Newton's law of viscosity, we can get that:

$$\frac{F}{A} \propto \frac{dv}{dy} \tag{1.3}$$

Then, if a coefficient μ, a function of temperature and pressure, is introduced, Eq. 1.3 becomes:

$$\mu = \frac{F/A}{dv/dy} \tag{1.4}$$

where

μ the viscosity of fluid, also called dynamic viscosity or absolute viscosity, Pa s;

F/A the shear stress or inner frictional force per unit area, N/m or Pa;

dv/dy velocity gradient, s^{-1}.

Fig. 1.1 Diagram of the viscous fluid flowing. **1** In cylinder tube and **2** between flat plates

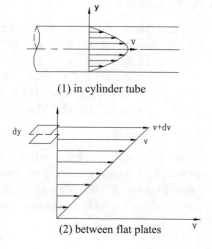

(1) in cylinder tube

(2) between flat plates

In the International System of Units (SI), the unit of viscosity is Pascal second or Pa s in short. But the commonly used units in petroleum engineering are poise, or mPa s, and centipoise (cP). The conversion relationships among these units are as follows:

$$1\,Pa\,s = 1000\,mPa\,s$$
$$1\,P = 100\,cP = 10^6\,\mu P$$
$$1\,mPa\,s = 1\,cP$$

The viscosity of fluid can also be indicated by kinematic viscosity, which is the ratio of the absolute viscosity to the density, subjected to the same pressure and temperature.

$$v = \frac{\mu}{\rho} \tag{1.5}$$

where

v kinematic viscosity, m^2/s;
μ dynamic viscosity, Pa s;
P the density of the fluid, kg/m^3.

Besides, there are some other standards for the description of the viscosity of crude oil, for example, Saybolt viscosity, Redwood viscosity, and Engler viscosity. They are all conditional viscosity and can only be measured under specified conditions and with specified instruments. The viscosity values, units, and expressions of the same oil vary according to the different viscosity measuring method.

Some properties of the crude oil from several oil fields in China are listed in Table 1.3. The specific gravity is all above 0.86. So, these oils are quite close to heavy oil.

5. Flash Point

Flash point is the lowest temperature under specific conditions at which a combustible liquid will give off sufficient vapor to form a flammable mixture with air but will not continue to burn when the flame is removed. The atmosphere pressure has impact on the flash point to some extent, so those casually measured flash points are usually converted to the corresponding values measured under the standard pressure, 1.01 kPa. The flash point of crude oil ranges from 30 to 180 °C.

6. Fluorescence

Besides heating or bombarding with fast-moving particles, an atom can also be excited by the absorption of photons of light. Many substances undergo excitation, when illuminated with photons of ultraviolet light, and then emit visible light upon the excitations. Such substances, crude oil, for example, are called fluorescent

Table 1.4 Fluorescent light colors of the oil with different densities	Oil density (g/cm^3)	Fluorescent light colors
	>0.9659	Brown
	0.9659–0.9042	Red-orange
	0.9042–0.8498	Yellow-milky white
	0.8498–0.8251	White
	<0.8251	White-violet

substances, and the phenomenon is called fluorescence. Fluorescence is a luminescence that depends upon the substance's chemical constructure, and it is one of the characters possessed by the aromatic compounds with ring structures. Additionally, what should be mentioned is that saturated hydrocarbon compounds do not have this property. Different components of the crude oil will lead it to emit different colors with fluorescent light: Light colloid leads to green light, heavy colloid leads to yellow light, and asphaltene leads to brown light. The variety in oil density can also make obvious differences in fluorescent light colors as shown in Table 1.4.

7. Optical Activity

Optical activity is a property caused by asymmetrical molecular or crystal structure that enables a compound or crystal to rotate the plane of incident polarized light. The angle that the plane of the polarized light rotates is called deflection angle. Generally speaking, aside from extreme cases, the deflection angle of crude oil is less than 1.

8. Conductivity

Crude oil or hydrocarbon compounds is nonpolar and a nonconductor of electricity. The electrical resistivity of crude oil is about 10^{11}–10^{18} Ω m. What's more, the specific inductive capacity (SIC) value of the water-free crude oil is usually within a quite slight variation, from 1.86 to 2.38.

1.2.2 Classification of Tank-Oil

According to the properties of oil, the tank-oil is classified and evaluated from the commercial points of view.

1. According to the sulfur content, oil can be classed and named as follows:

Low sulfur crude oil—sulfur content below 0.5 %;
Sulfur-bearing crude oil—sulfur content between 0.5 and 2.0 %; and
High sulfur crude oil—sulfur content above 2.0 %.
The sulfur element causes the crude oil to give off an unpleasant odor and become a liability in oil refining. What's more, the sulfur dioxide yielded on combustion seriously pollutes the air. The sulfuret compounds in oil products are

Table 1.5 Oil classification by density

Classification	Specific gravity	
	γ_0	γ_{API}, °API
Light oil	<0.855	>34
Medium-density oil	0.855–0.934	34–20
Heavy oil	>0.934	<20

the most unexpected impurities, so it is required to eliminate sulfur before the oil products are sold. Most of the crude oil in China has low sulfur content.

2. According to colloid and asphaltene content, oil can be classed and named as follows:

Low colloid and asphaltene oil—colloid and asphaltene content below 8 %;

Colloid- and asphaltene-bearing crude oil—colloid and asphaltene content between 8 % and 25 %;

High colloid and asphaltene oil—colloid and asphaltene content above 25 %.

The colloid and asphaltene form colloform texture in crude oil, and they have great impact on the fluidness of the crude oil, as well as other properties. In China, the oil from most oil fields belongs to colloid- and asphaltene-bearing or low colloid and asphaltene crude oil.

3. According to wax content, oil can be classed and named as follows:

Low wax crude oil—wax content below 1 %;

Wax-bearing crude oil—wax content between 1 and 2 %;

High wax crude oil—wax content above 2 %.

The wax content in crude oil affects its freezing point. The more the wax has, the higher the freezing point it is. It brings big trouble to both the oil production and the transportation.

4. According to key components, oil can be classed and named as follows:

Condensate oil—density <0.82 g/cm^3 (20 °C);

Paraffin-base oil—density between 0.82 and 0.89 g/cm^3 (20 °C);

Mixed-base oil—density between 0.89 and 0.93 g/cm^3 (20 °C); and

Naphthene base crude oil l—density >0.93 g/cm^3 (20 °C).

According to the density of the degassed tank-oil, oil can be classed and named as shown in Table 1.5. This is the international classification standard according to the specific gravity. Most of the oil in China is medium density or heavy oil.

Tables 1.6 and 1.7 show the tank-oil properties of both home and abroad.

Table 1.6 Properties of oil form oil fields in China

Oilfield	Properties								Fractional composition (%)		
	Specific gravity (d_4^{20})	50 °C Viscosity (mPa s)	Freezing point (°C)	Wax (%)	Colloid (%)	Asphalt (%)	Sulfur (%)	Remanent charcoal (%)	Dropping point	<200 °C	<300 °C
S district, Daqing	0.8753	15.2	24	28.6	13.3	–	0.15	2.5	88	14	28
Shengli	0.8845	33.3	33	17.9	18.3	3.1	0.47	5.5	79.5	9	20
Gudao, Shengli	0.9547	408.1	–12	0	27.5	6.6	2.25	8.95	15.8	1.9	11.2
Dagang	0.9174	47.7	–12	6.17	13.98	6.27	0.13	4.81	97	40	20.5
Karamay	0.8699	16.7	–50	2.04	12.6	0.01	0.13	3.7	58	18	35
Yumen	0.858	11.1	–15.5	8.3	22.6	–	–	–	–	–	–
Jianhgan	0.9744	–	21	3.8	51	9.6	11.8	9.5	89	5	218
Liaohe	0.9037	33.8	–7	4.73	17.6	0.15	0.26	6.4	–	–	–
Chuanzhong	0.8394	10.3	30	18.1	3.4	–	–	–	–	–	–
Renqiu	0.8893	56.5	33	22.6	20.7	–	2.35	–	148	–	–

Table 1.7 Oil properties of some overseas oil fields

Crude oil	Property								
	Specific gravity (d_4^{20})	Viscosity (mPa s)	Freezing point (°C)	Wax (%)	Colloid (%)	Asphalt (%)	Sulfur (%)	Remanent charcoal (%)
Hass Messaoud (Algeria)	0.804	2.769/(20 °C)	−45.5	2.40	0	0.06	0.13	0.83
Kirkuk (Iraq)	0.844	4.61/(20 °C)	−36	3.9	–	1.5	1.95	3.8
Xinta (India)	0.856	23.7/(50 °C)	43.3	29.3	–	19.5	0.08	–
Bomu (Nigeria)	0.78–0.95	4.4/(37.7 °C)	17.78	5.1	–	–	0.16	–
Azadegan (Iran)	0.852	6.56/(21 °C)	–	Medium	–	0.60	1.42	–
Pembina (Canada)	0.8988	–	10.0	–	–	–	0.29	1.76
East Texas (America)	0.827–0.835	3.4/(15.6 °C)	−38.89	–	–	–	0.21	–
Romashkino (Russia)	0.868	7.3/(50 °C)		4.3	–	–	1.61	–

Table 1.8 Classification of heavy oil

Classification		Major index	Supplementary index	Development method
Name	Rank	Viscosity,mPa s	Specific gravity	
Normal heavy oil	I			
	I-1	50–150	>0.92	Water or steam injection
	I-2	>150–10,000[a]	>0.92	Steam injection
Extra heavy oil	II	>10,000[a]–50,000[a]	>0.95	Steam injection
Extraordinary heavy oil	III	>50,000[a]	>0.98	Steam injection

[a]Refers to land-surface viscosity and others refer to underground viscosity

1.2.3   The Classification of Reservoir Oil

1. By Viscosity

Viscosity is a major physical property of crude oil, because it is one of the factors control the well productivity, the level of complexity in developing, and the ultimate recovery. According to the viscosity, the reservoir oil can be classified as follows:

Low-viscosity crude oil—viscosity <5 mPa s under reservoir conditions;

Medium-viscosity crude oil—viscosity between 5 and 20 mPa s under reservoir conditions;

High-viscosity crude oil—viscosity between 20 and 50 mPa s under reservoir conditions; and

Heavy oil—viscosity >50 mPa s and specific gravity above 0.920 under reservoir conditions.

For heavy oil, it can be classified into three subtypes and four ranks as shown in Table 1.8.

2. By States

Condensate oil The hydrocarbons are in gaseous state under reservoir conditions. When the pressure is below due point pressure, during the production, it condenses to be a liquid, i.e., light oil, whose specific gravity is below 0.82. This kind oil is called condensate oil.

Volatile oil Under reservoir conditions, the hydrocarbons are in liquid state and its critical point is close the point of initial reservoir condition Pi and Ti. The oil exhibits great volatility and strong shrinkage during the development. The gas–oil ratio is usually in the range of 210–1200 m^3/m^3, the specific gravity of oil is less than 0.825, and the volume factor is greater than 1.75. This kind oil is called volatile oil.

3. By Freezing Point

High-freezing-point oil: The oil whose freezing point higher than 40 °C is named as high-freezing-point oil.

1.3   Chemical Composition of Natural Gas

Natural gas is a mixture of gaseous hydrocarbon alkanes. It is composed primarily of methane with varying amount of ethane, propane, and butanes. Usually, natural gas consists of 70–98 % methane, less than 10 % ethane, and small amounts of heavier gaseous hydrocarbon compounds such as propane, butane, and pentane (about a few percent).

There are also non-hydrocarbon presents in natural gas, including nitrogen (N_2), carbon dioxide (CO_2), hydrogen sulfide (H_2S), water vapor, and occasionally some rare gases such as helium and neon. In addition, sometimes, there are also poisonous gases such as thiol and thioether.

The chemical composition of natural gas exhibits considerable variation. The CO_2 content of the natural gas from some areas tends to be very high. For instance, the value of this parameter can be as high as 70 % in an oil field located in the oil- and gas-bearing Basin of Siberia, Russia, and the gas produced in Binnan gas field of Shandong, China, possesses a high CO_2 content up to 50 %. They can provide gas source for the tertiary recoveries.

The hydrogen sulfide usually makes a contribution no more than 5–6 %; however, there are exceptions: The gas from the Lac gas reservoir of France bears a high H_2S content of 17 %, and the gas from a Canadian gas reservoir even bears a percentage of more than 20 %. Hydrogen sulfide leads to metal instruments corrosion and brings hydrogen embitterment and the following fracturing. Even more serious, once its concentration reaches a safe value (10 mg/m^3) in the air, humans will be poisoned and suffocated to death. However, we can also benefit from it if the hydrogen sulfide is recycled and processed to produce sulfur.

1.4   Classification of Oil and Gas Reservoirs

Normally, the oil and gas reservoirs are classified and named in the following methods:

1. According to the geological features of the reservoirs;
2. According to the characteristics and distribution of the reservoir fluids; and
3. According to the reservoir natural drive energy and types.

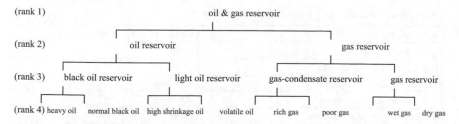

Fig. 1.2 Classification of petroleum reservoir

In the course of petrophysics, the object primarily studied is the characteristics of the reservoir fluids, so the second classification method listed above will be the focus to introduce.

1.4.1 Classification 1

The oil and gas reservoirs can be classified as shown in the Fig. 1.2.

1.4.2 Classification 2

With reference to fluid characteristics, phase behavior, field data of gas–oil ratio, tank-oil density, methane molar concentration, etc., the reservoirs can be approximately as follows:

1. Gas reservoir: primarily comprised of methane and a small amount of ethane, propane, and butane;
2. Gas-condensate reservoir: Comprised of alkanes from methane to octane (C_8), this type of reservoirs is in gaseous state under original underground conditions. As the pressure drops, the hydrocarbons will condense as liquid. The liquid hydrocarbons are light-colored, with a specific gravity of 0.72–0.80, and are named condensate oil. According to the concentration of the condensate oil, the gas-condensate reservoirs can be classified as in Table 1.9.
3. Volatile reservoir (also called critical reservoir): This type of reservoirs usually contains light oil, with specific gravity of 0.7–0.8, and its hydrocarbons composition containing heavier than C_8. In the geological structure of reservoirs, the fluid is like gas at the top, while it is like oil at the bottom, but no distinct gas/oil surface. In China, volatile reservoirs are found in Oilfield of Jinlin, Zhongyuan, and Xinjiang. Volatile reservoirs are also found in other countries such as UK and USA.

Table 1.9 Classification of gas-condensate reservoirs

Gas-condensate reservoirs type	Condensate oil content (g/m^3)
Extraordinary high condensate oil content	>600
High condensate oil content	250–600
Medium condensate oil content	100–250
Low condensate oil content	50–100
Meager condensate oil content	<50

4. Oil reservoir: It is primarily comprised of liquid hydrocarbons with a certain amount of natural gas dissolved in them. The tank-oil specific gravity is about 0.8–0.94. Crude oil in this type of reservoirs is usually called "black oil" in foreign countries.
5. Heavy-oil reservoir (also called thick-oil reservoir): It is featured by high viscosity and great specific gravity. According to the standards made in the 11th World Petroleum Congress held in London, 1983, this type is defined as the oil reservoirs which own a tank-oil specific gravity of 0.934–1.0 and a gas-free oil viscosity of 100–1000 mPa s under reservoir temperature.

1.4.3   Classification 3

Table 1.10 shows another representative classification.

The gas reservoirs can be further divided into to subtypes: dry gas reservoir and wet gas reservoir. In actual practice, the dry gas reservoir, also named poor gas reservoir, often refers to gas reservoirs whose production GOR (gas–oil ratio) is greater than 18,000 m^3/m^3, and the other group whose production GOR is smaller than 18,000 m^3/m^3 is called wet gas reservoir.

Table 1.11 shows the molar concentrations of several constituents in these typical reservoir fluids, as well as some more detailed information about their physical properties. There exists obvious difference among the molar concentrations

Table 1.10 Classification according to the properties of the reservoir fluids

Classification	Gas–oil ratio (m^3/m^3)	Methane content (%)	Condensate oil content (cm^3/m^3)	Tank-oil density (g/cm^3)
Natural gas	>18,000	>85	<55	0.70–0.80
Condensate gas	550–18,000	75–90	55–1800	0.72–0.82
Light oil	250–550	55–75	–	0.76–0.83
Black oil	<250	<60	–	0.83–1.0

Table 1.11 Molar composition and physical properties of typical reservoir fluids

Component	Black oil	Light oil	Condensate oil	Dry gas
C_1 (%)	48.83	64.36	87.01	95.85
C_2 (%)	2.75	7.52	4.39	2.67
C_3 (%)	1.93	4.47	2.29	0.34
C_4 (%)	1.60	4.12	1.08	0.52
C_5 (%)	1.15	2.97	0.83	0.08
C_6 (%)	1.59	1.38	0.60	0.12
C_7 (%)	4115	14.91	3.80	0.42
Sum (%)	100.00	100.00	100.00	100.00
C_7^+ molar weight	225	181	112	157
GOR (m^3/m^3)	111	356	3240	18,690
Tank-oil density (g/cm^3)	0.8534	0.7792	0.7358	0.7599
Color	Green	Orange-yellow	light straw	Colorless

of some constituents, especially the constituent C_7+, in the differently natured reservoir fluids. Furthermore, these differences result in disparities in some physical properties of the liquid, for example, the gas–oil ratio, the density and the color.

1.4.4   Other Classifications

According to the burial depths of the formations, the oil and gas reservoirs are classified as follows: shallow reservoir (burial depth <1500 m), medium-depth reservoir (burial depth 1500–2800 m), deep reservoir (burial depth 2800–4000 m), and ultra-deep reservoir (burial depth >4000 m).

The classifications introduced above are just several of their numerous fellow methods. For more convenience in application, the classification of oil and gas reservoirs needs further study and analysis. A rational classification method is of great significance to the development of reservoirs, calculation and assessment of reserves, and the establishment of reasonable developing indexes.

1.5   The Chemical Composition and Classification of Formation Water

1.5.1   The Chemical Composition of Formation Water

1. Chemical composition

Because of long-time contact with the reservoir rocks and crude oil, the formation water contains a considerable deal of metallic salts, such as sodium salt, potassium salt, calcium salt, and magnesium salt, and for this reason, the formation water is also called brine. Formation water has competitively higher salt concentration that makes the formation water distinguished from the land-surface water. The total salt concentration in formation water is called salinity.

Positive ions commonly encountered in formation water include Na^+, K^+, Ca^{2+}, and Mg^{2+}, and some other positive ions such as Ba^{2+}, Fe^{2+}, Sr^{2+}, and Li^+ also take a little share; the commonly encountered negative ions include Cl^-, SO_4^{2-}, HCO_3^- CO_3^{2-}, NO_3^-, and Br^-, as well as some trace ions.

Usually, in the formation water, there also exist diverse microorganisms, among which the most common one is the anaerobic sulfate-reducing bacterium which is rather stubborn to be eliminated. The microorganisms cause the corrosion of casings and build-up in pores of the formation. However, the origin of them is still rather vague for us. It is guessed that they may originally exist in the trapped and closed reservoirs or, perhaps, are brought to the underground formations during drilling operations.

Besides, some trace organic substances such as naphthenic acid, fatty acid, amino acid, and humic acid, and some other complicated organic compounds, are also present in formation water. Considering that these organic acids hold direct impact on the oil-washing ability of the injected water, extra attention should be paid during the selection of the injected water's quality.

2. The salinity and the milligram-equivalent concentration of the ions

Salinity is the total concentration of the mineral salts in water and is expressed with the unit mg/L or ppm (one millionth). It is the total concentration of both the positive and negative ions in it. Salinity varies within a wide range from several thousand to hundreds of thousand (unit mg/L) in different oil field. A high salinity, which sometimes can be up to 3×10^5 ml/L, is a typical value for formation water.

Under original reservoir conditions, the formation water is in a state of saturated solution, but when it flows up to the land surface, due to the pressure and temperature drop, the salts will fall of the solution. In severe cases, the wellbores can be salt-crusted and the production will be brought into trouble.

Because the positive and negative ions react with each other in equivalent amounts, the salt concentration in formation water is usually expressed by milligram-equivalent concentration which equals the ratio of a certain ion's concentration to its equivalent value. The compounded equivalent values of some common ions in oil formations are shown in Table 1.12. For example, the concentration of chloride ion (Cl^-) is 7896 mg/L, and the compounded equivalent value is 35.3; then, the milligram-equivalent concentration of the chloride ion can be obtained as follows: 7896/35.3 = 225.6 mg/L.

3. Water hardness

Hardness of water is a term used to indicate its content of calcium and magnesium ions. In the chemical flooding (e.g., injection of polymers or active agents), high water hardness tends to worsen the effect of the stimulations by causing

Table 1.12 Compounded equivalent values of common ions in oil formations

No.	Positive ion	Compounded equivalent values	No.	Negative ion	Compounded equivalent values
1	Ba^{2+}	68.7	1	CO_3^{2-}	30
2	Ca^{2+}	20	2	HCO_3^-	61
3	H^+	1	3	Cl^-	35.3
4	Fe^{3+}	27.9	4	OH^-	17
5	Fe^{2+}	18.6	5	O^{2-}	8
6	Mg^{2+}	12.2	6	SO_4^{2-}	48
7	Na^+	23	7	S^{2-}	16
8	Sr^{2+}	43.8	8	SO_3^{2-}	40

precipitation of the chemical agents. Therefore, clear and detailed knowledge about it is indispensable during the production.

In addition to the formation water's total salinity and water hardness, it is also necessary to know quantificationally the species and individual concentration of the ions, upon which the water type and the depositional environment it represents can be determined.

1.5.2   The Classification of Formation Water

Roughly, the total concentration of the formation water can be known in accordance with its total salinity. For further grasp of the individual concentration of every ion, either positive or negative, and their corresponding proportions, the formation water is classified to different types. Water type is named according to the tendency of a certain compound's appearance in the water. For example, the bicarbonate-sodium water refers to the water in which the bicarbonate-sodium is in a tendency to appear.

1. The classification of water—Sulin's system

   According to the chemical components in water or, in other words, their milligram-equivalent concentration: $\dfrac{Na^+}{Cl^-}, \dfrac{Na^+-Cl^-}{SO_4^{2-}}, \dfrac{Cl^--Na^+}{Mg^{2+}}, \dfrac{SO_4^{2-}}{Cl^-}$ and $\dfrac{Ca^{2+}}{Mg^{2+}}$, and water is classified to four types.

   (1) Sulfate-sodium (Na_2SO_4) type: It represents an environment of continental washing. Generally speaking, this type reflects a weak enclosed condition on which the oil and gas could not be gathered or stored easily. Most of the land-surface water belongs to this type.

   (2) Bicarbonate-sodium ($NaHCO_3$) type: It represents an environment of continental deposit. This type is widely distributed in oil fields and can be utilized as a sign of a good oil and gas bearing.

Table 1.13 Chemical composition and water type of the formation water in oil fields

Oil Field	Total salinity mg/L	Na^+ (K^+)	Mg^{2+}	Ca^{2+}	Cl	SO_4^2	HCO_3	CO_3^2	Water type
Daqang	16,316	5917	11	95	7896	18	2334	45	$CaCl_2$
Gudao, Shengli	3228	1038	L3	25	1036	0	1116	0	$NaHCO_3$
Shengli	17,960	4952	836	620	10,402	961	187	0	$CaCl_2$
Renqiu	178	21	9	20	43	18	67	0	$CaCl_2$
East Texas (America)	64,725	23,029	536	1360	39,000	216	578	0	$CaCl_2$
Gachsaran (Iran)	95,313	33,600	30	1470	55,000	4920	–	293	$CaCl_2$
Furrial-Musipan (Venezuela)	5643	1739	59	54	1780	–	2001	–	$MgCl_2$

(3) Chloride-magnesium ($MgCl_2$) type: It represents an environment of marine deposit. This type usually exists in the interior of the oil and gas fields.

(4) Chloride-calcium ($CaCl_2$) type: It represents an environment of deep enclosed construction. This type reflects a good enclosed condition on which the oil and gas can be gathered and stored easily. This is also a sign of fine oil and gas bearing.

This method discussed above is called Sulin's water classification system. Representing each geologic environment with a special water type, it has associated the chemical components in formation water with their corresponding forming conditions. Table 1.13 shows the chemical compounds and types of the formation water exist in some oil fields at home and abroad.

2. The determination of water types

As illustrated in Fig. 1.3, the order in which the positive and negative ions combine with each other is in accordance with the ions' affinity for their opposite ones. Moreover, this also offers the principle on which the water types are determined:

(1) When $Na^+/Cl^- > 1$, the redundant Na^+ will combine with the SO_4^{2-} or HCO_3^- ions and forms Na_2SO_4 type or $NaHCO_3$ type. Further judgment can be continued as follows:

Fig. 1.3 Diagram of the order in which the ions combine with each other

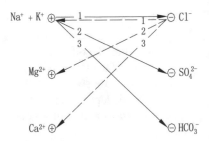

(a) When $\dfrac{Na^+ - Cl^-}{SO_4^{2-}} < 1$, it suggests that after balancing the Cl^-, the redundant Na^+ will combine with the SO_4^{2-}, and since the ratio is less than 1, there will be no Na^+ left for HCO_3^-. As a result, when the whole combination reaction ends what is formed finally is Na_2SO_4 and that is why this kind of water belongs to the Na_2SO_4 type.

(b) When $\dfrac{Na^+ - Cl^-}{SO_4^{2-}} > 1$, it suggests that after balancing the Cl^- and SO_4^{2-}, the redundant Na^+ will combine with the HCO_3^- to form $NaHCO_3$, whereupon the water is named $NaHCO_3$ type.

(2) When $Na^+/Cl^- < 1$, the redundant Cl^- will combine with the Mg^{2+} or Ca^{2+} ions and forms $MgCl_2$ type or $CaCl_2$ type. Further judgment is as follows:

(a) When $\dfrac{Cl^- - Na^+}{Mg^{2+}} < 1$, it suggests that after balancing the Na^+, the redundant Cl^- will combine with the Mg^{2+}, and since the ratio is less than 1, there will be no Cl^- left for Ca^{2+}. As a result, the $MgCl_2$ type is formed.

(b) When $\dfrac{Cl^- - Na^+}{Mg^{2+}} < 1$, it suggests that after balancing the Na^+ and Mg^{2+}, the redundant Cl^- will combine with the Ca^{2+} and forms $CaCl_2$, upon which the water is named $CaCl_2$ type.

The procedure for water-type determination is summarized in Table 1.14.

Table 1.14 Sulin's system of water classification

Equivalent proportion	Parameter for judgment	Water type	Environment
$\frac{Na^+}{Cl^-} > 1$	$\frac{Na^+ - Cl^-}{SO_4^{2-}} < 1$	Na_2SO_4 type	Continental washing (land-surface water)
	$\frac{Na^+ - Cl^-}{SO_4^{2-}} > 1$	$NaHCO_3$ type	Continental deposit (water in Oil and gas fields)
$\frac{Na^+}{Cl^-} < 1$	$\frac{Cl^- - Na^+}{Mg^{2+}} < 1$	$MgCl_2$ type	Marine deposit (water in Oil and gas fields)
	$\frac{Cl^- - Na^+}{Mg^{2+}} > 1$	$CaCl_2$ type	Deep enclosed construction (gas field water)

Table 1.15 Chemical composition of the formation water in a certain oil field

Content	Ion type						
	$Na^+ + K^+$	Mg^{2+}	Ca^{2+}	Cl^-	SO_4^{2-}	HCO_3^{-1}	CO_3^{2-}
mg/L	23,029	536	1360	39,000	216	518	0
Ionic equivalent ratio	Na^+23.00 K^+39.10	12.15	20.04	35.45	48.03	61.01	30.00
Equivalent value/L	1001.26	44.12	68.16	1100.14	4.50	8.49	0.00

[Example 1.1] Table 1.15 shows the chemical composition of the formation water present in a certain oil field. Please determine its water type.

Solution:

(1) Judge the equivalent proportion of the ions whose valence is 1: $\frac{Na^+ + K^+}{Cl^-} = \frac{1001.26}{1100.14} < 1$, As an initial judgment, the water can be $MgCl_2$- or $CaCl_2$-typed.

(2) Because that $\frac{Cl^- - Na^+}{Mg^{2+}} = \frac{1100 \cdot 14 - 1001 \cdot 26}{44 \cdot 12} = \frac{98 \cdot 88}{44 \cdot 12} > 1$, the given water belongs to the $CaCl_2$ type.

Chapter 2
Natural Gas Physical Properties Under High Pressure

It is commonly known that natural gas owns compressibility. When subjected to the reservoir pressure and temperature, the natural gas is in compression. However, various parameters of it undergo gradual changes at any stage of its flowing course which involves the underground percolation in formation and the wellbore flowing, resulting from the pressure drop and volume expansion. State equations are commonly used for gas to indicate its change in state (pressure P, volume V, and temperature T), and some high-pressure physical parameters which are quite practical in engineering are also introduced, for example, volume factor, isothermal compressibility, and viscosity ratio. This chapter will focus on these parameters that are commonly used and play a role of prerequisite in the oil and gas field development.

2.1 Apparent Molecular Weight and Density of Natural Gas

2.1.1 Composition of Natural Gas

1. The Definition and Classification of Natural Gas

Generally speaking, natural gas is flammable or inflammable gaseous mixtures of hydrocarbons and non-hydrocarbons, extracted from the underground, at normal temperature and pressure.

Narrowly speaking, natural gas can be classified into different types:

(1) According to the essential features of the deposits: The natural gas can be ranged into two types: oil reservoir associated gas and gas from gas reservoir. The former one is dissolved in crude oil at the reservoir condition and released from oil as pressure drops. The other one, gas from gas reservoir, is in a

© Petroleum Industry Press and Springer-Verlag GmbH Germany 2017
S. Yang, *Fundamentals of Petrophysics*, Springer Geophysics,
DOI 10.1007/978-3-662-55029-8_2

gaseous state in the underground formations and includes two types named pure gas and condense gas, respectively. The principal constituent of the gas in pure gas reservoirs is methane, with some others such as ethane, propane, and butane. The gas from gas-condensate reservoir, on analysis of the effluent at the oil wellhead, contains not only methane, ethane, propane and butane, but also some liquid C_5+ hydrocarbons between C_7 and C_{11}.

(2) According to the composition, natural gas can be described in two ways: dry gas and wet gas, or poor gas and rich gas.

Dry gas: The content of liquid C_5+ heavy hydrocarbons in wellhead effluent is less than 13.5 cm^3/m^3 at standard temperature and pressure.

Wet gas: On analysis of the wellhead effluents which are subjected to standard temperature and pressure, the content of liquid C_5+ heavy hydrocarbons is more than 13.5 cm^3/m^3.

Poor gas: On analysis of the wellhead effluents which are subjected to standard temperature and pressure, the content of liquid C_3+ heavy hydrocarbons is less than 94 cm^3/m^3.

Rich gas: On analysis of the wellhead effluents which are subjected to standard temperature and pressure, the content of liquid C_3+ heavy hydrocarbons is more than 94 cm^3/m^3.

(3) According to the content of the acidic gases such as H_2S and CO_2:

Sour natural gas contains significant amount of acidic gases such as H_2S and CO_2. This kind of natural gases is required to be processed and clarified to reach the standards for pipeline transportation.

Clean natural gas is also named as sweet gas. This kind of natural gas contains very minor amount of acidic gases, and the cleaning (purification) treatment is not required for it.

2. The Composition of Natural Gas

Natural gases are mixtures of hydrocarbons which, as stated earlier, may be characterized by composition. Some analytical instruments such as gas chromatograph can help us to know the components contained in the gaseous mixture to some extent. The composition of natural gas can be reported in terms of mole fraction (mole percentage), mass fraction (weight percentage), or volume fraction (volume percentage).

(1) Mole fraction (mole percentage)

The mole per cent of the ith component, expressed with the symbol y_i, is the percentage ratio of the mole number of this component to the total mole number of the entire gas:

$$y_i = \frac{n_i}{\sum_{i=1}^{N} n_i} \qquad (2.1)$$

In the equation:

n_i the mole number of the ith component;
$\sum_{i=1}^{N} n_i$ the total mole number of the gas;
N the number of different components.

(2) Volume fraction (volume percentage)

The volume percentage φ_i is:

$$\varphi_i = \frac{V_i}{\sum_{i=1}^{N} V_i} \tag{2.2}$$

In the equation:

V_i the volume of the ith component;
$\sum_{i=1}^{N} V_i$ the total volume of the gas.

Provided that the mole number of the ith component is n_i, it is easy to obtain its volume, 22.4 n_i, and thence, the total volume of the gas present is 22.4 $\sum_{i=1}^{N} n_i$. According to Eqs. (2.1) and (2.2), we can get that $\varphi_i = y_i$.

Within gases, the mole percentage of a component equals the volume percentage of it, because when subjected to the same conditions of pressure and temperature, every 1 mol of any gas occupies the same volume. In particular, upon the standard conditions, the volume of 1 mol gas is 22.4 L.

(3) Mass fraction (Weight percentage)

The weight percentage G_i is:

$$G_i = \frac{w_i}{\sum_{i=1}^{N} w_i} \tag{2.3}$$

In the equation:

w_i the mass of the ith component;
$\sum_{i=1}^{N} w_i$ the total mass of the gas.

It is easy to convert the mass fraction to mole fraction using the following equation:

$$y_i = \frac{G_i/M_i}{\sum_{i=1}^{N} G_i/M_i} \tag{2.4}$$

where

M_i the molecular weight of the ith component.

Table 2.1 The composition of natural gas and the mass fraction

Component	Mass fraction	Molecule weight, M_i	Mass fraction/molecule weight	Mole fraction, y_i
Methane	0.71	16.0	0.044	0.85
Ethane	0.14	30.1	0.005	0.09
Propane	0.09	44.1	0.002	0.04
Butane	0.06	58.1	0.001	0.02
Summation	1.00		0.052	1.00

Solution: the conversion and results can be seen in Table 2.1

Actually, the composition of the liquid crude oil can also be expressed by the three methods talked above.

[Example 2.1] The first two columns of Table 2.1 shows the natural gas composition and the weight percentage of each component. Please convert their weight percentage to mole percentage.

2.1.2　The Molecular Weight of Natural Gas

Since natural gas is a mixture composed of molecules of various sizes, not the same as the pure substance, it cannot be expressed by a definite chemical formula upon which the molecular weight can be calculated and gained easily. To solve this problem, the concept of "apparent molecular weight" is introduced.

The apparent molecular weight is defined as the mass of every 22.4 L natural gas when it is subjected to a condition of 0 °C and 760 mmHg. In other words, the numerical value of the apparent molecular weight is equal to the mass per mole of the natural gas. Obviously, this parameter is a hypothetical one that does not exist really, and this is why it is named "apparent molecular weight." In practical application, it is directly called "molecular weight" for short.

If the mole fraction and molecular weight of each component are given, the following equation can be used to calculate the apparent molecular weight of natural gas:

$$M = \sum_{i=1}^{N} (y_i M_i) \tag{2.5}$$

where

M　the molecular weight of the natural gas present;
y_i　the mole per cent of the ith component;
M_i　the molecular weight of the ith component.

Table 2.2 Calculation of the molecular weight of the natural gas

Component	Mole fraction, y_i	Molecular weight, M_i	$y_i M_i$
Methane	0.85	16.0	13.60
Ethane	0.09	30.1	2.71
Propane	0.04	44.1	1.76
Butane	0.02	58.1	1.16
Summation	1.000		$M = 19.23$

Result: the molecular weight of the natural gas is 19.23

Obviously, the different compositions of natural gas will lead to their different numerical values in apparent molecular weight. Generally speaking, the apparent molecular weight of dry gas is about 16.82–17.98 usually.

[Example 2.2] The mole fraction is shown in Table 2.1 to calculate the molecular weight of the natural gas.

Solution: The conversion and results are shown in Table 2.2.

2.1.3   The Density and Specific Gravity of Natural Gas

1. Density

The density of natural gas is defined as the mass of per unit volume of the natural gas.

$$\rho_g = m/V \tag{2.6}$$

where

ρ_g the density of natural gas, g/cm^3 or kg/m^3;
m the mass of natural gas, g or kg;
V the volume of natural gas, cm^3 or m^3.

The density of natural gas at given temperature and pressure can be gained through the state equation of gas (for details, see the lesson 2 of this chapter):

$$\rho_g = \frac{PM}{ZRT} \tag{2.7}$$

where

ρ_g the density of natural gas, kg/m^3;
P the pressure to which the natural gas is subjected, MPa;
M the molecular weight of natural gas, kg/kmol;

T the absolute temperature of natural gas, K;
Z the compressibility factor of natural gas;
R universal gas constant, $R = 0.008134$ MPa m^3/(kmol K).

2. Specific gravity

The specific gravity of natural gas is defined as the ratio of the density of the natural gas to the density of dry air, both subjected to standard conditions adopted in oil industry (293 K, 0.101 MPa). The specific gravity is dimensionless.

$$\gamma_g = \rho_g / \rho_a \tag{2.8}$$

where

ρ_g the density of natural gas;
ρ_a the density of air.

To integrate Eqs. (2.7) and (2.8), with consideration that the molecular weight of dry air is $28.96 \approx 29$, we can get:

$$\gamma_g = M/29 \tag{2.9}$$

When the specific gravity of a natural gas is known, its apparent molecular weight can be gained through the equation written above.

Generally speaking, the specific gravity of natural gas ranges from 0.55–0.8 (lighter than air). But when the content of heavy hydrocarbons or non-hydrocarbon components is comparatively high, the value of specific gravity may be larger than 1.

2.2   Equation of State for Natural Gas and Principle of Corresponding State

2.2.1   Equation of State (EOS) for Ideal Gas

Three limiting assumptions have to be made to define the ideal gas:

(1) The volume occupied by the molecules is insignificant with respect to the total volume occupied by the gas.
(2) There are no attractive or repulsive forces between the molecules.
(3) All collisions of molecules are perfectly elastic; that is, there is no loss of internal energy upon collision.

The molecules of an ideal gas must have the above-mentioned properties.

Table 2.3 Values of R for various unit systems

n	T	P	V	R
Kmol	K	MPa	m^3	0.008314 MPa m^3/(kmol K)
Kmol	K	Pa	m^3	8314 J/(kmol K)
Mol	K	Atmospheric pressure	L	0.08205 atm L/(mol K)
Mol	K	Pa	m^3	8.3145 Pa m^3/(mol K)
Mol	R	Psi	Cu ft	10.732 Psi Cu ft/(mol R)

The equation of state for ideal gas is:

$$PV = nRT \qquad (2.10)$$

where

P the absolute pressure of the gas, MPa;
V the volume occupied by the gas, m^3;
T absolute temperature, K;
n the number of moles of the gas present, kmol;
R universal gas constant, $R = 0.008134$ MPa m^3/(kmol K).

The universal gas constant is considered to be independent of the type of gas, but its numerical value depends on the system of units used. Table 2.3 shows the values of R for various unit systems.

Realizing that it is not correct to neglect the volume occupied by the natural gas molecules as well as the mutual force among them, the natural gas is not an ideal gas but a real one. At very low pressure, the molecules are relatively far apart, and the conditions of ideal gas behavior are more likely to be met. However, the situation is quite different for the real gases buried in underground reservoirs with high temperature and pressure; that is to say, the behavior of natural gas deviates drastically from the behavior predicted by the state equation for ideal gas. Therefore, based on the state equation for ideal gas, correction methods are adopted to describe the real gas state equation.

2.2.2 Equation of State (EOS) for Real Gas

The best way of writing an equation of state for real gas is to insert a correction factor to the ideal gas equation. This results in the equation of state for real gas:

$$PV = ZnRT \qquad (2.11)$$

It is also known as the compressibility equation of state.

In this equation, the correction factor Z is known as the compressibility factor or deviation factor. As for its physical meaning, it is the ratio of the volume actually occupied by a gas at given pressure and temperature to the volume of the gas occupied at the same pressure and temperature if it behaved like an ideal gas.

$$Z = \frac{V_{actual}}{V_{ideal}}$$

Compared with the ideal gas, the molecules of real gas, on the one hand, are big enough to occupy significant volume and make it harder to be compressed; on the other hand, the molecules are close enough with each other to exert some attraction between them, which makes the real gas easier to be compressed. The compressibility factor Z just reflects the integral effect of these two opposite interactions: When Z equals 1, it indicates that the real gas behaves like ideal gas; when Z is larger than 1, it indicates that the real gas occupies greater volume and is harder to be compressed than ideal gas; when Z is smaller than 1, it indicates that the real gas occupied less volume and is easier to be compressed than ideal gas.

The Z-factor is not a constant and varies with changes in gas composition, temperature, and pressure. It must be determined experimentally. The results of our predecessors' experimental achievements, including the relationship between the Z-factor and T or P, have been made into the form of charts for reference. The Z-factor charts for methane and ethane have been, respectively, shown in Figs. 2.1 and 2.2. But in order to get the compressibility of natural gas, principle of corresponding state has to be introduced in advance.

2.2.3 Principle of Corresponding State

1. The principle of corresponding state and the method to get the Z value of natural gas

Although real gases exhibit different properties (including compressibility factor) and various critical parameters upon different conditions of temperature and pressure, they possess similar characteristics when they are at their own critical points. Taking the critical state of gas as the datum mark, all the pure-hydrocarbon gases which are subjected to the same conditions of reduced pressure P_r and reduced temperature T_r have the same compressibility factors; in other words, they obey principle of corresponding state.

Reduced temperature and reduced pressure are defined as follows:

$$P_r = \frac{P}{P_c} \quad T_r = \frac{T}{T_c} \tag{2.12}$$

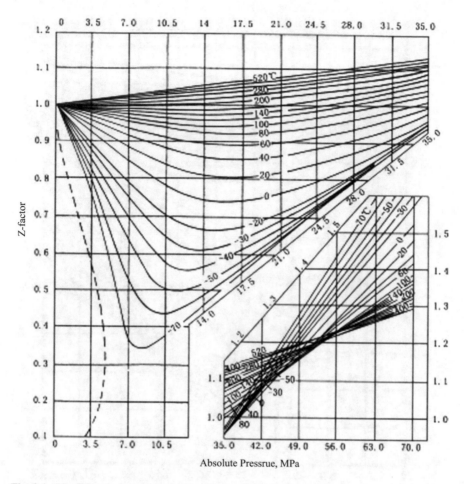

Fig. 2.1 The Z-factor charts for methane (Brown 1948)

where

P_r, T_r the reduced pressure and reduced temperature, dimensionless;
P, P_c the absolute pressure and critical temperature to which the gas is subjected (MPa);
T, T_c the absolute temperature and critical temperature of the gas (K).

Principle of corresponding state: When two gases are in the same corresponding state, many of their intensive properties (i.e., properties that are independent of volume) are approximately uniform. This law is highly accurate if the gases have similar molecular characteristics and pretty nearly the same values of critical temperature. The critical parameters of some hydrocarbon or non-hydrocarbon gases have been presented, respectively, in Table 2.4 and Table 2.5.

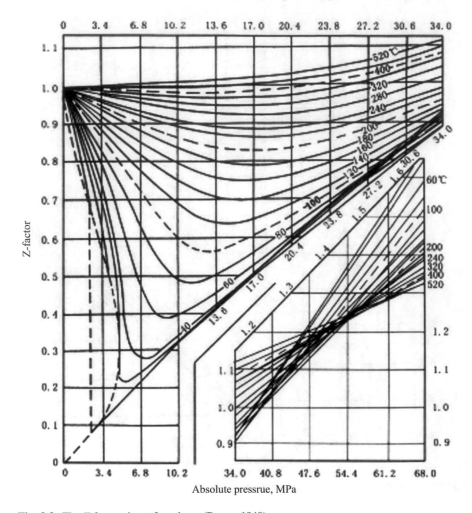

Fig. 2.2 The Z-factor charts for ethane (Brown 1948)

At present, the biparametric compressibility factor method (i.e., $Z = f(P_r, T_r)$), taking principle of corresponding state as theoretical basis, is used commonly to describe the state change of real gases. Figure 2.3 shows the compressibility factors of methane, ethane, and propane after a treatment according to the principle of corresponding state.

Natural gases, as multicomponent mixtures of hydrocarbons, have similar chemical composition and therefore can be covered by principle of corresponding state. That is to say, the creditability of the results got from Z-charts is good enough for engineering.

Table 2.4 Physical-property-related constants of some hydrocarbon gases

Component	Molecular formula	Molecular weight	Boiling point °C (at 0.1 MPa)	Critical pressure P_c (MPa)	Critical temperature T_c (K)	Density of liquid (at standard conditions) (g/cm^3)	Eccentric factor ω
Methane	CH_4	16.043	−161.50	4.6408	190.67	0.3	0.0115
Ethane	C_2H_6	30.070	−88.61	4.8835	303.50	0.3564	0.0908
Propane	C_3H_8	44.097	−42.06	4.2568	370.00	0.5077	0.1454
Isobutane	$i\text{-}C_4H_{10}$	58.124	−11.72	3.6480	408.11	0.5631	0.1756
Normal butane	$n\text{-}C_4H_{10}$	58.124	−0.50	3.7928	425.39	0.5844	0.1928
Isopentane	$i\text{-}C_5H_{12}$	72.151	27.83	3.3336	460.89	0.6247	0.2273
Normal pentane	$n\text{-}C_5H_{12}$	72.151	36.06	3.3770	470.11	0.6310	0.2510
Isohexane	$n\text{-}C_6H_{14}$	86.178	68.72	3.0344	507.89	0.6640	0.2957
Normal heptane	$n\text{-}C_7H_{16}$	100.205	98.44	2.7296	540.22	0.6882	0.3506
Normal Octane	$n\text{-}C_8H_{18}$	114.232	125.67	2.4973	569.39	0.7068	0.3978
Normal nonane	$n\text{-}C_9H_{20}$	128.259	150.78	2.3028	596.11	0.7217	0.4437
Normal decane	$n\text{-}C_{10}H_{22}$	142.286	174.11	2.1511	619.44	0.7342	0.4502

The meaning and application of the eccentric factor can be seen in the lesson 4 of this chapter

Table 2.5 Common physical-property-related constants of some non-hydrocarbon gases

Component	Molecular formula	Molecular weight	Boiling point °C (at 0.1 MPa)	Critical pressure P_c (MPa)	Critical temperature T_c (K)	Density of liquid (at standard conditions) (g/cm^3)	Eccentric factor ω
Air	N_2, O_2	28.964	−194.28	3.7714	132.78	0.856	
Carbon dioxide	CO_2	44.010	−78.51	7.3787	304.17	0.827	0.2250
Helium	He	4.003	−268.93	0.2289	5.278	–	
Hydrogen	H_2	2.016	−252.87	1.3031	33.22	0.07	−0.2234
Sulfureted hydrogen	H_2S	34.076	−60.31	9.0080	373.56	0.79	0.0949
Nitrogen	N_2	28.013	−195.80	3.3936	126.11	0.808	0.0355
Oxygen	O_2	31.999	−182.96	5.0807	154.78	1.14	0.0196
Water	H_2O	18.015	100	22.1286	647.33	1.0	0.3210

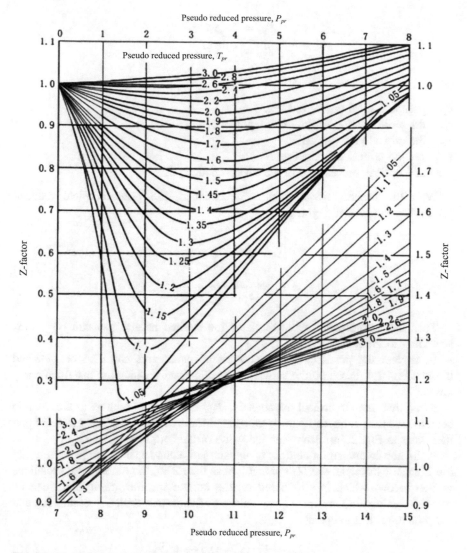

Fig. 2.3 The Z-factor charts for natural gas (Standing and Katz 1942)

In light of the impurity of natural gases, Kay introduced the concept of "pseudo-critical" and "pseudo-critical reduced" properties for treating mixtures of gases.

The pseudo-critical pressure and pseudo-critical temperature are defined mathematically:

$$P_{pc} = \sum y_i P_{ci}$$
$$T_{pc} = \sum y_i T_{ci}$$

$$(2.13)$$

where

P_{pc} pseudo-critical pressure, MPa;
y_i the mole fraction of the ith component;
P_{ci} the critical pressure, MPa;
T_{pc} pseudo-critical temperature, K;
T_{ci} and critical temperature of the ith component, K.

With the pseudo-critical parameters obtained, the pseudo-critical reduced parameters can be easily got, too:

$$P_{pr} = \frac{P}{P_{pc}} = \frac{P}{\sum y_i P_{ci}}$$
$$T_{pr} = \frac{T}{T_{pc}} = \frac{T}{\sum y_i T_{ci}}$$

$$(2.14)$$

Then, the compressibility factor Z of the studied natural gas can be gained according to the Z-charts (Fig. 2.3).

In engineering practice, the pseudo-critical parameters can also be obtained through Fig. 2.4, based on the values of specific gravity which are usually already known.

After that pseudo-critical parameters, P_{pc} and T_{pc}, are known, the pseudo correspondence parameters can be calculated through the equations in (2.14) and then refer to Fig. 2.3 to determine the value of the compressibility Z.

If the non-hydrocarbon content of the studied natural gas is comparatively not too high, for example, the N_2 content is less than 2 %, the empirical correlations written below, which is also based on the known specific gravity (γ_g), can be employed to obtain the approximate values of the pseudo-critical pressure P_{pc} and the pseudo-critical temperature T_{pc}.

$$T_{pc} = 171(\gamma_g - 0.5) + 182 \tag{2.15}$$

$$P_{pc} = [46.7 - 32.1(\gamma_g - 0.5)] \times 0.09869 \tag{2.16}$$

$$P_{pc} = \sum y_i P_{ci}$$
$$T_{pc} = \sum y_i T_{ci}$$

$$(2.13)$$

where

P_{pc}, T_{pc} The pseudo-critical pressure (MPa) and pseudo-critical temperature (K);
y_i The mole per cent of the ith component;

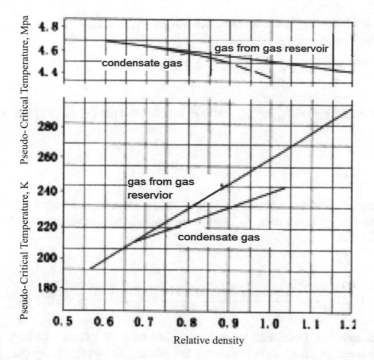

Fig. 2.4 Diagram about the natural gas special gravity and pseudo-critical parameters (Brown 1948)

Then, calculate the P_{pr} and T_{pr} according to the p and T conditions to which the studied gas is subjected to, and then, determine the value of Z through the Z-charts (Fig. 2.3).

[Example 2.3] Table 2.6 shows the composition of a certain natural gas. Please determine its Z value at 65 °C and 12 MPa, and give the volume of 1 mol of this gas.

Solution:

(1) List related data and figure out the pseudo-critical parameters, P_{pc} and T_{pc}, as shown in able 2.6.

Table 2.6 Calculation of the parameters of natural gas

Component	Mole fraction y_i	Critical temperature, T_{ci} (K)	$y_i T_{ci}$	Critical pressure, P_{ci} (Mpa)	$y_i P_{ci}$
CH₄	0.85	190.5	162.0	4.6408	3.9447
C₂H₆	0.09	306.0	27.5	4.8835	0.4395
C₃H₈	0.04	369.6	14.8	4.2568	0.1703
n-C₄H₁₀	0.02	425.0	8.5	3.7928	0.0759

(2) Determine the pseudo-reduced parameters, P_{pr} and T_{pr}.

Pseudo-reduced T_{pr}: $T_{pr} = \frac{T}{T_{pc}} = \frac{273+65}{212.8} = 1.59$

Pseudo-reduced P_{pr}: $P_{pr} = \frac{P}{P_{pc}} = \frac{12}{4.6303} = 2.59$

(3) With the gained P_{pr} and T_{pr}, refer to Fig. 2.3 and get the value of the Z-factor:

$$Z = 0.849$$

(4) According to the compressibility equation of state, determine the volume occupied by 1 mol natural gas which is subjected to the condition of 65 °C and 12 MPa:

$$V = \frac{ZnRT}{P} = \frac{0.849 \times 1.0 \times 0.008314 \times (273+65)}{12} = 0.1988 \,(\mathrm{m^3})$$

[Example 2.4] Determine the value of Z-factor at 65 °C and 12 MPa for the given natural gas (from a gas reservoir). It is already known that its specific gravity (γ_g) is 0.80.

Solution:

(1) With the known $\gamma_g = 0.80$, refer to Fig. 2.4 to get the pseudo-critical temperature $T_{pc} = 220$ K and the pseudo-critical pressure $P_{pc} = 4.58$ MPa.
(2) Determine the pseudo-reduced parameters:

$$T_{pr} = T/T_{pc} = (273+65)/228 = 1.48$$

$$P_{pr} = P/P_{pc} = 12/4.58 = 2.68$$

(3) With the obtained T_p and P_{pr}, refer to Fig. 2.3 and get the value of the Z-factor:

$$Z = 0.783$$

Sometimes, natural gases contain a small quantity of components heavier than heptane (C_7+). Since the lump of the heptane plus components should be seen as a mixture, too, the Z-factor values for this mixture can be obtained only by empirical

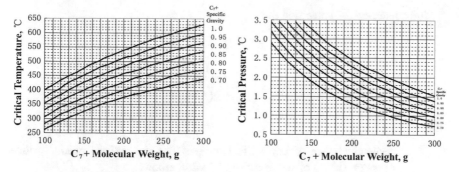

Fig. 2.5 Critical parameters of C$_7$+ components

methods. Figure 2.5 shows an empirical chart for the C$_7$+ components to determine their critical parameters, and what's more, once these parameters are got, the Z-factor can be gained by the way mentioned above.

2. Effect of Non-hydrocarbon Components

Natural gases commonly contain hydrocarbon sulfide, carbon dioxide, and nitrogen. The presence of them does affect the Z-factor obtained by the previously described methods. Hence, limits have to be put up for the usage of those methods: The content of the non-hydrocarbons in the studied natural gas must be less than 5 % by volume (the volume content of N$_2$ is less than 2 % and the volume content of CO$_2$ is less than 1 %), and what's more, the volume content of methane should not be less than 50 %. Exceeding these limits means great error (more than 3 %). For this reason, when the non-hydrocarbon or C$_5$+ components of a natural gas (e.g., condensate gas) make a comparatively high contribution, the methods should undergo a correction by referring to some other charts and equations. To remedy this problem, methods have been developed to eliminate the error caused by non-hydrocarbons, and three of them will be introduced below.

(1) Method of Chart Correction (Wichert Method)

In order to get a corrected Z-factor, this method adjusts the pseudo-critical parameters and the pseudo-reduced parameters, through an adjustment of the curve on the reference chart, to account for the unusual behavior of these gases containing impurities: Firstly, bring in an adjustment factor ε, a function of the concentrations of H$_2$S and CO$_2$, to the pseudo-critical temperature T_{pr}; Secondly, correct the pseudo-critical pressure P_{pc}; thirdly, with the corrected T_{pr} and P_{pc}, calculate the T'_{pc} and P'_{pc}; and finally get the value of Z-factor on basis of T_{pr} and P_{pr}, referring to their original Z-chart. Then, we can get the compressibility factor for the acid natural gases that contain H$_2$S and CO$_2$.

The equations for correction are:

$$T'_{pc} = T_{pc} - \varepsilon \qquad (2.17)$$

$$P'_{pc} = \frac{P_{pc}T'_{pc}}{[T_{pc} + n(1 - n)\varepsilon]}$$

where

T_{pc}, P_{pc} respectively the pseudo-critical temperature (K) and the pseudo-critical pressure (MPa) of the mixture of hydrocarbons;

T'_{pc}, P'_{pc} the pseudo-critical temperature (K) and the pseudo-critical pressure (MPa) after correction;

n the mole fraction of the H_2S in natural gas;

ε the adjustment factor for the pseudo-critical temperature (K) as a function of the concentrations of H_2S and CO_2.

The value of ε can be obtained from Fig. 2.6.

(2) Method of Empirical Formula (Wichert and Aziz Method)

The steps of this method are the same as the method introduced above, but there lies difference in their empirical formulas upon which the ε is determined. This method is more convenient because the reference to the charts can be omitted.

$$\varepsilon = \frac{120(A^{0.9} - A^{1.6}) + 15(B^{0.5} - B^4)}{1.8} \qquad (2.18)$$

where

ε the adjustment factor for pseudo-critical temperature, K;

A the sum of the mole fractions of H_2S and CO_2 in natural gas;

B the mole fraction of H_2S in natural gas.

(3) Method of Weighting Treatment (Eilerts, 1948)

Firstly, extract the Z-factors for both the hydrocarbon and non-hydrocarbon components from their corresponding charts and secondly, add them up, through a weighting treatment, to get the final value of the Z-factor for the natural gas mixture. Details can be seen in Example 2.6.

[Example 2.5] The composition of a natural gas which contains H_2S and CO_2 is tabulated below (Table 2.7). Determine its Z value at 65 °C and 11 MPa.

Solution:

(1) Calculate the pseudo-critical parameters in accordance with the mole composition of the given gas. The steps and results of calculation are listed in Table 2.7.

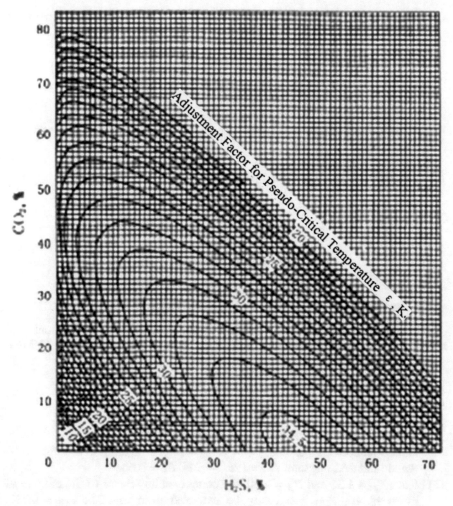

Fig. 2.6 The relationship between the adjustment factor for pseudo-critical temperature and the non-hydrocarbon concentration (White and Brown 1942)

Table 2.7 Calculation of the parameters of natural gas

Component	y_i	P_{ci} (MPa)	$y_i P_{ci}$	T_{ci} (K)	$y_i T_{ci}$
CO_2	0.10	7.3787	0.73787	304.17	30.42
H_2S	0.20	9.008	1.8016	373.56	74.71
CH_4	0.60	4.6408	2.838	190.3	114.18
C_2H_6	0.10	4.8835	0.48835	306.0	30.6
		$P_{pc} = \sum y_i P_{ci} = 5.8123$		$T_{pc} = \sum y_i T_{ci} = 249.91$	

(2) With the known concentrations of H_2S and CO_2, refer to Fig. 2.6 to get $\varepsilon = 28$ K, so

$$T'_{pc} = T_{pc} - \varepsilon = 249.91 - 28 = 221.91 \text{ K}$$

$$P'_{pc} = \frac{P_{pc}T'_{pc}}{[T_{pc} + n(1-n)\varepsilon]} = \frac{5.8123 \times 221.91}{249.91 + 0.2 \times (1 - 0.2) \times 28} = 5.070 \text{ MPa}$$

(3) Calculate the pseudo-reduced parameters:

$$T'_{pr} = T/T'_{pc} = 338/221.91 = 1.523$$

$$P'_{pr} = P/P'_{pc} = 11/5.070 = 2.169$$

(4) Refer to Fig. 2.3 and get the value of the Z-factor: $Z = 0.833$.

[Example 2.6] Determine the compressibility factor Z for the N_2-contained natural gas described in Table 2.8. The designated condition is 35 °C and 12 MPa.
Solution:

(1) Figure out y'_i, see Table 2.8.
(2) Calculate and schedule the pseudo-parameters:
  Pseudo-critical temperature: $T_{pc} = \sum y'_i T_{ci} = 222.68$ K
  Pseudo-critical pressure: $P_{pc} = \sum y'_i P_{ci} = 4.62$ MPa;
  Pseudo-reduced temperature: $T_{pr} = T/T_{Pc} = (273 + 35)/222.68 = 1.38$;
  Pseudo-reduced pressure: $P_{pr} = P/P_{pc} = 12/4.62 = 2.60$.
(3) Since $T_{pr} = 1.38$ and $P_{pr} = 2.60$, the compressibility factor of the mixture of hydrocarbons, Z_{CH}, can easily be extracted from Fig. 2.3: $Z_{CH} = 0.712$. Furthermore, the compressibility of nitrogen at 35 °C and 12 MPa can be obtained through related handbooks: $Z_N = 1.03$.

Table 2.8 Calculation of the parameters of nitrogen-contained natural gas

Component	Mole fraction y_i	Mole fraction of hydrocarbon $y'_i = \frac{y_i}{1-y_{N2}}$	T_{ci}, (K)	$y'_i T_{ci}$	P_c (MPa)$_i$	$y'_i P_{ci}$
N_2	0.435	–	–	–	–	–
CH_4	0.445	0.790	190.67	150.63	4.641	3.67
C_2H_6	0.062	0.110	303.56	33.39	4.884	0.54
C_3H_8	0.039	0.070	370.00	25.90	4.257	0.30
C_4H_{10}	0.019	0.030	425.39	12.76	3.793	0.11
Sum	1.000	1.000	–	222.68	–	4.62

(4) Setting the symbol y_N as the mole fraction of non-hydrocarbon components, the compressibility factor for the nitrogen-contained natural gas can be obtained through the equation written below:

$$Z = y_N Z_N + (1 - y_N)Z_{CH}$$
$$= 0.435 \times 1.03 + (1 - 0.435) \times 0.712$$
$$= 0.8503$$

Based on this example, let us compare the Z-factor values of the nitrogen-contained and nitrogen-free natural gases.

Nitrogen-contained compressibility factor: $Z = 0.8503$

Nitrogen-free compressibility factor: $Z_{CH} = 0.712$

This result indicates that great error would occur if no correction is taken. As for the source of the error, it can be explained by the applicability of principle of corresponding state: The curves on the Z-chart (Fig. 2.3) are made on the basis of principle of corresponding state; however, this law can only apply nicely to the gases with similar molecular characteristics. It is also suggested that as a correction aimed to get more accurate results, the gained Z-factor could be multiplied by an adjustment factor C, numerically 1–1.04.

Finally, another two points should be mentioned:

(1) The higher the pressure in gas reservoirs, the greater the compressibility factor they have. For ultradeep gas reservoirs, they have high Z-factor value. An accurate determination of the Z-factor is essential in the calculation of reserves. If the obtained Z-factor is smaller than its actual value, the original reserve figured out is greater than it really is.

(2) It is found that when the heavy hydrocarbons make a comparatively big contribution in natural gases (e.g., the condensate gas), the value of Z-factor obtained through the charts suffers significant error, so the compressibility equation of state as stated above could not be used to describe their p–V–T relations. Therefore, in recent years, other state equations are developed for the phase equilibrium calculations of condensate gases.

2.2.4   Other Equations of State for Natural Gas

The equation of state for real gas (2.11) is established on the basis of experimental research, and the Z-factor is actually an adjustment coefficient born from experimental tests.

Another way to establish the equations is through theoretical analyses: Proceeding from the microscopic structure of the substance, research on the movement of the gas molecules and their interaction between each other and then conclude their influence on the macroscopic properties of natural gases. In this way, corresponding corrections are added into the equation of state for ideal gas; thus, the citation of the ideal equation by real gases can be ensured.

1. VDW State Equation

One of the earliest attempts to represent the behavior of real gases by an equation was that of van der Waals (1873). Based on the kinetic theory of gases, he took the microscopic structure of gases in consideration and proposed two correction terms which were concerned essentially with the two fundamental assumptions in the definition of the ideal gas model.

(1) The correction in consideration of the volume occupied by the molecules: When confined in a limited space, since the gas molecules take up a certain amount of volume themselves, the room for their free action or movement is reduced. Therefore, the real gas molecules strike the walls of the container more frequently, leading to an increase in the pressure exerted on the container walls, from $P = \frac{RT}{V_m}$ to $P = \frac{RT}{V_m - b}$. In the formula, the parameter b is associated with the molecules themselves, and its value is four times of the molecular volume.

(2) The correction in consideration of the interaction between the gas molecules: Within the gases, each molecule has to expose itself to the attraction forces from the other molecules around it, and an equilibrium state will be reached because of the counteraction of their interactions. However, the molecules nearby the container walls, also called outer-layer molecules, are only acted upon by the forces from the inner part of the container, resulting in fewer strikes and lower pressure on the walls. This decrease is in direct proportion with the number of the molecules that strike on the container walls in every unit interval of time and also in direction proportion with the number of the other molecules that attract the outer-layer ones. Thus, the decrease of pressure is in direct proportion with the square of the gas density. The term $\frac{a}{V_m^2}$ is used to represent it.

From the knowledge mentioned above, we can get the corrected equation of state for real gases, the VDW equation:

$$P = \frac{RT}{V_m - b} - \frac{a}{V_m^2} \tag{2.19}$$

where

a constant, MPa m^3/(kmol K);
b constant, m^3/kmol;
P the pressure, MPa;
T the temperature, K;
V_m the specific volume (the volume of every 1000 mol gas), m^3/kmol;
R universal gas constant, $R = 0.008134$ MPa m^3/(kmol K).

According to the index of the specific volume, the equation above can be written in another form with descending indexes:

$$PV_m^3 - (bP + RT)V_m^2 + aV_m - ab = 0 \tag{2.20}$$

This is the VDW equation in cubic form, and when the temperature is fixed, V_m has three roots in the coordinate system of V_m–P.

2. RK (Redlich–Kwong) State Equation

Soon after the appearance of the van der Waals equation, other researchers began attempts to improve it. The RK equation is one of the improved:

$$P = \frac{RT}{V - b} - \frac{a}{V(V + b)T^{0.5}} \tag{2.21}$$

Among the parameters: $a = \frac{C_a R^2 T_c^{2..5}}{P_c}$ $b = \frac{C_b RT_c}{P_c}$

$$C_a = 0.42747; \quad C_b = 0.08664;$$

where

T_c the critical temperature, K;
T temperature, K;
P_c the critical pressure, MPa;
P pressure, MPa;
V mole volume, m³/kmol;
R universal gas constant, $R = 0.008134$ MPa m³/(kmol K).

In Eq. (2.21), which is comparatively simple, only two empirical constants are involved in. But it is able to offer reasonable accuracy for ordinary gases.

3. PR(Peng–Robinson) State Equation

In 1975, Peng and Robison proposed another state equation in corrected form.

$$P = \frac{RT}{V_m - b} - \frac{a(T)}{V_m(V_m + b) + b(V - b)} \tag{2.22}$$

Upon critical conditions:

$$a(T_c) = 0.45724R^2 T_c^2 / P_c;$$
$$b(T_c) = 0.07780RT_c / P_c;$$

At other temperatures:

$$a(T) = a(T_c)\beta(T_r, \omega);$$
$$b(T) = b(T_c);$$
$$\beta(T_r, \omega) = [1 + m(1 - T_c^{0.5})]^2;$$
$$m = 0.37464 + 1.542260 - 0.26992\omega^2;$$

where

ω eccentric factor, refer to Table 2.1 and Table 2.2;
P_c critical pressure, Pa;
T_c critical temperature, K;

Equation (2.22) can be written in a cubic form expressed by the compressibility factor:

$$Z^3 - (1 - B)Z^2 + (A - 3B^2 - 2B)Z - (AB - B^2 - B^3) = 0 \qquad (2.23)$$

$$A = \frac{aP}{R^2T^2} \quad B = \frac{bP}{RT}$$

4. SRK State Equation

To make use of the state equations in the numerical simulation of gas-condensate reservoirs, Soave made a correction upon the PK state equation and suggested that $1/T^{0.5}$ be replaced with a temperature-dependent term α, regarding the term of attraction force as a function of the eccentric factor and temperature.

$$P = \frac{RT}{V_m - b} - \frac{a\alpha}{V_m(V_m + b)} \qquad (2.24)$$

To set that:

$$A = aP/(R^2T^2)$$

$$B = bP/(RT)$$

$$V_m = ZRT/P$$

Then, the Eq. (2.24) can be transformed into a cubic equation expressed by the compressibility factor:

$$Z^3 - Z^2 + Z(A - B - B^2) - AB = 0 \qquad (2.25)$$

where
  For pure components:

$$A = \frac{C_a \alpha_i P T_{ci}^2}{P_{ci} T^2} \tag{2.26}$$

$$B = \frac{C_b P T_{ci}}{P_{ci} T} \tag{2.27}$$

For mixed gases:

$$A_m = C_a \frac{P}{T^2} \left(\sum y_i \frac{T_{ci} \alpha^{0.5}}{P_{ci}^{0.5}} \right)^2 \tag{2.28}$$

$$B_m = C_b \frac{P}{T} \sum y_i \frac{T_{ci}}{P_{ci}} \tag{2.29}$$

$$\alpha_i = \left[1 + m_i \left(1 - T_{ri}^{0.5} \right) \right]^2 \tag{2.30}$$

$$m_i = 0.480 + 1.574 \omega_i - 0.176 \omega_i^2 \tag{2.31}$$

$$b = \sum_i^c x_i b_i \tag{2.32}$$

$$b_i = 0.08664 \frac{R T_{ci}}{P_{ci}} \tag{2.33}$$

$$a = \sum_i^c \sum_j^c x_i x_j \left(a_i a_j \right)^{0.5} \left(1 - \bar{K}_{ij} \right) \tag{2.34}$$

$$a_i = a_{ci} \alpha_i \tag{2.35}$$

$$a_{ci} = 0.42748 \frac{(R T_{ci})^2}{P_{ci}} \tag{2.36}$$

where

ω_i the eccentric factor of the ith component, refer to Table 2.1 and Table 2.2;
T_{ri} the reduced temperature of the ith component;
$\bar{K}_{ij}$ a coefficient for the binary interactive effect;
i, j, c represent the ith and jth component and the number of components, respectively .

The equations of state have various forms, and each one has their own rules. Owing to the limitation of space, not all these equations can be introduced here, so it is better for the interested readers to get some reference books for more details.

[Example 2.7] Calculate the pressure exerted by the CO (3.7 kg, 215 K) which is confined in a vessel of 0.03 m^3, respectively, using (1) the equation of state for real gas and (2) the van der Waals equation. It is already known that the pressure in actual measurement is 6.998 MPa.

Solution: Refer to related charts and get molecular weight of CO which is $\mu = 28.01$ kg/kmol, and thence, the mole number of it is $n = \frac{3.7}{28.01} = 0.13209$ (kmol).

The volume every 1000 mol of CO is $v_m = 0.030/0.13209 = 0.227$ (m^3/kmol).

(1) Using the equation of state for real gas:

$$P = \frac{RT}{V_m} = \frac{0.008311 \times 215}{0.227} = 7.874 \,(\text{MPa})$$

The error is: $\frac{7.874-6.998}{6.998} \times 100\% = 12.52\%$

(2) Using the van der Waals equation:

Refer to related charts and get the van der Waals constants:

$$a = 0.14739 \,\text{Pa}\,\text{m}^6/\text{mol}^2 = 0.14739 \,\text{MPa}\,\text{m}^6/\text{kmol}^2$$

$$b = 0.03949 \times 10^{-3} \,\text{m}^3/\text{mol} = 0.03949 \,\text{m}^3/\text{kmol}$$

Substitute a and b into the Eq. (2.19):

$$P = \frac{R_m T}{V_m - b} - \frac{a}{V_m^2} = \frac{0.008314 \times 215}{0.227 - 0.03949} - \frac{0.14739}{0.227^2} = 6.672 \,(\text{MPa})$$

The error is: $\frac{6.672-6.998}{6.998} \times 100\% = -4.65\%$

Comparing the two results, in this example, the result obtained through the van der Waals equation is more close to the real situation.

2.2.5 The Calculation of Z-Factor Using State Equation Correlations

Apart from the methods based on preprepared charts as previously mentioned, another way used to calculate the Z-factor is a computational method, which can be further divided into two branch ways. The first one is called "fitting method," using the algebraic expressions to fit the Z-factor charts. Because the Z-curves are non-monotonic ones, piecewise fitting is required. The second method is to calculate the compressibility factor through state equations (e.g., the SRK equation and the PR equation). The improvements in computer technology have also promoted the development in the analytical computation of the Z-factor. Two of the methods that are used to make a straightforward calculation of Z-factor will be presented here.

1. The calculation formula for Z-factor at low pressure ($p < 35$ MPa)

$$Z = 1 + \left(0.31506 - \frac{1.0467}{T_{pr}} - \frac{0.5783}{T_{pr}^3}\right)\rho_{pr} + \left(0.5353 - \frac{0.6123}{T_{pr}}\right)\rho_{pr}^2 + 0.6815\frac{\rho_{pr}^2}{T_{pr}^3}$$

(2.37)

$$\rho_{pr} = 0.27P_{pr}/(ZT_{pr})$$

(2.38)

where

ρ_{pr} the dimensionless reduced density;
the others   the same as mentioned before.

When the p and T conditions are already given, only by the trial calculation approach can the value of Z-factor be obtained. Usually, an alternation approach is used on computer to obtain the result, and the computation flow process is listed here:

(1) With the given conditions, calculate the T_{pr}, P_{p}, or T'_{pr}, P'_{pr};
(2) With the given p and T, calculate the T_{pr} and P_{pr};
(3) For initialization, set $Z^{(0)} = 1$, and resort to Eq. (2.38) to get ρ_{pr};
(4) Substitute the value of ρ_{pr} into Eq. (2.37) and get $Z^{(1)}$;
(5) Go back to the step (3) using $Z^{(1)}$. By controlling the cycle number in the program execution, output the result after 5 iterations. The accuracy is already fine enough to meet engineering need.

2. The DPR Method

In 1974, Dranchuk, Purvis, and Robinson used the BWR state equation to fit the Z-factor charts and proposed the equation below:

$$Z = 1 + \left(A_1 + A_2/T_{pr} + A_3/T_{pr}^3\right)\rho_{pr} + \left(A_4 + A_5/T_{pr}\right)\rho_{pr}^2 + \left(A_5 A_6 \rho_{pr}^5\right)/T_{pr} + \left(A_7 \rho_{pr}^2/T_{pr}^3\right)\left(1 + A_8 \rho_{pr}^2\right)\exp\left(-A_8 \rho_{pr}^2\right)$$

(2.39)

$$\rho_{pr} = 0.27 P_{pr}/(Z T_{pr}) \tag{2.40}$$

In the equation, A_1–A_8 are all coefficients:

$$
\begin{aligned}
A_1 &= 0.31506237; & A_2 &= -1.0467099; \\
A_3 &= -0.57832729; & A_4 &= 0.53530771; \\
A_5 &= -0.61232032; & A_6 &= -0.10488813; \\
A_7 &= 0.68157001; & A_8 &= 0.68446549;
\end{aligned}
$$

When the p and T conditions are already given, the simultaneous equation of the two can be used to obtain ρ_{pr}. Then, substitute the ρ_{pr} back to one of the two equations to get the value of the Z-factor.

The new function F is constructed by the two equations:

$$
\begin{aligned}
F(\rho_{pr}) &= \rho_{pr} - 0.27 P_{pr}/T_{pr} + \left(A_1 + A_2/T_{pr} + A_3/T_{pr}^3\right)\rho_{pr}^2 \\
&\quad + \left(A_4 + A_5/T_{pr}\right)\rho_{pr}^3 + \left(A_5 A_6 \rho_{pr}^6\right)/T_{pr} + \left(A_7 \rho_{pr}^3/T_{pr}^3\right)\left(1 + A_8 \rho_{pr}^2\right)\exp\left(-A_8 \rho_{pr}^2\right) \\
&= 0
\end{aligned}
\tag{2.41}
$$

Equation (2.41) is a nonlinear equation, and the value of ρ_{pr} can be obtained through the Newton's alternation method. The computation steps are listed hereunder:

(1) Initialization: $\rho_{pr}^{(0)} = 0.27 P_{pr}/T_{pr}$;
(2) Calculate $F(\rho_{pr})$ using Eq. (2.41);
(3) Calculate $F'(\rho_{pr})$: figure out the differential coefficient of Eq. (2.41);

$$
\begin{aligned}
F'(\rho_{pr}) &= 1 + \left(A_1 + A_2/T_{pr} + A_3/T_{pr}^3\right)(2\rho_{pr}) + \left(A_4 + A_5/T_{pr}\right)\left(3\rho_{pr}^2\right) \\
&\quad + \left(A_5 A_6/T_{pr}\right)/\left(6P_{pr}^5\right) + \left(A_7/T_{pr}^3\right)\left[3\rho_{pr}^2 + A_8\left(3\rho_{pr}^4\right) - A_8^2\left(2\rho_{pr}^6\right)\right]e^{-A_8\rho_{pr}^2}
\end{aligned}
\tag{2.42}
$$

(4) Obtain the value of ρ_{pr} through the Newton's alternation method;

$$\rho_{pr}^{(k+1)} = \rho_{pr}^{(k)} - \frac{F(\rho_{pr})}{F'(\rho_{pr})} \tag{2.43}$$

(5) Estimate the accuracy, and when $\left|\rho_{pr}^{(k+1)} - \rho_{pr}^{(k)}\right| < 0.0001$, the accuracy is fine enough; then, calculate the value of Z through Eq. (2.39);

(6) Go back to the step (2) and continue with the alternation until $F\left(\rho_{\mathrm{pr}}\right) \approx 0$;
(7) Substitute the ρ_{pr} which is already accurate enough for engineering requirement back to Eqs. (2.39) or (2.40) and get the value of the Z-factor.

2.3 Physical Properties of Natural Gas Under High Pressure

2.3.1 Natural Gas Formation Volume Factor (FVF)

In reservoir engineering calculations, sometimes it is necessary to convert the gas volume upon reservoir conditions (high temperature and high pressure) to standard gas volume, which also called surface volume. Hereupon, the concept of the volume factor of natural gas can be introduced.

The natural gas formation volume factor is defined as the ratio of the volume of a certain amount of natural gas under reservoir conditions to the volume of it under surface standard conditions,that is the volume of gas under reservoir conditions required to produce one unit volume of oil or gas at the surface:

$$B_{\mathrm{g}} = \frac{V}{V_{\mathrm{sc}}} \qquad (2.44)$$

where

B_{g} the volume factor of natural gas, $\mathrm{m}^3/\mathrm{m}^3$;
V_{sc} the volume of a certain amount of natural gas under reservoir conditions, m^3;
V the volume of a certain amount of natural gas under surface standard conditions, m^3.

Under surface standard conditions, the natural gas is approximately regarded as ideal gas; that is to say, it is compressibility is $Z = 1$. Then, the volume of gas can be calculated according to the state equation:

$$V_{\mathrm{sc}} = \frac{nRT_{\mathrm{sc}}}{P_{\mathrm{sc}}} \qquad (2.45)$$

In the equation, P_{sc}, V_{sc}, and T_{sc}, respectively, represent the pressure, volume, and temperature under standard conditions.

At reservoir conditions with a pressure of p and a temperature of T, the volume occupied by the same amount of natural gas can also be obtained through the compressibility equation of state:

Table 2.9 Definitions of standard state in different countries and organizations

Country or organization		Temperature and pressure set as standards	Volume of 1 mol ideal gas
Standard state		0 °C, 760 mmHg (0.101325 MPa)	22.4 L
China	Petroleum industry	20 °C (293 K), 0.101325 MPa	24.04 L
Russian	Petroleum industry	20 °C (293 K), 0.101325 MPa	24.04 L
USA	Petroleum industry	20.0 °C (68°F), 0.101325 MPa	24.04 L
USA	High-pressure gas association	20.0 °C (68°F), 1 atm	24.04 L
USA	Gas association	60°F (15.6 °C), 30 in Hg (760 mmHg)	23.68 L
Britain		60°F (15.6 °C), 30 in Hg (760 mmHg)	23.68 L

$$V = \frac{ZnRT}{P} \tag{2.46}$$

Substitute Eqs. (2.45) and (2.46) into Eq. (2.44):

$$B_g = \frac{V}{V_{sc}} = \frac{ZTP_{sc}}{T_{sc}P} = \frac{273 + t}{293}\frac{P_{sc}Z}{P} \tag{2.47}$$

In the equation, the unit for B_g is m^3/(standard) m^3; t is the reservoir temperature, °C.

In fact, the natural gas volume factor B_g describes the conversion coefficient of the gas volumetric change from the reservoir state to the land-surface state.

Table 2.9 shows the details about the standard state.

For real reservoirs, their pressure is much higher than that on the land surface (about dozens or hundreds of times), while the reservoir temperature is about several times of the temperature on the surface. So, when the underground gases are subjected to surface conditions, their volume is ten times greater, resulting in a B_g far less than 1. The definition of the swelling coefficient is the ratio of the surface gas volume to the underground gas volume, symbolically $E_g = 1/B_g$, and it is the swelling coefficient of the gas from underground to the surface.

Along with the development of the reservoirs, the gases are continuously depleted out and the reservoir pressure decreases in companion, while the reservoir temperature is comparatively stable and can be seen as a constant. Therefore, in such a situation, it is reasonable to regard the B_g as a function only of the reservoir pressure. That is, $B_g = CZ/P$ (C is a coefficient).

An example curve about the P–B_g relationship is shown in Fig. 2.7.

Fig. 2.7 The P–B_g relationship of a studied natural gas

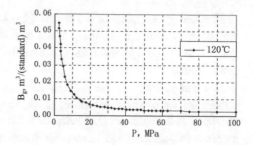

2.3.2 Isothermal Compressibility of Natural Gas

In reservoir engineering calculations, especially when considering the elastic reserve of the reservoirs, it is necessary to calculate the volumetric change rate of gas along with the change of pressure. Symbolically:

$$C_g = -\frac{1}{V}\left(\frac{\partial V}{\partial P}\right)_T \tag{2.48}$$

In companion with the increase of pressure, the volume of gas drops, and hence, the $\left(\frac{\partial V}{\partial P}\right)_T$ is negative. To ensure C_g to have a positive value, a minus should be added before the right formula of Eq. (2.48).

By equation of state for real gas, the volumetric change rate of gas in Eq. (2.48) can be obtained.

$$V = nRT\frac{Z}{P} \tag{2.49}$$

To calculate the differential of it:

$$\left(\frac{\partial V}{\partial P}\right)_T = nRT\frac{P\frac{\partial Z}{\partial P} - Z}{P^2} \tag{2.50}$$

Substitute Eqs. (2.50) and (2.49) into Eq. (2.48), and get the expression of C_g:

$$C_g = -\frac{1}{V}\left(\frac{\partial V}{\partial P}\right)_T = -\frac{P}{ZnRT}\left[\frac{nRT}{P^2}\left(P\frac{\partial Z}{\partial P} - Z\right)\right]$$
$$= \frac{1}{P} - \frac{1}{Z}\frac{\partial Z}{\partial P} \tag{2.51}$$

The $\frac{\partial Z}{\partial P}$ in Eq. (2.51) can be determined by the Z-charts: Resort to the p–Z curve at the corresponding temperature, and find the point in accordance with the given Z; then, extract value of the Z-factor and figure out the slope of the tangent line at this point,$\Delta Z/\Delta P$; finally, substitute them into the equation presented above and get the value of C_g.

For real gases, since $Z = 1.0$ and $\frac{\partial Z}{\partial P} = 0$, we can know that $C_g = \frac{1}{P}$; in other words, the value of C_g is proportional to the reciprocal of pressure. For real gases, since $Z \neq 1$, the $\frac{1}{Z}\frac{\partial Z}{\partial P}$ must have a certain value which sometimes can be quite considerable.

We can know from the Z-charts that the value of $\frac{\partial Z}{\partial P}$ is not uniform when the conditions of pressure are different. It can be either positive or negative. For example, at low pressure, the value of the compressibility factor decreases with the increase in pressure, and hence, the value of $\frac{\partial Z}{\partial P}$ is negative. As a result, the corresponding C_g is greater than that of the ideal gas. On the contrary, under high pressure, the value of the compressibility factor increases with the increase in pressure, and hence, the value of $\frac{\partial Z}{\partial P}$ is positive. As a result, the corresponding C_g is smaller than that of the ideal gas.

[Example 2.8] Calculate the isothermal compressibility C_g of the natural gas at 20 °C and 6.8 MPa.

Solution:

(1) Refer to the P–Z-charts of methane (Fig. 2.1) and get $Z = 0.89$;
(2) Calculate the slope at this point: $\frac{\partial Z}{\partial P} = -0.01551$ MPa^{-1};
(3) Calculate the value of C_g through Eq. (2.51):

$$C_g = \frac{1}{P} - \frac{1}{Z}\frac{\partial Z}{\partial P} = \frac{1}{6.8} - \left(\frac{1}{0.89}\right)(-0.0155)$$
$$= 1.645 \times 10^{-4}\,\text{MPa}^{-1}$$

It is not convenient to calculate by Eq. (2.51) for multicomponent natural gases. With the reference of principle of corresponding states, therefore, Eq. (2.51) can be converted into another form, i.e., to change the pressure p to the pseudo-reduced pressure P_{pr}. The details of the derivation are presented here:

From the relational equation, $P_{pr} = P/P_p$, we can get that:

$$P = P_{pc} \cdot P_{pr} \tag{2.52}$$

where

P_{pc}, P_{pr} the pseudo-critical pressure and the pseudo-reduced pressure of natural gas, respectively.

Because that:

$$\frac{\partial Z}{\partial P} = \frac{\partial Z}{\partial P_{pr}} \cdot \frac{\partial P_{pr}}{\partial P} \tag{2.53}$$

From Eq. (2.52), we can get that:

$$\frac{\partial P_{pr}}{\partial P} = \frac{1}{P_{pc}}$$

(2.54)

Substitute Eq. (2.54) into the Eq. (2.53):

$$\frac{\partial Z}{\partial P} = \frac{1}{P_{pc}} \cdot \frac{\partial Z}{\partial P_{pr}}$$

(2.55)

Substitute Eqs. (2.52) and (2.55) into Eq. (2.51):

$$C_g = \frac{1}{P} - \frac{1}{Z}\frac{\partial Z}{\partial P} = \frac{1}{P_{pc} \cdot P_{pr}} - \frac{1}{Z P_{pc}}\frac{\partial Z}{\partial P_{pr}}$$

$$= \frac{1}{P_{pc}}\left(\frac{1}{P_{pr}} - \frac{1}{Z}\frac{\partial Z}{\partial P_{pr}}\right)$$

(2.56)

Equation (2.56) is commonly used to calculate the compressibility factor C_g of natural gas. Given the composition of a certain gas, the pseudo-critical parameter (P_{pc}) and the pseudo-reduced parameter (P_{pr}) can be figured out. Then, make use of the Z-charts (Fig. 2.3) and get the value of Z, as well as the $\frac{\partial Z}{\partial P_{pr}}$. Finally, refer to Eq. (2.56) and calculate the value of C_g.

Sometimes, people prefer to cancel out the dimension of C_g and define the pseudo-reduced compressibility factor of natural gas as follows:

$$C_{pr} = C_g \times P_{pc}$$

(2.57)

where

C_{pr} the pseudo-reduced compressibility factor of natural gas, dimensionless;
C_g the compressibility factor of natural gas, MPa^{-1}.

Substitute Eq. (2.56) into Eq. (2.57):

$$C_{pr} = P_{pc} \cdot \left[\frac{1}{P_{pc}}\left(\frac{1}{P_{pr}} - \frac{1}{Z}\frac{\partial Z}{\partial P_{pr}}\right)\right]$$

$$= \frac{1}{P_{pr}} - \frac{1}{Z}\frac{\partial Z}{\partial P_{pr}}$$

(2.58)

With the known values of the pseudo-reduced parameters (P_{pr}, T_{pr}), the pseudo-reduced compressibility factor of natural gas, C_{pr}, can be obtained by referring to Fig. 2.8.

[Example 2.9] Using the gas composition given in Example 2.3, calculate the isothermal compressibility C_g of the natural gas at 49 °C and 10.2 MPa (102 atm).

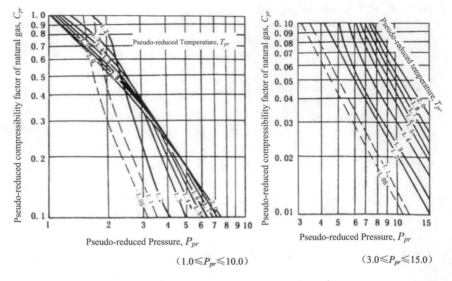

Fig. 2.8 C_{pr}–T_{pr}, P_{pr} charts (Mattar 1975) $(1.0 \leq P_{pr} \leq 10.0)$ $(3.0 \leq P_{pr} \leq 15.0)$

Solution:

(1) Calculate pseudo-critical parameters and the pseudo-reduced parameters (already known from the Example 2.3):

$$P_{pc} = 4.54 \, \text{MPa}; \quad T_{pc} = 212.8 \, \text{K};$$
$$P_{pr} = 2.25; \quad T_{pr} = 1.51.$$

(2) Refer to the Z-charts (Fig. 2.3), extract $Z = 0.813$, and figure out $\frac{\partial Z}{\partial P} = -0.053$.

(3) Calculate the value of C_g through Eq. (2.56):

$$
\begin{aligned}
C_g &= \frac{1}{P_{pc}} \left(\frac{1}{P_{pr}} - \frac{1}{Z} \frac{\partial Z}{\partial P_{pr}} \right) \\
&= \frac{1}{4.54} \left[\frac{1}{2.25} - \frac{1}{0.813} (-0.053) \right] \\
&= \frac{1}{4.54} \times 0.5096 = 1122 \times 10^{-4} \, \text{MPa}^{-1}
\end{aligned}
$$

[Example 2.10] Using the gas composition given in Example 2.3, calculate the isothermal compressibility C_g of the natural gas at 49 °C and 10.2 MPa (102 atm) by referring to Fig. 2.8.

Solution:

(1) Calculate the values of:

$$P_{pc} = 4.54; \quad T_{pc} = 212.9K;$$
$$P_{pr} = 2.25; \quad T_{pr} = 1.51.$$

(2) Refer to Fig. 2.8 and determine that when $P_{pr} = 2.25$ and $T_{pr} = 1.51$, $C_{pr}T_{pr} = 0.77$ and $C_{pr} = (C_{pr}T_{pr})/T_{pr} = 0.77/1.51 = 0.51$;
(3) Calculate the value of C_g through Eq. (2.57):

$$C_g = \frac{C_{pr}}{P_{pc}} = \frac{0.51}{4.54} = 1123 \times 10^{-4}\,\text{MPa}^{-1}$$

2.3.3 Viscosity of Natural Gas

The natural gas is a viscous fluid. Its viscosity is much lower than that of water and oil; hence, the unit used for it is μP.

The viscosity of gas depends on its composition and the temperature and pressure conditions to which it is subjected. The changing regularities of it are not uniform at high- and low-pressure conditions. The partition between high and low pressure is shown in Fig. 2.9.

1. Viscosity of Natural Gas at Low Pressure

When the natural gas is put under pressure condition as low as atmospheric pressure, its viscosity almost has nothing to do with the pressure, but the viscosity increase with increase in temperature.

Fig. 2.9 The division of high- and low-pressure regions in gas viscosity calculation

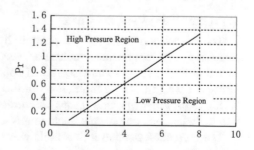

According the molecular dynamics of gas:

$$\mu = \frac{1}{3}\rho\bar{\upsilon}\bar{\lambda} \tag{2.59}$$

where

μ the viscosity of gas, g/cm s;
ρ the density of gas, g/cm^3;
$\bar{\upsilon}$ the average velocity of gas molecules, cm/s;
$\bar{\lambda}$ the average free-travel length of gas molecules, cm.

Equation (2.59) indicates that the viscosity of gas has a connection with the quantities of ρ, $\bar{\upsilon}$, and $\bar{\lambda}$. Among the three parameters, the velocity of gas molecules $\bar{\upsilon}$ has nothing to do with pressure; the density ρ is proportional to pressure; and the free-travel length $\bar{\lambda}$ is inversely proportional to pressure. On the whole, we can conceive that the product of the three quantities is indispensable of pressure. What needs to be stressed is that this conclusion is feasible only for low-pressure conditions that are near the atmospheric pressure.

As the temperature is increased, the kinetic energy of molecules increases in companion with accelerated movement of the molecules and more collisions occur between each other, and thus, the internal friction of the fluid is raised, resulting in greater viscosity.

Figure 2.10 shows the viscosity of single-component natural gases at atmospheric pressure. From the chart, it is obvious that for hydrocarbon gases, an increase in temperature causes an increase in viscosity, and the viscosity decreases with the increase in molecular weight.

2. Viscosity of Natural Gas under High Pressure

The viscosity of natural gas under high pressure tends to increase with the increase in pressure, decrease with the increase in temperature, and increase with the increase in molecular weight. That is to say, the gases under high pressure possess characteristics that typically belong to liquid fluids.

3. Methods for Determining the Viscosity of Natural Gases

(1) Carr (1954) Viscosity Chart

The viscosity retaliations of paraffin hydrocarbons at atmosphere pressure, correlated with temperature, molecular weight, and gas specific gravity, are presented in Fig. 2.11: (1) The viscosity of natural gas decreases with the increase in molecular weight; (2) the inset charts in the figure, at the top right corner and the bottom left corner, provide means of correcting the viscosity for the presence of non-hydrocarbon components, including N_2, CO_2, and H_2S.

The correction method follows this way: Firstly, refer to the corner charts and extract the corresponding value of the adjustment factor in accordance with the mole fraction shared by the non-hydrocarbon components; secondly, add the

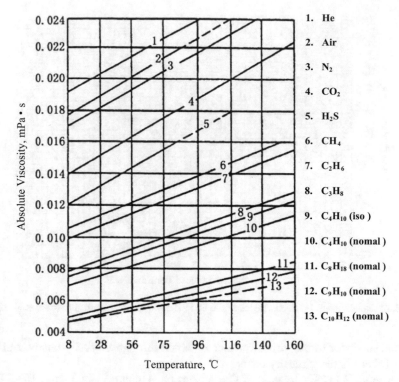

Fig. 2.10   The viscosity–temperature relation curve at atmospheric pressure (Carr 1954)

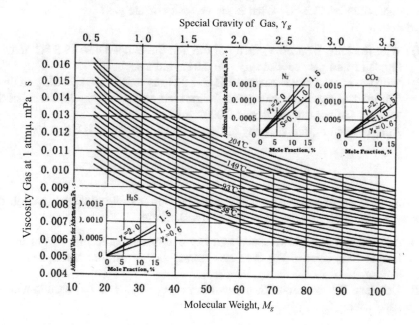

Fig. 2.11   Viscosity of paraffin hydrocarbon gases at 1 atm (Carr et al. 1954)

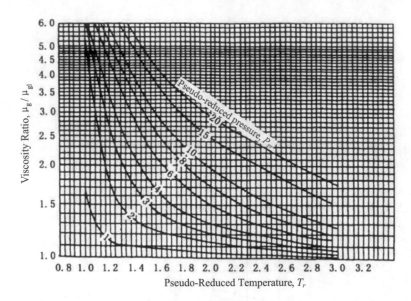

Fig. 2.12 Relationship between viscosity ratio and pseudo-reduced parameters (Carr et al. 1954)

adjustment factor to the viscosity value obtained through viscosity charts, and the sum of them is the viscosity result.

[Example 2.11] Calculation of Gas Viscosity. A natural gas having 10 % H_2S (mole fraction) and a special gas gravity of 0.6 exist at the pressure of 0.1 MPa and the temperature of 43.3 °C. What is the viscosity of the gas?

Solution:

(1) With the known specific gravity of 0.6 and the temperature 43.3 °C, refer to Fig. 2.11 and get the viscosity $\mu = 0.0113$ mPa s.
(2) With the known mole fraction of H_2S, 10 %, as well as the specific gravity of 0.6, refer to the inset chart at the bottom left corner of Fig. 2.11 and get the value of the adjustment factor: 0.0002 mPa s.
(3) After correction, the viscosity of the given natural gas at 0.1 MPa is:

$$\mu_g = 0.0113 + 0.0002 = 0.0115 \, \text{mPa s}$$

The viscosity of natural gas under high pressure can be obtained through the charts in Fig. 2.12, and the steps to use this chart are:

(1) According to the known specific gravity or apparent molecular weight and the temperature, refer to Fig. 2.11 and extract the viscosity at atmospheric pressure, μ_{g1};
(2) Calculate the pseudo-critical parameters (P_{pc}, T_{pc}) and the pseudo-reduced parameters (P_{pr}, T_{pr});

(3)   Refer to Fig. 2.12 and get the ratio of the viscosity values at high and low pressure, symbolically $\left(\frac{\mu_g}{\mu_{g1}}\right)$.

(4)   Finally, figure out the natural gas viscosity at the designated temperature and pressure:

$$\mu_g = \left(\frac{\mu_g}{\mu_{g1}}\right) \times \mu_{g1} \tag{2.59}$$

2)   Method Based on the Composition of Natural Gas

Besides the charts, some computational methods have been developed to obtain the viscosity of natural gas. For example, when the composition of a gas is already known, its viscosity at 1 atm and varied temperatures can be obtained through the equation below:

$$\mu_{g1} = \frac{\sum y_i \mu_{gi} M_i^{0.5}}{\sum y-i M_i^{0.5}} \tag{2.60}$$

where

μ_{gi} the viscosity of the ith component (refer to Fig. 2.10 for its value);
M_i the molecular weight of the ith component;

[Example 2.12] Calculation of Gas Viscosity. Determine the viscosity of the gas described in Table 2.10. The designated condition is 93.3 °C and 0.1 MPa.
Solution:

(1)   Refer to Fig. 2.10 to get the values of viscosity at 93.3 °C for each component present in the gas, and list them in Table 2.10.
(2)   According to the Eq. (2.60), calculate the natural gas viscosity at 93.3 °C and 0.1 MPa:

Table 2.10 Calculation of the viscosity of natural gas

Composition	y_i	M_{gi}	μ_{gi}, cP	$M_i^{0.5}$	$y_i M_i^{0.5}$	$\mu_{gi} y_i M_i^{0.5}$
CH_4	0.85	16.0	0.0130	4.0	3.40	0.0442
C_2H_6	0.09	30.1	0.0112	5.48	0.493	0.0055
C_3H_8	0.04	44.1	0.0098	6.64	0.266	0.0026
n-C_4H_{10}	0.02	58.1	0.009 1	7.62	0.152	0.0014
Sum					4.311	0.0537

$$\mu_g = \frac{\sum y_i \mu_{gi} M_i^{0.5}}{\sum y_i M_i^{0.5}} = \frac{0.0537}{4.311} = 0.0125 (\text{mPa s})$$

3) Method of Empirical Formula—Residual Viscosity Method

When it is necessary to determine the gas viscosity under different pressures, many empirical formulas are applicable. The residual viscosity method is one of them:

$$(\mu - \mu_0)\eta = 1.08[\exp(1.439\rho_{pr}) - \exp(-1.111\rho_{pr}^{1.858})] \qquad (2.61)$$

$$\eta = T_{pc}^{1/6} / M^{1/2} P_{pc}^{2/3}; \ \rho_{pr} = \rho/\rho_{pc}$$

$$\rho_{PC} = P_{pc} / (ZRT_{pc}); \quad M = \sum y_i M_i$$

$$Z = \sum y_i Z_{ci}$$

where

μ the viscosity of natural gas under high pressure, 10^2 mPa s;
μ_o the viscosity of natural gas at low pressure, 10^2 mPa s;
η coefficient;
ρ_{pr} the pseudo-reduced density of natural gas;
ρ_{pc} the pseudo-critical density of natural gas;
M the molecular weight of natural gas;
Z compressibility factor.

4) Method of Empirical Formula—Lee–Gonzalez Semiempirical Method (1966)

On the analysis of the eight oil samples from four oil companies, a semiempirical equation can be derived from the laboratory measurements taken by Lee, Gonzalez, and Eakin:

$$\mu_g = 10^{-4} K \exp\left(X\rho_g^y\right) \qquad (2.62)$$

Among the parameters:

$$K = \frac{2.6832 \times 10^{-2}(470 + M)T^{1.5}}{116.1111 + 10.5556M + T} \qquad (2.63)$$

$$X = 0.01009\left(350 + \frac{54777.7}{T} + M\right) \qquad (2.64)$$

$$Y = 2.447 - 0.2224X \tag{2.65}$$

where

μ_g the viscosity of natural gas under the given condition of p and T, mPa s;
T the temperature to which the studied natural gas is subjected, K;
M the apparent molecular weight of natural gas, g/mol;
ρ_g the density of natural gas, g/cm^3.

Actually, the influence of pressure on natural gas is implicit in the computing formula for the gas density ρ_g.

This method is exclusive of the correction for non-hydrocarbon gases, and when it is applied to pure-hydrocarbon gases for viscosity determination, the standard calculation error is ± 3 %, and the maximum error allowed is 10 %. This degree of accuracy is good enough for the calculations in gas reservoir engineering.

2.4 Natural Gas with Water Vapor and the Gas Hydrate

2.4.1 Water Vapor Content in Natural Gas

Reservoir pressures are well in excess of the saturation pressure of water at prevailing reservoir temperatures. Therefore, due to longtime connection with underground gases, a part of the reservoir water tends to be volatilized into the gas, in which it exists as water vapor, and at the same time, the gases can be dissolved in the reservoir water, too. The term "natural gas with water vapor" refers to the natural gases containing water vapor. The water vapor content (use "water content" for short) depends on the factors written below:

(1) Pressure and temperature: The water content decreases with the increase in pressure and increases with the increase in temperature.
(2) The salinity of reservoir water: The water content decreases with the increase in salinity;
(3) The specific gravity of natural gas: The greater the specific gravity of natural gas, the less the water vapor contained in it.

Generally speaking, absolute humidity and relative humidity are used to describe water content in natural gas.

1. Absolute Humidity (Absolute Content)

The absolute humidity refers to the mass of the water vapor contained in 1 m^3 wet natural gas. The relational expression is:

$$X = \frac{W}{V} = \frac{P_{vw}}{R_w T} \tag{2.66}$$

where

X absolute humidity, kg/m^3;
W the mass of water vapor, kg;
V the volume of the wet natural gas, m^3;
P_{sw} the partial pressure of water vapor, Pa;
T the absolute temperature of wet natural gas, K;
R_w the gas constant of water vapor, $R_w = 461.53$ kg m^3/(kg K).

If the partial pressure of water vapor reaches as high as the saturation vapor pressure, the saturation absolute humidity can be written like this:

$$X_s = \frac{P_{sw}}{R_w T} \tag{2.67}$$

where

X_s the saturation absolute humidity, kg/m^3;
P_{sw} the saturation vapor pressure of water vapor, Pa.

The saturation absolute humidity reflects the maximum weight of water that can be hold in natural gas at a specific temperature.

2. Relative Humidity (Relative Content)

The relative humidity is defined as the ratio of the absolute humidity to the saturation absolute humidity, both subjected to the same specific temperature. Symbolically:

$$\phi = \frac{X}{X_s} = \frac{P_{vw}}{P_{sw}} \tag{2.68}$$

For absolute dry natural gases, $P_{vw} = 0$; thence, $\phi = 0$; when the natural gas studied reaches its state of saturation, $P_{vw} = P_{sw}$; thence $\phi = 1$. For ordinary natural gases: $0 < \phi < 1$.

3. Determination of the Water Content in Natural Gas

The method commonly used to calculate the water content in natural gas relies on experimental curves. A brief introduction will be given here. In 1958, Mcketta and Wehe proposed a correlation diagram of water content on the basis of analyzing the testing data (Fig. 2.13). This diagram, also named dew-point diagram, indicates the saturation content of water vapor in natural gases upon different conditions of pressure and temperature. The dashed line in the diagram signifies that the hydrates begin to form. When the pressure condition is below the temperature point at which the hydrate formation can start, the gas reach an equilibrium with the hydrates;

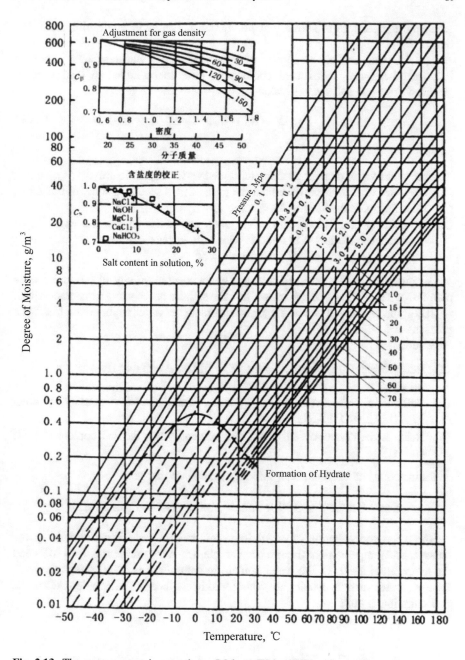

Fig. 2.13 The water content in natural gas (Mcketta-Wehe 1958)

when the temperature is above that point, the gas stays in equilibrium with the liquids.

This diagram can only be applied directly to the gases whose specific gravity is 0.6; for other gases, adjustment coefficients have to be added to their specific gravity values and salt content values:

$$X = X_\text{o} \times C_\text{s} \times C_\beta \tag{2.69}$$

where

X the water vapor content of the natural gases with a specific gravity of r_g, g/m^3;
X_o the water vapor content of the natural gases with a specific gravity of 0.6 g/m^3;
C_β the adjustment coefficient for specific gravity equals X/X_o;
C_S the adjustment coefficient for salt content equals the ratio of the water vapor content when the gas is in contact with salty water to the water vapor content when the gas is in contact with pure water.

[Example 2.13] The given natural gas has a specific gravity of 0.6 and a temperature of 311 K. It is already known that the salinity of the water in contact with it is 30,000 mg/L (about 3 % NaCl); please calculate the water vapor content in it at a pressure of 3.44 MPa.

Solution:

(1) Refer to Fig. 2.13, when $\gamma_\text{g} = 0.6, T = 311\,\text{K}, P = 3.44\,\text{MPa}$:
     $Xo = 1.8 \times 10^{-3}$ kg/(standard) m^3
(2) Refer to the correction curve for specific gravity, when $\gamma_\text{g} = 0.6,\ T = 311\,\text{K}{:}C_\beta = 0.99$
(3) Refer to the correction curve for salt content, when the salt content is 3 %: $C_\text{s} = 0.93$

Substituting the obtained data into Eq. (2.69), we can get:

$$X = 1.8 \times 10^{-3} \times 0.99 \times 0.93 = 1.66 \times 10^{-3}\,\text{kg/(standard)m}^3$$

The water concentration in the underground natural gases is determined by the pressure and temperature of them. For example, when the pressure is 14.0 MPa and the temperature is 60 °C, the water content in natural gas is 1.740 g/(standard)m^3. Besides, water vapor is usually not considered in engineering calculations when the concentration is quite low.

2.4.2 Natural Gas Hydrate

At specific conditions of pressure and temperature, hydrocarbon gas and liquid water combine to form crystal solids. The solids are called "natural gas hydrates" or "hydrates" for short. The hydrates are one of the forms of complexes known as

Fig. 2.14 Diagram of the crystalline clathrate complexes

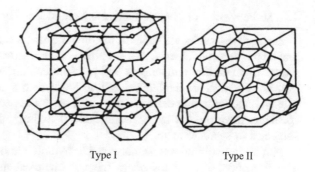

Type I   Type II

crystalline clathrates, evolving water and low molecular weight hydrocarbon or non-hydrocarbon components. The main framework of the hydrate crystal is formed with water molecules, and the hydrocarbon molecules occupy void spaces within the water molecules (see Fig. 2.14). The structure of the water framework complies with the shape and size of the gas molecules that are in contact with it, while the level to which the void spaces are filled depends on the temperature and pressure in the studied system.

A natural gas hydrate is a solid solution of water and natural gas with a "freezing point" which depends on the gas composition, the available water, the pressure, and the temperature. In the solid solution, the solvent is the crystalline framework formed by the water molecules, and the dissolved gas molecules distribute within it. Obviously, that is the reason why the natural gas hydrate is classified into the category of clathrate complexes. Hydrocarbon formation is physical rather than chemical in nature. Apparently, no strong chemical bonds are formed between the molecules: The water molecules form the clathrate crystal lattice by H-bonds between each other, while the hydrocarbon molecules are held within the void spaces in the lattice by van der Waals forces. The gas components that take part in the formation of the hydrates (e.g., N_2, CH_4, C2H4, C_3H_8, i-C_4H_{10}, CO_2, and H_2S) decide the size of the gas hydrate molecules. The mentioned gases, which can form hydrates when conditions allow, all have a molecular diameter less than 6.7 Å. For the gases whose molecular diameter is greater than 6.7 Å, such as the normal butane, hydrates cannot form. In addition, when the size of the molecules is too small (e.g., hydrogen), their van der Waals forces with water molecules are too weak to ensure the stability of the lattice, which make it impossible for them to form hydrates. However, it is also found that when a great amount of methane is present, the hydrogen and normal butane can also join in the formation of natural gas hydrates.

In petroleum engineering, the significance of the research on natural gas hydrates lies in three aspects: (1) The hydrates is a kind of resources, it can exist in underground formations under specific conditions; (2) during natural gas production, hydrates may occur in wellbores and at the surface chokes making great difficulties for gas production; (3) on the surface, the gaseous natural gas may change to hydrates, which improves the efficiency of storage and transportation.

1. The Formation of Hydrates and the Natural Gas Hydrates Resource

Since all reservoirs are believed to contain connate water, it is generally assumed that all mixtures which exist as a gas phase in the reservoirs are saturated with water vapor. Upon certain temperature and pressure conditions, the hydrate starts forming. The hydrates are ice-like crystalline solids with a density of 0.88–0.90 g/cm^3. The gas amount in a hydrate is determined by the gas compositions, and, generally speaking, there are 0.9 m^3 of water and 70–240 m^3 of gas in every 1 m^3 when subjected to standard state.

Not all the void spaces within a stable hydrate are occupied by gas molecules. The formation of hydrates obeys the non-chemical quantitative relation written below:

$$M + nH_2O = M(H_2O)_n \tag{2.70}$$

where

M the gas molecules to form hydrates;
n the ratio of the water molecular number to the gas molecular number, depending on the pressure and temperature in the studied system.

With the exception of methane, nitrogen, and inert gases, the other gas components contained in nature gases all possess their own critical temperature, a threshold above which no hydrates can start forming. Figure 2.15 shows the hydrate—forming conditions for several component gases, and Fig. 2.16 shows the conditions for natural gases. We can know from the figures that only when appropriate pressure and temperature are present can the gas hydrates form.

Fig. 2.15 The forming conditions for pure component gases (Katz 1945)

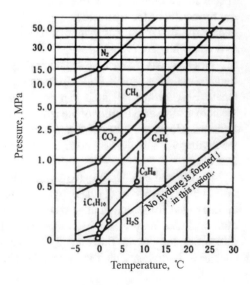

Fig. 2.16 The forming
conditions for natural gases
(Katz 1945)

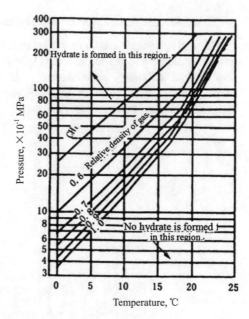

In the continental frozen crust or the sedimentary structures beneath the ocean bed, thermodynamic conditions are ready for the hydrate forming upon reservoir conditions. In the 1920s, the first discovery of the hydrate resource is found in Siberia, Russia. It was also found in Alaska, USA. In China, researches and study on the gas hydrate are in progress, and hydrate resources have been found in the rock formations beneath the bottom of South China Sea.

2. Prevention of Gas Hydrates in Wellbores During Gas Well Production

During the production of natural gases, the temperature and pressure of the mixture are reduced as it flows from the formation to wellbore and pipe. In particular, resulting from the pressure drop caused by choking restriction, when the gases pass through chokes and needle valves, their volume will expand and their temperature will drop, which creates conditions for the water-containing gases to form hydrates. Once the gas hydrates are formed, pipelines may be plugged and the production can be suspended.

The most important consideration in hydrate formation is the presence of liquid water in the natural gases. Thus, the fundamental approach to prevent hydrate formation is to remove the water from the gases. Besides, some other methods of prevention include raising the temperature of the gases before and after they encounter the chokes and injecting inhibiting agent into the gas flow ahead of the chokes.

3. Transform the Natural Gases into Hydrates for Storage and Transportation

Normally, natural gases are transported by pipelines. But when the pipeline is not convenient, it is needed to resort to high-pressure vessel transportation, which implies low efficiency and insecurity. If the natural gases are transformed to hydrates before transportation, high efficiency and security can be warranted. This technology is still under study.

Chapter 3
Phase State of Reservoir Hydrocarbons and Gas–Liquid Equilibrium

As we all know, oil and gas are naturally existing mixtures of a variety of hydrocarbons plus non-hydrocarbons. Reservoirs may be radically different from each other in fluid composition. Subjected to original reservoir conditions, these underground hydrocarbon resources are in diversified forms: gas reservoir (in an exclusive gaseous form), oil reservoir (in an exclusive oil form), oil reservoir with gas-cap (where gas and oil coexist), etc.

What's more, a complex succession of state change usually takes place along with the extraction of crude oil out of the ground. State changes include the liberation of soluted natural gas from oil, the retrograde condensation in which the consendate gas is changed from a gas into a liquid, etc.

What petroleum engineer concerned are the predevelopment state of reservoir, the succession of state changes occurring along with development, and the regulations governing these changes.

As a general rule, intrinsic reasons trigger changes under suitable external conditions. The chemical composition of reservoir hydrocarbons is the intrinsic reason for their state changes while the conditions of temperature and pressure serve as external reasons. Meanwhile, as the phase state changes, the hydrocarbon composition of the individual phases changes concurrently, too. This chapter will go into the phase-conversion behavior of reservoir fluids subjected to changes of pressure and temperature.

© Petroleum Industry Press and Springer-Verlag GmbH Germany 2017
S. Yang, *Fundamentals of Petrophysics*, Springer Geophysics,
DOI 10.1007/978-3-662-55029-8_3

3.1   Phase Behavior of Reservoir Hydrocarbon Fluids

3.1.1   Phase and the Descriptive Approaches of Phase

Theoretically based on the thermodynamics and the physical chemistry, the theory of reservoir fluid phase usually takes a concept named "system" as the object of study. The term "system" refers to the object which is artificially divided from the others for study and conceived as a space enveloped by boundaries. The boundaries, being static or dynamic, take the form of objectively existing solid interface or hypothetical ones.

A "phase" is defined as a definite part of a system which is homogeneous (having the same physical and chemical properties) throughout and physically separated from other phases by distinct boundaries. Paralleled with the three states in which materials can exist in, the materials that homogenously stay in one particular state can be called, respectively, the "gas phase," "liquid phase," and "solid phase." A phase may comprise of more than one component; for instance, a gas phase may contain methane, ethane, and so on.

In oil and natural gas, hundreds of kinds of molecules are contained, and each kind of these molecules can be referred to as a component. Sometimes, a component involves only a single and distinct kind of molecule, but at other times several kinds of molecules can be incorporated to one component, named pseudo-component, as long as their properties are similar to each other and the contents they shared are quite small. For example, we can regard the hydrocarbon fluids in reservoirs as a mixture of seven components: C_1, C_2, C_3, C_4, C_5, C_6, and C_7+, in which the C_1–C_6 are all pure components with only one kind of molecules, while the C_7+ is an incorporated one with the C_7 and all the molecules heavier than C_7 involved in. Another example, the mixture can also be divided into three components: C_1, C_{2-6}, and C_7+, in which all the molecules between C_2H_6 and C_6H_{14} are incorporated to be a single component and the molecules of C_7H_{16} and those heavier than them are treated as a single liquid-hydrocarbon component. The first principle of division is the contents, that is to say, if the content of a constituent is quite great, it can be exclusively treated as a component; and if the content of a constituent is quite small, it can be incorporated with other small-content constituents and form a component. The second principle is the need and the purpose of study.

The "composition" of a system refers to the components contained in a system and the proportions they, respectively, take up in the total.

For a system whose composition remains the same, the relationship among its state parameters, the pressure, the temperature, and the specific volume, can be expressed by the equation of state:

$$F(P, v, T) = 0 \qquad (3.1)$$

The equation of state is the mathematical method to describe the phase state in a system, and if we illustrated the equation of state by a corresponding diagram, we can get a "phase diagram." More exactly, a "phase diagram" is a graph used to show the conditions of state parameters, under which the various phases of a substance will be present. The phase state of a system depends on not only its composition but also the temperature and pressure to which it is subjected.

1. Three-Dimensional Phase Diagram

Drawing the relational curves in a coordinate system which takes the three variable parameters (p, V, and T) as a set of regular axes, we can obtain a three-dimensional phase diagram (Fig. 3.1). The three-dimensional phase diagram enables the illumination of the relationship among the variable parameters to be more detailed and exhaustive.

2. Two-Dimensional Phase Diagram

If one of the three parameters in the equation of state is held a constant, the relationship of the other two can be reflected in a two-dimensional diagram, which can be more intuitively clear and more convenient to realize. In petroleum engineering, the p–V (pressure–volume) phase diagram shown in Fig. 3.2 and the p–T (pressure–temperature) phase diagram shown in Fig. 3.3 are most commonly adopted. The later one of the two is the most convenient phase diagram in the study of the phase changes when the system is definite in composition.

3. Ternary Diagram

Compositional phase diagrams for three-component mixtures must be plotted in such a way that the composition of all three components can be displayed. As

Fig. 3.1 The 3-dimensional phase diagram of a single-component system and its 2-dimensional projection (From Zollzn Gyulay 1977)

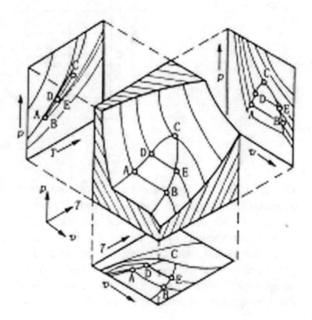

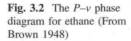

Fig. 3.2 The P–v phase diagram for ethane (From Brown 1948)

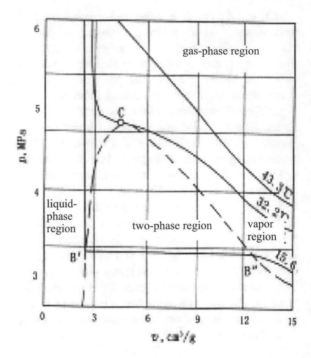

Fig. 3.3 p–T phase diagram for ethane

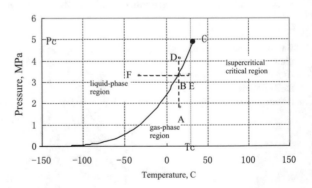

shown in Fig. 3.4, diagrams formed in equilateral triangles are called "ternary diagrams" and they are convenient to indicate the system in which the pressure and temperature are constants and only the composition changes. Each apex of the triangle corresponds to 100 percent of a single component (component 1, 2, and 3). Each side of the triangle represents two-component mixtures. For example, the bottom side of the triangle represents all possible mixtures of the component 1 and component 2 with no component 3 present. Points within the triangle represent three-component mixtures. For a mixture with three components, any composition

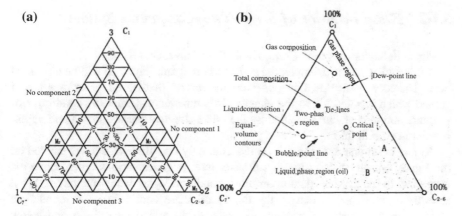

Fig. 3.4 Ternary diagram

pattern of it can fall within the ternary diagram and upon a corresponding point. The sum of the three concentrations is always 100 %.

For instance, point M_1 in Fig. 3.4a represents 100 % pure component 2; point M_2 represents a mixture of 70 % component 1 and 30 % component 3; and point M3 represents a mixture which consists of 20 % component 1, 50 % component 2, and 30 % component 3.

The straight lines parallel with the sides are called "equal-volume contours." For example, the straight lines which run parallel to the side 1–2 indicate that the points exactly on this line possess the same concentration of component 3.

Figure 3.4b shows how to use the ternary diagram to describe the phase state of a system. The curved line within the triangle is named "saturation envelope," and the area enclosed by it is the two-phase region. The straight lines which lie inside the saturation envelope are "equilibrium tie-lines," and the composition of any point upon these lines is determined by its coordinate values. The crossing points at which the bubble-point line and the dew-point line intersect with the equilibrium tie-lines can be used to determine the gas–liquid composition of the system. The equilibrium tie-lines are straight but not entirely parallel to the equal-volume contours. To the right of the cortical point, it is the "supercritical critical region," within which the area A is the miscible gas region and the area B is the miscible oil region.

A common use of three-component phase diagram is in analysis of miscible displacement of gas injection under the formation condition. Reference [36–38] about the application of the ternary diagram can be resorted to for details.

Which phase diagram should be adopted depends on the practical situation and the purpose established for which the phase diagram is to be used.

3.1.2   Phase Behavior of Single/Two–Component System

1. Phase Behavior of Single-Component (Pure Substance) System

Taking ethane as an example, the $p–T$ phase diagram (Fig. 3.3) will be discussed here. The curve plotted is a "vapor–pressure curve," that is, the vapor pressure of ethane which is subjected to the changes in temperature. Under the conditions of pressure and temperature specified by the curve, the two phases, liquid and vapor, coexist in equilibrium.

At any point that does not fall on the line, only one phase exists. As labeled on the figure, ethane exists in the liquid state upon those conditions lying above the curve and in the gaseous state upon those conditions lying below the curve. What should be mentioned is that in the theory of phase equilibrium the terms "gas phase" and "vapor phase," respectively, refer to the temperature above and below the critical temperature to which the substance is subjected; but in practice the two words, vapor and gas, are used interchangeably. Holding the temperature as a constant, when the pressure decreases from P_D to the point of the saturation vapor pressure P_B, which has a fixed value at the given temperature, gas begins to form as bubbles and leaves the liquid. Conversely, when the pressure increases from P_A to the point of the saturation vapor pressure, the vapor condenses into liquid droplets, and hence the formation of a liquid phase is readily identified. Obviously, for ethane or any single-component system, the vapor–pressure curve represents the trace of both the bubble-point and dew-point curves on the pressure–temperature plane.

The bubble-point pressure is the pressure at which the first bubble of gas is formed and leaves from the liquid when decreasing the pressure while holding the temperature as constant; the dew-point pressure is the pressure at which the first droplet of liquid is formed and appears in the vapor when increasing the pressure while holding the temperature as constant.

For the single-component system, the vapor–pressure curve is the only realm where the liquid and gas can coexist. The upper limit of the vapor–pressure line, the point labeled C, is called "the critical point," and, what's more, the temperature and pressure at this point are, respectively, called "the critical temperature" and "the critical pressure."

For a single-component system, the critical point is the highest value of pressure and temperature at which two phases can coexist. The area above the critical point is known as the supercritical region, wherein there is no distinction between the gaseous and liquid states—that is to say, no matter how great a pressure is put on the system, it is impossible for it to be in a situation of two phase.

As shown in Fig. 3.3, the point A is in a vapor state. With a constant temperature, when the pressure is raised to 3.4 MPa, the value that is corresponding to the point B on the vapor–pressure curve, the condensation of a liquid from the ethane vapor is observed, leading to a situation of two-phase; then, continue to raise the pressure until it passes the vapor–pressure curve and reaches a value higher than the dew-point pressure (e.g., the point D), so that we can see the ethane is completely in liquid state. It should be noted that when reflecting the point B in the $p–T$ phase

Fig. 3.5 Saturated vapor–pressure curve of several alkanes and nonhydrocarbons

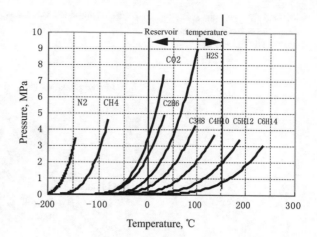

diagram to the p–V phase diagram (Fig. 3.2), there appears two points: point B′ and point B″. That is to say, the p–V phase diagram illustrates the two phases or, in other words, shows the course of vaporization or liquefaction, whereas the p–T phase diagram fails to do. In fact, different phase diagrams can reflect different aspects of a course, and which of them should be chosen and adopted depends on purposes.

Figure 3.5 shows the vapor–pressure curves of several pure substances. During the development of oil & gas reservoirs, the temperature of the hydrocarbon fluids ranges from the original reservoir temperature (can be as high as 120 °C) to the land-surface temperature. As inspection of this figure, the critical temperatures of N_2 and CH_4 are much lower than reservoir temperature, while the critical points of CO_2 and C_3H_8 fall within this range. This is the main reason why it is the CO_2 and C_3H_8 but not the N_2 or CH_4 that are used as the miscibility agents in the EOR practice of miscible displacement.

2. Phase Behavior of Two-Component System

When a second component is added to a hydrocarbon system, the phase behavior becomes more complex. Figure 3.6 is a typical phase diagram of a two-component hydrocarbon mixture. The opening loop curve that is shaped long and narrow is the saturation envelope of the binary system. The characteristics of this phase diagram are:

(1) The diagram of a two-component system is no longer a monotonically increasing curve but an opening loop one. CE is the dew-point line; CAF is the bubble-point line; the point at which the two lines join is the critical point. Rigorously speaking, the critical point is the state of pressure and temperature at which the intensive properties of the gas and liquid phases are continuously identical. The intensive properties are the properties that are independent of the quantity of material present, for example, the density, viscosity, and compressibility factor. Above and to the left of the bubble-point line, the

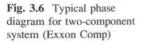

Fig. 3.6 Typical phase diagram for two-component system (Exxon Comp)

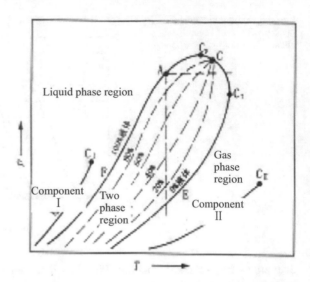

mixture exists as a liquid. Below and to the right of the dew-point line, the mixture exists as a gas. The area enclosed by these curves is the two-phase region. The dashed lines in the two-phase section are the equal-volume contours.

(2) The conditions represented by the critical point are no longer the highest temperature or pressure at which the two phases can coexist, because a particular area above the critical point also allows the occurrence of a two-phase state. Actually, the highest temperature and highest pressure for the two-phase coexistence, respectively, stand on the points labeled by C_T and C_P. The particular area mentioned refers to the scope of $T_c < T < T_{CT}$ and the scope of $P_c < P < P_{CP}$. When the temperature of the system is above the highest temperature T_{CT}, it is impossible for the system to undergo liquefaction no matter what a high pressure it is subjected to. Regarding the reasons above, T_{CT} is called the critical condensate temperature. Likewise, when pressure of the system is above the highest pressure P_{CP}, the system will undergo vaporization by no matter how high a temperature it is subjected to, remaining in a single phase of liquid. And in similar manner, P_{CP} is called the critical condensate pressure.

Furthermore, for each hydrocarbon system, a distinct phase diagram exists, and different proportions of each component presented in binary system lead to different shapes of the diagrams. Taking a methane–ethane mixture as example, the phase diagrams of a variety of studies proportions of the two components are shown in Fig. 3.7. Comparing the diagrams, we can conclude that:

(1) The two-phase region of any binary mixtures must lie between the vapor–pressure curves of the two pure substances.

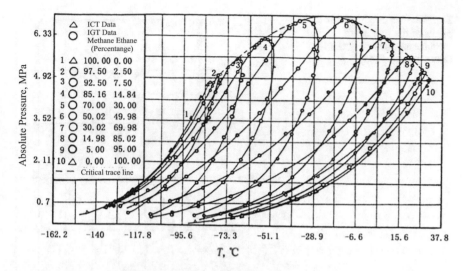

Fig. 3.7 Phase diagram for the mixture of methane and ethane

(2) The critical pressure of the system will always be higher than the critical pressure of any component in the system; the critical temperature lies between the critical temperatures of the two components present in the mixture.

(3) When the proportion of the heavier component is increased, the critical point of the mixture is shifted to the right, or, in other words, to the direction closer to the vapor–pressure curve of the heavier component.

(4) As one component becomes more predominant, the dew-point line or the bubble-point line of the mixture tends to approach the vapor–pressure curve of the major pure component.

(5) As the composition of the mixture becomes more evenly distributed between the components, the two-phase region increases in size, whereas when one component becomes predominant, the two-phase region tends to shrink in size.

Figure 3.8 shows the phase diagrams of a C_1–C_{10} hydrocarbon mixture. Comparing Fig. 3.8 with Fig. 3.7, we can know that:

(1) The critical pressure of the mixture is much higher than the critical pressure of any component in the system. In fact, the variety of the molecular size present in the mixture tends to increase the critical pressure of it.

(2) The area enclosed by the trace of the critical points tends to be larger when the properties of the two components (e.g., the molecular weight and the volatility) are more similar to each other.

(3) As long as one of the two components becomes absolutely predominant, the two-phase region of the mixture will be narrowed largely.

Fig. 3.8 Traces of the critical points of the mixture of methane and other alkanes (From Brown 1948)

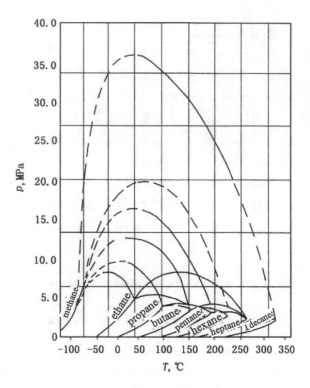

3.1.3 Phase Behavior of Multicomponent System

Usually, mixtures consisting of many components are encountered because naturally occurring hydrocarbon systems are composed of a wide range of constituents. The phase behavior of a hydrocarbon mixture is dependent on the composition of the mixture as well as the properties of the individual constituents. The phase state of the hydrocarbons is also associated with the conditions of pressure and temperature to which they are subjected. Therefore, the phase diagrams of different oil & gas reservoir fluids, although sharing some common characteristics of multicomponent system and exhibiting some similar features, are obviously different from each other. In the following section of this lesson, they will be introduced in detail.

1. Phase Diagram of Multicomponent System

A typical phase diagram for a multicomponent system is shown in Fig. 3.9. The envelope line aC_PC_Tb separates the whole plane into two parts: the two-phase region and the single-phase region. The area enclosed by the envelope is the two-phase region, and the dashed lines within it are the iso-vol lines (quality lines), the loci of points of equal liquid volume percent. The area outside the envelope is the single-phase region. The bubble-point line aC_PC represents 100 % liquid by volume and is the demarcation line between the liquid-phase region and the

Fig. 3.9 Phase diagram of multicomponent hydrocarbon system

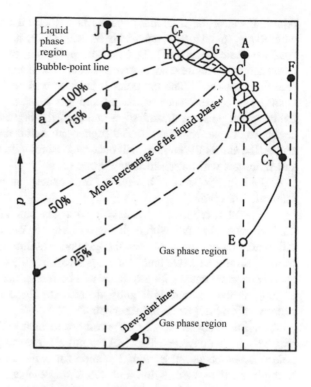

two-phase region. When the pressure is dropped to the bubble-point pressure, the first bubble of gas is formed, and therefore the bubble-point line is also called saturation pressure line.

The dew-point line CC_Tb represents 100 % vapor by volume and is the demarcation line between the vapor–phase region and the two-phase region. When the pressure is raised to equal the dew-point pressure, the first droplet of liquid is formed in the system.

The bubble-point line coincides with the dew-point line at the critical point, marked by C. On the envelope line, the point C_P is named "cricondenbar," meaning the highest pressure at which the liquid and vapor can coexist in equilibrium; the point C_T is called "cricondentherm," meaning the highest temperature at which the liquid and vapor can coexist.

2. Phase Diagrams of Petroleum Reservoir Fluids

The phase diagram of a petroleum reservoir fluid can be gained according to its composition. Then, given that the information about the original reservoir conditions (temperature and pressure) is readily available, the type of the reservoir can be determined by the critical point and is relative position with the point identified corresponding to the original reservoir conditions.

As shown in Fig. 3.9, the point labeled J represents a pure oil reservoir. Upon original reservoir conditions, the hydrocarbon mixture exists in a single state of liquid. Because the reservoir pressure is higher than the saturation pressure, the natural gas in oil could not reach its saturation, and thence this type is called "undersaturated reservoir." As fluids are produced, the reservoir pressure declines while the temperature comparatively remains constant. When the decreasing pressure becomes lower than the point I, bubbles will be formed and leave the liquid, resulting in a coexistence of the oil and gas phases. The pressure at point is the bubble-point pressure or saturation pressure, the highest pressure at which the gas begins to escape from the oil. The higher saturation pressure an oil reservoir possesses, the earlier the bubbles will occur. Continuing to decrease the pressure, more and more gas will escape from the crude oil.

If the initial pressure of a reservoir falls exactly at the point I, it can be called a "saturated reservoir."

The point L represents a reservoir with gas-cap. The initial conditions of this kind of reservoirs fall within the two-phase region, and because of gravitative differentiation of the two phases, the gas phase at initial state accumulates at the top of the reservoir structure and forms a gas-cap. As the crude oil in the reservoir has already been saturated with gas, there will be further escaping of gas from the liquid as long as the pressure is slightly decreased. These reservoirs with gas-cap are usually called "oversaturated reservoirs."

If the initial conditions of a reservoir are located to the right of the critical point and outside the envelope line (e.g., point F), the reservoir is in a gaseous state before development. The point F stands for a gas reservoir, which never passes through the two-phase region and keeps single-phase state during the isothermal pressure decreasing along with the gas production.

A typical phase diagram for gas-condensate reservoir is presented in Fig. 3.9. Its significant feature is that between the critical temperature and the cricondentherm, the initial reservoir pressure is greater than the critical pressure. Obviously, point A lies above the retrograde region (the shaded area in Fig. 3.9). It can be determined that the hydrocarbon mixture in this kind of this reservoir originally exists as a single phase of gas. The system maintains a state of gas before the pressure decreases to the point B, but as long as the pressure becomes lower than the dew-point pressure, liquid phase begins to form and appears in the gas phase; as the pressure continues to drop, more condensed liquid will emerge and comes into being as condensate oil in the formations.

3. The Retrograde Region of Gas-condensate Reservoir (The Region of Retrograde Condensation)

As shown in Fig. 3.9, the two shaded areas, CBC_TDC and CGC_PHC, are, respectively, called the region of isothermal retrograde condensation and the region of isobaric retrograde condensation. Reservoirs featuring retrograde regions like this in phase diagram are named retrograde gas-condensate reservoirs, gas-condensate reservoirs for short. In practical gas production, pressure decline occurs along with the fluid depletion while the reservoir temperature can be regarded as a constant, that is to say, the path of pressure dropping travels through the region of isothermal retrograde condensation.

Fig. 3.10 Phase diagram of condensate gas (by Muskat 1949)

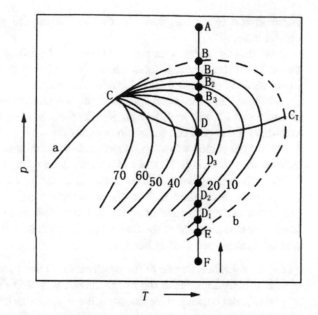

Let's suppose that a reservoir has initial conditions which fall upon the point labeled A in Fig. 3.10, and it is in a single state of gas. When the pressure decreases from A to B, there will be no change in phase state; from B to D, the liquid content in the system will gradually increase from 0 to 10, 20, 30, and 40 %; from D to E, as the pressure continues dropping, the liquid content in the system will gradually decrease from 40 to 30, 20, 10, and 0 %.

It is normal for liquid to evaporate as the pressure in system declines along the path D–E, but it is irregular that more and more condensed liquid occurs as the pressure drops along the path B–D. Therefore, the later phenomenon is called retrograde condensation.

The retrograde condensation can be explained from a perspective of the molecular dynamics: A pressure decline within the studied system tends to cause the distance between the hydrocarbon molecules to extend and thence to decrease the attraction force interacted between each other. In particular, when the pressure decreases to the point B, the weakened force acted upon heavy hydrocarbon molecules by the light the gaseous ones will lead the heavy molecules isolate from the light hydrocarbons and form the first few droplets of liquid. As the pressure continues to decrease, the attraction force between molecules becomes smaller and smaller, and more and more liquid will be condensed out. By the time that the pressure decreases to the point D, most of the heavy molecules have already isolated out and the quantity of the condensed liquid reached its maximum.

It should be mentioned that this irregular phenomenon of retrograde condensation only occurs under special conditions which fall upon a particular region nearby the critical point on the phase diagram. It will never take place when the conditions fall upon far away from the critical point. In a word, the retrograde condensation is an irregular phenomenon when the system comes close to its critical state.

Note that there exists an upper dew-point B (also called the second dew-point) and a lower dew-point E (also called the first dew-point). When the pressure is lower than E, all the condensed liquid will evaporate into the vapor phase.

Likewise, the maximum quantities of the condensed liquid at changing temperature can be got, and they all join to form the region enclosed by CDC_TBC, known as the region of retrograde condensation.

As for the region of isobaric retrograde condensation, it is hard to use isobaric method to deal with field problems; so no deep research on it is needed.

Regarding the gas-condensate reservoirs, due to the condensation of the heavy components, the heavy hydrocarbons, which are the most valuable part of oil, precipitate and attach onto the reservoir rocks. This part of oil (the condensate oil) is hard to be extracted out of the formations and will be left in the reservoir, leading to a considerable loss. This is because:

(1) The saturation degree of the condensate oil that precipitates from gas reservoir is usually lower than the lowest degree at which the liquid hydrocarbons can flow, and consequently the condensate oil forms on the rock surface oil films or exists as oil droplets that are sequestered obstinately in the corners of the porosities.

(2) Based upon the phase diagrams, we can imagine that when the pressure continues to drop and becomes lower than the dew-point E, also known as the first dew-point, the condensate oil will evaporate again and join in the gas phase, which thus can be extracted from the underground. However, in practice, this does not take place readily. The reason why that process cannot occur stems from the fact that when the gas reservoir pressure drops below the upper dew-point B, as the heavy hydrocarbons condensate from the gas phase and the light hydrocarbons are extracted from the formations, the total molecular weight of the residual hydrocarbons in gas reservoir increases considerably. That is to say, along with the long-term production, the composition of the residual hydrocarbon system is already different from that under the original reservoir conditions, and increased molecular total molecular weight tends to lead the position of the phase envelope of the residual hydrocarbon system to move downwards and to the right, and the shape of the phase diagram will change. Therefore, the revaporization of the condensate oil is prevented to take place.

(3) The precipitation of the condensate liquid results in an increased drag to the gas flow and a change of the seepage-flow characteristics, because of which the gas wells may suffer a decline in their capacity of production and even a quit of the gas flowing caused by "hydrocarbon lock."

So, in order to prevent the heavy hydrocarbons from compensating from the natural gas, methods such as cyclic gas injection should be adopted to maintain the

Fig. 3.11 Oil & gas reservoir types determined by reservoir. Temperature and critical temperature

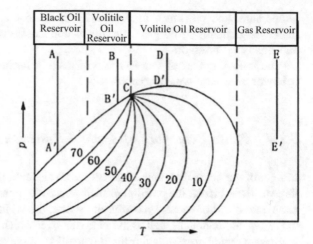

reservoir pressure and to make sure that the reservoir pressure condition can be kept above the dew-point during the production.

4. Judgment of Reservoir Types

The judgment of reservoir types is preceded essentially on the basis of the relative position relationship between the point of the original reservoir conditions, including the pressure and the temperature, and the critical point. As indicated in Fig. 3.11, the relative position relationship between the two points in the discussed reservoir types is different from each other.

In fact, different hydrocarbon systems with different composition will have different shapes and positions for phase diagrams. Figure 3.12 shows the position and shape of 4 kinds of reservoirs. As the heavy components make a greater contribution in the hydrocarbon system, the phase envelope tends to move downwards and to the right. The reservoir fluid type is named according to the relative position of critical point on the phase envelope and the reservoir conditions (i.e.,

Fig. 3.12 Relative position of the phase diagrams of various oil & gas reservoirs (from Exxon Comp 1987)

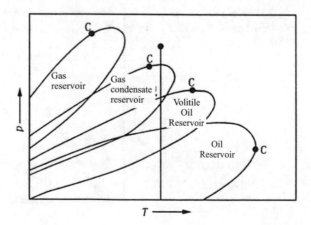

temperature and pressure). In Fig. 3.12, the molecular weight of reservoir fluid increases gradually, from gas reservoir to condensate reservoir, volatile reservoir, and finally oil reservoir

The following part of this lesson will make a further analysis about the phase behavior of several typical reservoirs.

3.1.4 Phase Behaviors of Typical Reservoirs

In light of the phase diagrams, different reservoirs exhibit their difference mainly in three aspects: Firstly, the position of their phase diagrams, that is to say the scope of temperature and pressure; secondly, the scope, the width, and the area confined in the two-phase region, as well as the intervals over which the iso-vol lines within the two-phase region distribute; thirdly, the position of the critical point upon the phase envelope and the relative position relationship (relationship of left and right, upper and lower) between it and the point corresponding with the original reservoir conditions, including the temperature and pressure.

Natural gas escapes from the liquid instantly and the oil is extracted from the underground and lifted up to the surface, which leads to the shrinkage of the crude oil. The oil that undergoes a high degree of shrinkage in volume is named high-shrinkage oil, and in a similar way, the oil that undergoes a low degree of shrinkage in volume is named low-shrinkage oil.

1. The Phase Diagram of Conventional Heavy Oil Reservoir (Low-Shrinkage Oil Reservoir)

As shown in Fig. 3.13, the point 1 which represents the reservoir conditions lies above the bubble-point line and to the left of the critical point, and the hydrocarbons exist as a single phase of liquid, therefore it indicates an oil reservoir. As the formation pressure decreases along with the depletion of oil, the first gas bubbles are formed directly and the pressure drops upon the bubble-point (point 2), and they

Fig. 3.13 Phase diagram of low-shrinkage oil (from Clark 1953)

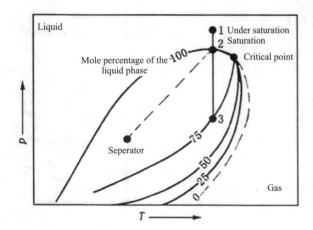

will leave the liquid, so the system began to contain two phases. With the continuing decrease in formation pressure, more and more bubbles separate from the liquid.

The phase diagram of reservoir type features that its comparatively closely spaced iso-vol lines within the two-phase region cluster near the side of the dew-point line. When the reservoir pressure decreases to the bubble-point pressure, although there is a small amount of gas separating from the oil, the system is still predominantly liquid-phased. The gas–oil ratio of this kind of reservoirs, usually lower than 90 m^3/m^3, is the lowest among all the mentioned types. The specific gravity of low-shrinkage oil can be higher than 0.876, and the tank-oil on the land surface often exhibits the color of black or dark brown. The vertical line on the diagram represents the route of the pressure dropping under constant reservoir temperature. We can imagine that when the pressure drops to the point 3, the residual liquid in the reservoir contains about 75 % oil (molar percent) and 25 % natural gas. The oblique broken line on the diagram represents the route of pressure dropping and temperature dropping when the fluids flow from the wellbore to the land-surface separator. It indicates that under the condition of separator about 85 % (molar percent) of the well effluents is oil, which is a considerable high percentage and could be owned by the low-shrinkage oil.

2. The Phase Diagram of Light Oil Reservoir (High-shrinkage oil)

The phase diagram of light oil is shown in Fig. 3.14. It is quite similar to that of conventional heavy oil, but the iso-vol lines within the two-phase region are more widely spacing, as a result of its comparatively more light components contained. Once the pressure drops lower than the bubble-point pressure, a large amount of gas will escape, and this type is therefore also called high-shrinkage oil reservoir.

The vertical line and the oblique broken line have the same significance as before, and under the condition of separator, about 65 % of the liquid is crude oil, which owns a specific gravity smaller than 0.78, a dark color, and a gas–oil ratio lower than 1800 m^3/m^3.

Fig. 3.14 Phase diagram of high-shrinkage oil (from Clark 1953)

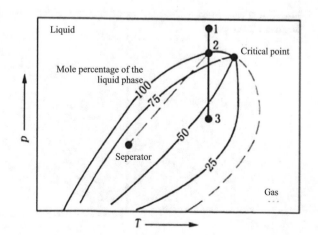

Fig. 3.15 Phase diagram of
gas-condensate reservoir

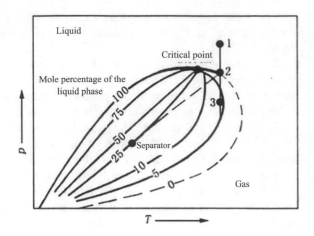

3. The Phase Diagram of Retrograde Gas-condensate Reservoir

As mentioned previously, the gas-condensate reservoirs are also called retro-grade gas-condensate reservoirs (Fig. 3.15). The reservoir temperature of them is between the critical temperature and the cricondentherm, and the reservoir pressure is outside the envelope line. Under initial reservoir conditions (the point 1), the hydrocarbon system stays in a single gaseous state, existing as a gas reservoir.

Under the conditions of separator, about 25 % of the yield on the surface is liquid, more accurately, condensed oil; the yielded gas is condensate gas. The gas–oil ratio of a gas-condensate reservoir can be up to 12,600 m^3/m^3, and the specific gravity of the condensed oil, which is lightly colored and transparent, can be as small as 0.74. Along with the depletion, the gas–oil ratio tends to increase, resulting from the condensation of the heavy components in the gas reservoir.

4. The Phase Diagram of Wet Gas Reservoir

Fig. 3.16 shows the phase diagram of a wet gas reservoir. Its features are: The reservoir temperature is much higher than the critical temperature, and as the

Fig. 3.16 Typical phase
diagram of wet gas

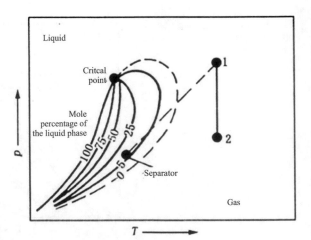

Fig. 3.17 Typical phase diagram of dry gas

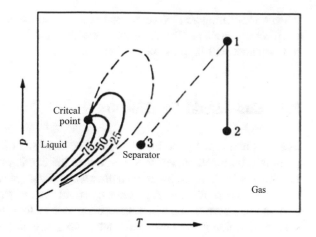

reservoir pressure decreases, for instance, from the point 1 to the point 2, the fluid keeps in gaseous state throughout the process. However, when subjected to the conditions of surface separators, the system lies in the two-phase region, and there will be some condensed hydrocarbon liquid within the separators. This liquid is transparent light oil which is colorless like water and possesses a specific gravity less than 0.78 and a surface gas–oil ratio lower than 18,000 m^3/m^3.

5. The Phase Diagram of Dry Gas Reservoir

The dry gas is the kind of natural gas which is comprised largely of methane (70–98 %). A typical phase diagram of dry gas is shown in Fig. 3.17. The separator conditions as well as the reservoir conditions lie in the single-phase region, and both the path of the changing of the reservoir conditions (from point 1 to point 2) and the path from the wellbore to the separator are outside the two-phase region. Hydrocarbon liquid is not condensed from the mixture either in the reservoir or at the surface, and theoretically, and the gas–oil ratio of the mixture is infinite. If there is a minute amount of condensed oil appearing at the surface separator, the gas–oil ratio gained is usually higher than 18,000 m^3/m^3.

Nowadays, the petroleum exploration and development are on their way to the deeper formations. In the ultradeep formations that bear high temperature and pressure, most of the hydrocarbon systems encountered belong to high-volatile oil reservoirs or gas-condensate reservoirs. Therefore, the phase behaviors of these kinds of reservoirs should be well known to ensure a better development of them.

The phase state changes, taking place when the reservoir hydrocarbons are subjected to the changing pressure and temperature, can be reflected by the separation or solution of the gas in the oil; hence the composition and properties of the system also follow to change. By resorting to arguments from the thermodynamic phase equilibrium theory, the phase-state and phase equilibrium processes will be introduced in the lesson 2 of this chapter. What's more, it will also discuss the methods that are used to figure out the quantities and composition of the two phases when a multicomponent system reaches its equilibrium. Then, in lesson 3, the two engineering methods, the separation and the solution, and their application in oil development will be presented.

3.2 Gas–Liquid Equilibrium

The area bounded by the bubble-point and dew-point lines on the p–T phase diagram of a multicomponent mixture defines the conditions for the gas and liquid to exist in equilibrium. To put it in another way, any point within the two-phase region on the p–T phase diagram represents a certain state of gas–liquid equilibrium of the system, and each equilibrium state corresponds to a unique composition of the system in which every component takes up a one and only content.

In the oil reservoir engineering, the equilibrium calculation of the two phases is to figure out the unknown parameters listed below, in the case that the composition of the underground fluids is already known:

(1) The bubble-point pressure or bubble-point temperature of the system;
(2) The dew-point pressure or dew-point temperature of the system;
(3) The quantities and composition of the gas and liquid at conditions within the two-phase region.

Using the calculation method mentioned above, plus the analysis results of the underground fluid samples that are taken from the initial stage of development, the goals that can be achieved include:

(1) Forecast the phase-state characteristics and change of the reservoir fluids during the development of a reservoir.
(2) Forecast the quantity of the reservoir fluids at the surface and provide the basis for the design of their processing and delivery.

This lesson will begin with an introduction of the behavior of a hypothetical fluid known as an ideal solution and then discuss the theory of gas–liquid equilibrium as well as its application.

3.2.1 Ideal Solution

When two or more liquids are mixed to form a true solution, thermal effect happens as a result of the association between different kinds of molecules. In the discussion of the phase-state changes of a multicomponent system, we will neglect the thermal effect and the volume changes of the true solution to simplify the problem. An ideal solution should satisfy the following conditions for hypothesis:

(1) Absolute mutual solubility happens when the components are mixed;
(2) No chemical interaction occurs upon mixing;
(3) The molecular diameters of the components are the same; that is, the inter-molecular forces of attraction and repulsion are the same between unlike and like molecules.

A solution that possesses these features is called an ideal solution. When the components are mixed, no exchange of heat occurs between the ideal solution system and the environment, the temperature and pressure hold the same, and the volume of the ideal solution equals the sum of the volumes the components would occupy as pure liquids at the same temperature and pressure.

In fact, ideal solutions do not exist. Actually, the only solutions which approach ideal solution behavior are gas mixtures at low pressure and liquid mixtures of similar molecular-structured components of the same series. However, studies of the phase behavior of ideal solutions help us understand the behavior of real solutions. The laws and formulas for ideal solutions can be applied to solve engineering problems after proper modification.

1. Raoult's Law

Raoult's law states that the partial pressure of a component in the gas is equal to the mole fraction of that component in the liquid multiplied by the vapor pressure of the pure component. The mathematical statement is:

$$P_j = x_j P_{vj} \tag{3.2}$$

where

P_j the partial pressure of component j in the gas phase;
x_j the mole fraction of component j in the liquid phase;
P_{vj} the vapor pressure of component j at the given temperature.

2. Dalton's Law

Dalton's law is used to figure out the partial pressure of the components in the gas phase of an ideal solution. It states that the total pressure of a confined mixture of gas is equal to the sum of the partial pressure of the individual components each taken alone in the same volume, or in other words, the partial pressure of each component equals the mole fraction of it multiplied by the total pressure of the system. Expressed mathematically:

$$P_j = y_j P \tag{3.3}$$

where

y_j the mole fraction of component j in the gas phase;
P the total pressure of the system.

3. Gas–Liquid Equilibrium Calculation of an Ideal Solution

The two equations may be combined by eliminating partial pressure p_j, and the resulting equation relates the composition of the gas and liquid phases in equilibrium to the pressure and temperature at which the gas–liquid equilibrium exists:

$$y_j P = x_j P_{vj} \tag{3.4}$$

$$\frac{y_j}{x_j} = \frac{P_{vj}}{P} \tag{3.5}$$

or

$$y_j = \frac{P_{vj}}{P} x_j$$

In case of ideal solution, Eq. (3.15) indicates that for a component in the system the ratio of its proportion in the liquid to its proportion in the gas is equal to the ratio of the vapor pressure of this component to the total pressure of the studied system, which is subjected to the same temperature.

Then, with the consideration of material balance, the formula for the calculation of the gas–liquid composition of an ideal solution can be derived from Eq. (3.5).

First of all, the significance of the following symbols should be presented:

n—the total mole number of the gas and liquid phases in the studied ideal system;

n_g—the total mole number of the gas phase in the studied ideal system;

n_L—the total mole number of the liquid phase in the studied ideal system;

$\bar{n}_g$—the mole fraction of the gas phase;

$\bar{n}_L$—the mole fraction of the liquid phase;

x_j—the mole fraction of the component j in the liquid phase;

z_j—the mole fraction of the component j in the studied system which includes both the liquid and the gas phases;

y_j—the mole fraction of the component j in the gas phase;

m—the number of the components present in the system.

Under the stipulations above, we can know:

$z_j n$—the total mole fraction of the component j in the studied system;

$x_j n_L$—the mole fraction of the component j in the liquid phase;

$y_j n_g$—the mole fraction of the component j in the gas phase.

The significance of the mentioned symbols as well as the relationship among them is shown in Fig. 3.18.

For a closed system, since it has no material exchange with its environment, it obeys the law of conservation of mass, no matter what a state it is at. There occurs no chemical reaction, so the component j also obeys the law of conservation of mass:

		Total Mole Number	Mole Fraction	Mole Fraction of the jth Component	Mole Number of the jth Component
	Entire System	N	1	z_j	$z_j n$
	Gas	n_g	$\overline{n}_g$	y_j	$y_j n_g$
	Liquid	n_L	$\overline{n}_L$	x_j	$x_j n_L$

Fig. 3.18 Compositions of the whole system and the phases

$$z_j n = x_j n_L + y_j n_g \tag{3.6}$$

Substitute Eq. (3.5) into Eq. (3.6), eliminating y_j, we can get that:

$$z_j n = x_j n_L + \frac{P_{vj}}{P} n_g x_j \tag{3.7}$$

By a simple transformation of it, we can get:

$$x_j = \frac{z_j n}{n_L + \frac{P_{vj}}{P} n_g} \tag{3.8}$$

In consideration of the whole liquid phase, the sum of the mole fractions of all the components should equal 1, that is to say, $\sum_{j=1}^{m} x_j = 1$. Written in another form:

$$x_j = \frac{z_j n}{n_l + \frac{P_{vj}}{p} n_g}$$

$$\sum_{j=i}^{m} x_j = \sum_{j=1}^{m} \frac{z_j n}{n_l + \frac{P_{vj}}{p} n_g} = 1 \tag{3.9}$$

In a similar way, for the gas phase, $\sum_{j=1}^{m} y_j = 1$, and

$$y_j = \frac{z_j n}{n_g + \frac{P}{P_{vj}} n_L}$$

$$\sum_{j=1}^{m} y_j = \sum_{j=1}^{m} \frac{z_j n}{n_g + \frac{P}{P_{vj}} n_L} = 1 \tag{3.10}$$

With Eqs. (3.9) and (3.10), dealing with a multicomponent mixture system at any equilibrium state of it, we can calculate the quantities of the component j, respectively, in the liquid and gas phases.

Since $\overline{n_L} = n_L/n$, $\overline{n_g} = n_g/n$, $\overline{n_L} + \overline{n_g} = 1$, and $n = n_g + n_L$, it is clear that $\overline{n_L} + \overline{n_g} = 1$, withal.

So as to result in a simplification, if both of the numerator and the denominator of the right-hand side expression of Eq. (3.9) are divided by n, with Eq. (3.10) dealt with in the same way, we can get:

The composition equation for the liquid phase:

$$\sum_{j=1}^{m} x_j = \sum_{j-1}^{m} \frac{z_j}{1 + \overline{n_g}\left(\frac{p_{vj}}{p} - 1\right)} = 1 \qquad (3.11)$$

The composition equation for the gas phase:

$$\sum_{j=1}^{m} y_j = \sum_{j-1}^{m} \frac{z_j}{1 + \overline{n_L}\left(\frac{p}{p_{vj}} - 1\right)} = 1 \qquad (3.12)$$

As the value of z_j and n is already known, it is only needed to determine the value of n_g or n_L by adopting the trial-and-error method. Through solving the equations presented above, we can determine the composition of the liquid and the gas phases, respectively, and quantities of every component in each phase. Taking Eq. (3.12) as an example, the steps of the trial calculation are:

(1) Select a trial value of $\overline{n_L}$ between 0 and 1.0.
(2) Refer to related handbooks and obtain the vapor pressures, using the symbol P_{vj}, of the individual components at the given temperature.
(3) According to Eq. (3.12), figure out the values of y_j and $\sum_{j=1}^{m} y_j$; if the value of $\sum_{j=1}^{m} y_j$ is not equal to 1.0, reselect the value of $\overline{n_L}$ and repeat the step (2) and step (3) until the summation equals 1.0.
(4) Calculate the mole fraction of component j in the liquid phase: $y_j = z_j/[1 + \overline{n_L}(P/P_{vj} - 1)]$.
(5) If the total mole number n is already known, correspondingly, the mole number in the liquid phase is $n\,\overline{n_L}$, and the mole number in the gas phase is $n\,\overline{n_g}$.

[Example 3.1] Calculate the compositions and quantities of the gas and liquid when 1.0 mol of the following ideal mixture in Table 3.1 is brought to equilibrium at 65.5 °C (150 °K) and 1.378 MPa. Assume ideal solution behavior.

Table 3.1 The composition of an ideal mixture

Component	Composition, mole fraction	P_{vj}, Psia	Y_j, Mole fraction	X_j, Mole fraction
Propane	0.610	350	0.771	0.441
n-Butane	0.280	105	0.194	0.370
n-Pentane	0.110	37	0.035	0.189
Sum	1.0		$\sum y_j = 1.000$	$\sum x_j = 1.000$

Solution:

(1) Refer to Fig. 3.5 to obtain the vapor pressures (P_{vj}) of the individual components at the given 65.5 °C, and list them in Table 3.1 (line 3).
(2) Select $\overline{n_L} = 0.487$ to undergo the trial calculation, and figure out y_i through Eq. (3.12). List them in Table 3.1 (line 4).
(3) The result of the trial calculation satisfies the equation $\sum\limits_{j=1}^{3} y_j = 1.0$.
(4) Calculate the value of x_J through the Eq. (3.11), and list the composition of the liquid phase in Table 3.1 (line 5).

4. Calculation of the Bubble-point Pressure of Ideal Solutions

The bubble-point pressure, the pressure at which the first bubble of gas is formed in the system, represents the phase-change point at which a closed system is converted from a single phase of liquid into a two-phase of liquid and gas. For practical purposes, this formed bubble of gas is infinitely small and the quantity of it is negligible. Therefore, the composition of the liquid phase can be directly considered as the composition of the system, that is to say, $n_g \approx 0$, $n_L = n$. Then, the pressure of the system is exactly the bubble-point pressure, $P = P_b$, and Eq. (3.10) can be simplified:

$$\sum_{j=1}^{m} \frac{z_j}{P_b/P_{vj}} = 1 \tag{3.13}$$

After a transformation of it, we can get the bubble-point pressure:

$$P_b = \sum_{j=1}^{m} z_j P_{vj} \tag{3.14}$$

The equation above is the used to calculate the bubble-point pressure of an ideal solution. Furthermore, what it indicates is that, the bubble-point pressure of an ideal solution at a given temperature is the summation of the products of mole fractions and vapor pressure for each component.

[Example 3.2] Calculate the bubble-point pressure of the mixture in the example 3.1 at 65.5 °C (150 °K). Assume ideal solution behavior.

Solution: Solve Eq. (3.14), and show the results in Table 3.2.

Table 3.2 Calculation of bubble-point pressure

Component	Composition, mole fraction	Vapor pressure P_{vj}, psia	$z_j P_{vj}$
Propane	0.610	350	213.5
n-Butane	0.280	105	29.4
n-Pentane	0.110	37	4.1
Sum	1.0		$P_b = 247$psia

5. Calculation of the Dew-point Pressure of Ideal Solutions

The dew-point is the phase-change point at which a system is converted from a single phase of gas into a two-phase of gas and liquid. At this point, $n_L = 0$, $n_g = n$, and the total pressure of the system is equal to the dew-point pressure, $P = P_d$. From Eq. (3.9), we can get that:

$$\sum_{j=1}^{m} \frac{z_j}{P_{vj}/P_d} = 1 \tag{3.15}$$

Then, the dew-point pressure is:

$$P_d = \frac{1}{\sum\limits_{j=1}^{m} \frac{z_j}{P_{vj}}} \tag{3.16}$$

Therefore, the dew-point pressure of an ideal gas mixture at a given temperature is simply the reciprocal of the summation of the mole fraction divided by vapor pressure for each component.

[Example 3.3] Calculate the dew-point pressure of the mixture in the example 3.1 at 65.5 °C (150 °K). Assume ideal gas behavior.

Solution: Solve Eq. (3.16), and show the results in Table 3.3.

Finally,

$$p_d = \frac{1}{1.071} = 0.933 \text{ MPa.}$$

Table 3.3 Calculation of dew-point pressure

Component	Composition, mole fraction	Vapor pressure P_{vj}, MPa	z_j/p_{vj}
Propane	0.610	2.412	0.253
n-Butane	0.280	0.723	0.387
n-Pentane	0.110	0.255	0.431
Sum	1.0		1.071

3.2.2   Phase-State Equations for Real Gas–Liquid System

1. The Limitations of the Gas–Liquid Equilibrium Equation for Ideal System

With severely restricted conditions for validity, the composition equations for ideal solutions to predict the gas–liquid equilibrium can only be used as a theoretical guidance to help us understand the phase-state changes in real gas–liquid system. Generally, the use of the gas–liquid equilibrium equations for ideal system is limited in petroleum engineering because:

(1) Dalton's law is based on the assumption that the gas behaves as an ideal solution of the ideal gases, and furthermore, it can apply to a specific pressure of 0.69 MPa (100 psia) and a temperature range below the moderate.
(2) Raoult's law is applicable only when the studied liquid mixture is an ideal solution in which the components are very similar chemically and physically.
(3) Since a pure substance does not have a vapor pressure at temperatures above its critical temperature, these equations are limited to temperatures less than the critical temperature of the most volatile component in the mixture. For example, if methane is one of the components in the system, these equations can only be employed at temperatures below 190 K. Practically, due to the fact that there are always highly volatile hydrocarbon components existing in naturally occurring crude oil and gas, the engineering application of the gas–liquid equilibrium equations for ideal system is greatly limited.

2. The Definition of Equilibrium Ratio

Theoretically, it is impossible for a substance to liquidize at temperatures above its critical temperature, no matter what it is and no matter how high a pressure is put on it—in other words, it exists in a form of gas for ever. However, in fact, when a system exists as a mixture, it still has the problem of gas–liquid equilibrium although the temperature of it is already above the critical temperatures of its highly volatile components.

Various theoretical methods to overcome these problems have been tried, ended in failure. Good practice dictates that to determine experimentally the values of the compositions of the gas and liquid that exist at equilibrium is a way feasible to calculate the equilibrium. For this reason, a concept of "equilibrium ratio" is involved in the study. The equilibrium ratio K_j of component j is defined as:

$$K_j = \frac{y_j}{x_j} \qquad (3.17)$$

In the equation, y_j and x_j are the experimentally determined values of mole fractions of the component j, respectively, in the gas and liquid phases of a true solution that exist at equilibrium at a given pressure and temperature. Equilibrium ratios are sometimes also called vaporization equilibrium constants, equilibrium gas–liquid distribution ratios, distribution coefficients, or simply, K-values. K-values change with pressure, temperature, and type of mixture. The composition of a mixture has myriads of changes, and therefore the equilibrium ratio of a system varies by its changing composition and state.

3. The Gas–Liquid Equilibrium Calculation of Real Solutions

In this calculation, the composition equations both of the liquid and gas phases can be obtained by replacing the y_i/x_i in Eq. (3.5) with the experimentally determined K_j:

The composition equation for the liquid phase:

$$\sum_{j=1}^{m} x_j = \sum_{j=1}^{m} \frac{z_j n}{n_l + n_g k_j} = 1 \tag{3.18}$$

The composition equation for the gas phase:

$$\sum_{j=1}^{m} y_j = \sum_{j=i}^{m} \frac{z_j n}{n_g + \frac{n_l}{k_j}} = 1 \tag{3.19}$$

We can also simplify the two equations by expressing them with mole fraction:

$$\sum_{j=1}^{m} x_j = \sum_{j=1}^{m} \frac{z_j}{1 + \bar{n}_g(k_j - 1)} = 1 \tag{3.20}$$

$$\sum_{j=1}^{m} y_j = \sum_{j=1}^{m} \frac{z_j}{1 + \bar{n}_L(\frac{1}{k_j} - 1)} = 1 \tag{3.21}$$

[Example 3.4] Make a recalculation of the Example 3.1 on the assumption that the solution is a real one. The values of the equilibrium ratios have already been extracted from the charts in Appendix 3 and given in Table 3.4.

Solution:

(1) Select $\bar{n}_L = 0.5$ to undergo the trial calculation. Use Eq. (3.21) to figure out y_j, and obtain that $\sum y_j = 0.992$.
(2) Reselect $\bar{n}_L = 0.6$, and obtain that $\sum y_j = 1.010$.
(3) Then, chose a value between 0.5 and 0.6 for $\bar{n}_l$, for example, $\bar{n}_l = 0.547$, we can get that $\sum y_j = 1.0$. The calculations are listed in Table 3.4.

It can be noticed that when $\bar{n}_L = 0.547$ the system reaches its equilibrium. Obviously, the assumption of ideal solution behavior results in an error of about ten percent in $\bar{n}_L$.

Essentially, the composition calculation of a gas–liquid equilibrium is to figure out the quantitative proportions of the gas and liquid in the case that the composition of the studied system is definite and the conditions (pressure and temperature) of it are already known. So, the first thing to do is to determine the phase state of the system under the given conditions, and if the system is in a single phase of liquid or

Table 3.4 The compositions of the liquid and gas phases of a real solution

Component	Composition Z_j, mole fraction	Equilibrium ratio, K_j	y_j $(\bar{n}_L = 0.5)$ $y_j = z_j/[1 + \bar{n}_L(1/K_j - 1)]$	y_j $(\bar{n}_L = 0.6)$	y_j $(\bar{n}_L = 0.547)$	Composition of the liquid, x_j $(\bar{n}_L = 0.547)$ $x_j = y_j/K_j$
Propane	0.610	1.55	0.742	0.775	0.757	0.488
n-Butane	0.280	0.592	0.208	0.198	0.203	0.344
n-Pentane	0.110	0.236	0.042	0.037	0.040	0.168
Sum	1.000		$\sum y_j = 0.992$	$\sum y_j = 1.010$	$\sum y_j = 1.000$	1.000

gas, it is not necessary to go on with the calculation. Begin to select the first value for the trial calculation only when the system is in a two-phase state.

4. Calculation of the Bubble-point Pressure of Real Solutions

At the state of bubble-point, the mole number of gas phase is $n_g = 0$, and the mole number of liquid phase is $n_L = n$. By resorting to Eq. (3.19), we can get:

$$\sum z_j K_j = 1.0 \tag{3.22}$$

Pressure does not appear implicitly in Eq. (3.22) but is concealed in the equilibrium ratio K_j. Thus, for a system with a known composition, its bubble-point pressure at a given temperature may be determined by means of selecting a trial value of pressure, when the K_j obtained from the selected pressure can satisfy Eq. (3.22); this corresponding trial pressure is rightly the bubble-point pressure. The procedure to determine the bubble-point pressure for a real solution follows:

(1) Give the pressure a trial value $p_b^{(0)}$, or in another way, assume the solution an ideal one whose bubble-point pressure value can be assigned to $p_b^{(0)}$ as the first trial.

(2) Refer to related charts and obtain the value of K_j at the given pressure and $p_b^{(0)}$.

(3) Calculate $\sum z_j K_j$. If $\sum z_j K_j < 1.0$, choose another $p_b^{(1)}$ with a smaller value and determine a new K_j; If $\sum z_j K_j > 1.0$, choose another $p_b^{(1)}$ with a greater value.

(4) Calculate according to Eq. (3.22) and repeat the above steps until the result satisfies the required precision.

[Example 3.5] Assuming the mixture described in the example 3.1 as a real solution, calculate the bubble-point pressure of it. The values of K_j can be seen in Table 3.5.

Solution:

(1) Determine the bubble-point pressure of solution on the assumption that it was an ideal one, and take this pressure as the first trial value (Example 3.2).
$p_b^{(0)} = 1.702$ MPa

(2) Obtain the value of K_j when $T = 150\ °F$ and $P_b = 1.516$ MPa, and figure out $z_j K_j$. As it turns out, $\sum z_j K_j = 0.969 < 1.0$

Table 3.5 Calculation of the bubble-point pressure of a real solution

Component	z_j, mole fraction	The first trial		The second trial		The third trial	
		K_j (1.702 MPa)	$z_j K_j$	K_j (1.516 MPa)	$z_j K_j$	K_j (1.633 MPa)	$z_j K_j$
Propane	0.610	1.32	0.805	1.44	0.878	1.36	0.830
n-Butane	0.280	0.507	0.142	0.553	0.155	0.523	0.146
n-Pentane	0.110	0.204	0.022	0.218	0.024	0.208	0.023
Sum	1.000		0.969		1.057		0.999

Fig. 3.19 Determination of the bubble-point pressure by interpolation method

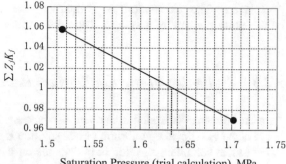

$\sum z_j K_j$

Saturation Pressure (trial calculation), MPa

(3) In the second circle of trial calculation, choose P_b = 220psia and redetermine the value of K_j, and then figure out $z_j K_j$. The summation is $\sum z_j K_j = 1.057 > 1.0$

(4) By adopting a linear interpolation method, it can be determined that P_b = 237psia when $\sum z_j K_j = 1.0$ (Fig. 3.19).

(5) Use P_b = 237psia to determine the value of K_j. The summation turns out to be $\sum z_j K_j = 0.999 \approx 1.0$, which results in an error of only 0.001 and satisfies the required precision in engineering. Now, the trial calculation can be stopped and the final result is P_b = 1.633 MPa.

5. Calculation of the Dew-Point Pressure of Real Solutions

When a mixture of gases is at a state of dew-point, we can also consider that $n_L = 0$ and $n_g = n$ although the system is two-phased. Then, Eq. (3.18) becomes:

$$\sum z_j / K_j = 1.0 \tag{3.23}$$

So, the dew-point pressure is the pressure that satisfies Eq. (3.23) at the given temperature. Similar to the calculation of the bubble-point pressure, a trial-and-error method should be used to obtain the dew-point pressure of a system. Firstly, give a trial value to the pressure using Eq. (3.16), then go on with the trial calculation step by step, approaching to the required precise result.

[Example 3.6] Assuming the mixture described in the example 3.1 as a system of real gas, calculate its dew-point at 65.5 °C. The values of K_j have already been extracted from related charts and shown in Table 3.6.

Table 3.6 Calculation of the bubble-point pressure of a real gas

Component	Composition z_j, mole fraction	K_j (0.971 MPa)	z_j/K_j
Propane	0.610	2.11	0.289
n-Butane	0.280	0.785	0.357
n-Pentane	0.110	0.311	0.354
Sum	1.000		1.000

Solution: The calculation is similar to the process of determining the bubble-point pressure of a system. The final result is: $P_d = 0.971$ MPa.

3.2.3   Evaluation of the Equilibrium Ratio K—By Charts

As seen from the discussion above and the presented examples, the key to calculation of the gas–liquid equilibrium of real solution is the evaluation of the K_j-value. For the ideal solutions, K_j-value only depends on the pressure and temperature of the studied system, mathematically $K_j = f(p, T)$. In the oil industry, since the oil & gas reservoir systems are quite complicated in compositions and usually subjected to high pressure, it is not reasonable to consider the fluids in them as ideal solutions. With regard to this, charts of experiential equilibrium ratios are made for practical reference. The charts are created on the basis of experimental approaches, in which the composition of the studied system is already known and the mole numbers of the individual component in each phase are measured at different equilibriums of the system. Then, figure out the values of K_j according to the equation $K_j = f(P, T)$, and make use of them to draw a chart. Obviously, the experiential equilibrium ratios are the values obtained from experimental approaches.

As proven by experiments, the composition of a mixture has influence on the equilibrium ratio of it when it is subjected to a high pressure. In other words, the equilibrium ratio is not only a function of temperature and pressure, but also that of the composition of the system, or $K_j = f(P, T,$ composition$)$.

1. The Effect of Composition on the Equilibrium Ratio

By making a K_j–P relationship curve under isothermal condition for every component in the studied system, we can note that there are two obvious characteristics of these curves as shown in Fig. 3.20:

(1) Almost all the curves have a slope of 1 when they are at low pressure. The pressure that exactly falls at the crossing point of every individual curve and the $K_j = 1$ line represent the vapor pressure of the corresponding component at the given temperature.

(2) Except CH_4, the slope-change trend of all the curves is similar to each other. When the pressure is low, the equilibrium decreases with the increase in pressure; but when the pressure rises up to a certain value, the equilibrium reaches its minimum value; then, after that lowest point, the equilibrium increases with the increase in pressure; finally, as the pressure continues increasing, the curves converge to $K_j = 1$ (i.e., $y_j = x_j$). The pressure at which the equilibrium converges to 1 is called the "pressure of convergence," referred to as P_{cv}.

In case that the given temperature is rightly the critical temperature of the mixture system, the convergence pressure is equal to the critical pressure of the system. If the given temperature is lower than its critical temperature, the dew-point

Fig. 3.20 The equilibrium
ratio with changing pressure

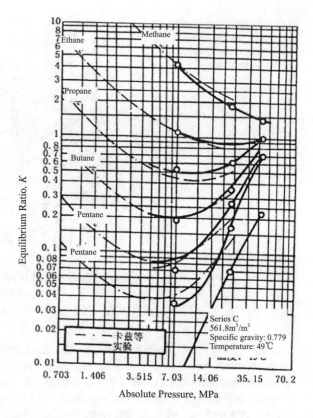

Absolute Pressure, MPa

or bubble-point may be encountered as the experimental pressure increases, which causes failure in the determination of the of convergence pressure of convergence. However, by a method of extrapolation, this problem can be solved. In accordance with their changing trend, extrapolate these curves to the points at which their equilibrium ratios reach 1. The pressure at a point like this is called "apparent pressure convergence." In fact, the system has already been converted into a single-phase state before it reaches its apparent pressure, for example, at the bubble-point and the dew-point, and consequently K_j makes no sense any more. Therefore, the pressure of convergence and the apparent pressure of convergence are not the same in significance although they are both called "pressure of convergence" in engineering.

Different mixtures have different critical parameters. With the introduction of the pressure of convergence, the relationship between the equilibrium ratio with the temperature, pressure, and composition, $K_j = (P, T, \text{composition})$ can now be replaced by the relationship between it with the temperature, pressure, and the pressure of convergence, $K_j = (P, T, P_{cv})$. According to the principle of corresponding state, if two systems are composed of homogeneous materials and have the same convergence pressure when they are subjected to the same temperature, for any involved individual component its equilibrium ratio in one system is equal

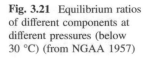

Fig. 3.21 Equilibrium ratios of different components at different pressures (below 30 °C) (from NGAA 1957)

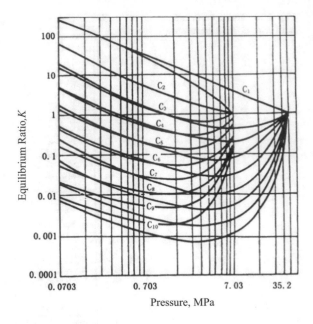

to that in the other one, no matter whether the composition and the quantity of each component are the same or not. The convergence pressure P_{cv} can be seen as the parameter that describes the effect of the composition on K_j. The two clusters of P–v curves in Fig. 3.21 represent two systems with different compositions, and the convergence pressure of them are, respectively, 7.0 and 35.0 MPa. Comparing the curves, we can see that when the pressure is lower than 0.7 MPa, the C_2–C_5 components from the two different systems overlap their curves quite closely, which means each of them has the same equilibrium ratio in the two systems. This phenomenon can be explained in this way: when the pressure of a system is low, it approaches an ideal solution whose composition affects little on the equilibrium ratio K. But when the pressure is higher than 0.7 MPa, the disparity between the two clusters of curves is much greater. In conclusion, in low-pressure situation (pressure lower than 0.7 MPa), the composition of a multicomponent system has little effect on the equilibrium ratios of its components; and in high-pressure situation, the composition of a multicomponent system has great effect on the equilibrium ratios of its components.

2. The Method to Determine K

In 1957, the NGAA published a set of K–p charts whose convergence pressures are 4.2, 5.6, 7.0, 21.0, 28.5, 35.0, 70.0, and 140.0 MPa. For reason of space, here we only show the K–p curve of propane, with a convergence pressure of 35.0 MPa, in Fig. 3.22, which illustrates the relationship among the convergence pressure, temperature, and equilibrium ratio.

Fig. 3.22 Equilibrium ratio of propane (the convergence pressure is 35 MPa)

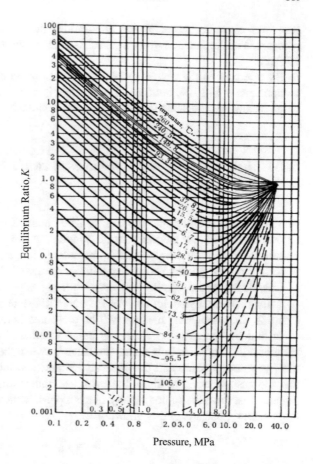

In use of a set of *K–p* curves, the first thing to do is determining the convergence pressure of the studied system, and then suitable charts are chosen according to the determined convergence pressure.

In engineering practice, the operating pressure is not high during the gas–oil separation on the surface and the liquefaction of natural gas, and the given charts with a convergence pressure of 35.0 MPa (Figs. 3.21, 3.22) enable the determination to satisfy the required precision.

In the calculation of phase equilibrium under reservoir temperature and pressure, the choosing of convergence pressure is much more complicated, and some special mathematical algorithms should be used to determine the convergence pressure of a multicomponent system. Some engineers conceive that the convergence pressure should be greater than the dew-point or bubble-point pressure by at least 10 %, and from this standpoint, we can estimate the value of the convergence pressure when the dew-point or bubble-point pressure is already known.

After the determination of the convergence pressure P_{cv}, or in other words, the determination of which *K–p* chart should be chosen, the *K* value can be obtained from the chart according to the conditions such as reservoir temperature and pressure. The Appendix 3 gives a set of $K-p$ charts with a convergence pressure of 75 MPa.

By referring to the charts to obtain equilibrium ratios, the error for the C_2-C_5 components is 5–10 %, but for CH_4 (the error is more than 30) and C_7+, it is quite great. Therefore, a calculation method is more suitable for CH_4 and C_7+.

3.2.4 Evaluation of the Equilibrium Ratio K—By Calculation

The gas–liquid equilibrium ratio can also be obtained through some mathematical algorithms based on thermodynamic models. A large number of models and algorithms have been developed at home and abroad, and the computer calculation which makes the problem solved much more conveniently has also been realized.

Calculating phase equilibrium ratio with a computer involves the concept of fugacity in thermodynamics and the PK and SRK state equations introduced in the Chap. 2. The calculating method will be presented here.

In thermodynamics, the criteria to judge whether a system reaches an equilibrium or not include not only the equivalence between the liquid-phase pressure and the gas-phase pressure and the equivalence between the liquid-phase temperature and the gas-phase temperature, but also, as another important criteria, each component must have the same fugacity in each phase which are subjected to the same temperature, pressure, and equilibrium:

$$f_{iV} = f_{iL} \qquad (3.24)$$

where f_{iV}, f_{iL}—the fugacity of component i, respectively, in the gas and liquid phases.

The physical significance of this equation is that the escaping tendency of the component i in the gas and liquid phases is the same. That is to say, only when the escaping tendency of the component i in the two phases is the same, the system can reach its equilibrium.

Besides existing as a criterion for the judgment of equilibrium, Eq. (3.24) can be used to calculate the equilibrium ratio K.

The fugacity of component i in the gas phase equals the product of the fugacity of component i as a pure gas-state substance times the mole fraction of it in the gas phase, subjected to the same pressure and temperature:

$$f_{iV} = f_{iV}^0 y_i \qquad (3.25)$$

The fugacity of component i in the liquid phase equals the product of the fugacity of component i as a pure liquid-state substance times the mole fraction of it in the liquid phase, subjected to the same pressure and temperature:

$$f_{iL} = f_{iL}^0 x_i \tag{3.26}$$

where

f_{iV}^0—the fugacity of pure gas-state component i at the equilibrium pressure and temperature;

f_{iL}^0—the fugacity of pure liquid-state component i at the equilibrium pressure and temperature;

y_i, x_i—the mole fractions of component i, respectively, in the gas and liquid phases

According to the mentioned equilibrium criteria, i.e., Eq. (3.24), we can get:

$$f_{iV}^0 y_i = f_{iL}^0 x_i \tag{3.27}$$

Then, the equilibrium ratio can be expressed with the fugacity, and all the following forms can be used:

$$K_i = \frac{y_i}{x_i} = \frac{f_{iL}^0}{f_{iV}^0} = \frac{f_{iL}/x_i}{f_{iV}/y_i} = \frac{f_{iL}/(x_i P)}{f_{iV}/(y_i P)}$$

$$= \frac{\phi_{iL}}{\phi_{iV}} \tag{3.28}$$

where

P	the equilibrium pressure of the system;
$\frac{f_{iL}}{x_i P} = \phi_{iL}$	the liquid-phase fugacity coefficient of the component i;
$\frac{f_{iV}}{y_i P} = \phi_{iV}$	the gas-phase fugacity coefficient of the component i.

Before using Eq. (3.28) to calculate K_i, the gas-phase and liquid-phase fugacity coefficients φ_i should be figured out through a suitable phase-state equation. As discussed above, the equilibrium ratio K_i expressed by Eq. (3.28) depends on not only the temperature T and the pressure p but also the compositions of the equilibrium phases. Therefore, the calculation is a procedure of iteration. Taking the SRK state equation as an example, the following part will illustrate how to calculate the equilibrium ratio K by adopting single-model method.

The SRK state equation $Z^3 - Z^2 + Z(A - B - C) - AB = 0$ has three roots of Z-factor, among which the greatest positive one Z_V is the compressibility factor of the gas-phase mixture while the smallest positive one Z_L is the compressibility factor of the liquid-phase mixture.

The equations to calculate fugacity coefficients:

For the gas phase:

$$\ln \phi_{iV} = (Z_V - 1)\frac{b_i}{b} - \ln(Z_V - B)$$
$$- \frac{A}{B}\left\{\frac{1}{a}\left[2a_i^0 \sum_j x_j a_j^{0.5}\left(1 - \overline{K_{ij}}\right) - \frac{b_i}{b}\right]\right\}\ln\left(1 + \frac{B}{Z_V}\right) \qquad (3.29)$$

For the liquid phase:

$$\ln \phi_{iL} = (Z_L - 1)\frac{b_i}{b} - \ln(Z_L - B)$$
$$- \frac{A}{B}\left\{\frac{1}{a}\left[2a_i^{0.5} \sum_j x_j a_j^{0.5}\left(1 - \overline{K_{ij}}\right) - \frac{b_i}{b}\right]\right\}\ln\left(1 + \frac{B}{Z_L}\right) \qquad (3.30)$$

The parameters such as b_i, b, a_i, a, A, and B in Eqs. (3.29) and (3.30) are parameters in the SRK state equation, and you can refer to the Chap. 2 for more details.

To obtain the phase equilibrium ratio K_i by utilizing the SRK or PR state equations, the first thing to do is figure out the gas-phase and liquid-phase

Fig. 3.23 Program graph of the calculation of K

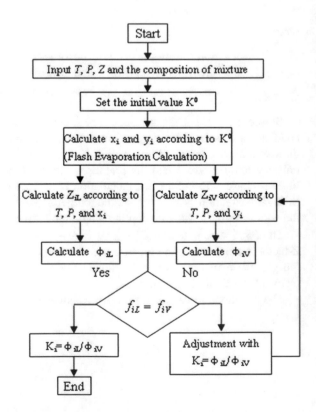

compressibility factors, Z_V and Z_L, through solving corresponding equations; then, substitute the compressibility factors into the equations for the fugacity calculation, and get the fugacity coefficients φ_{iv} and φ_{iL}; and finally, calculate the phase equilibrium ratio K_i. The gas-phase composition x_i and liquid-phase composition y_i are unknown in a system in which the pressure p, temperature T, and the composition Z_i are designed, and however, they are needed to figure out the Z-factors in the state equations; so, this calculation is a procedure of repeated iteration for which only a computer can be competent. The procedure chart to calculate the equilibrium ratio K_i is shown in Fig. 3.23. In engineering practice, the charts and calculation methods are usually combined for application.

3.3 Solution and Separation of the Gas in an Oil–Gas System

This lesson will focus on the process of the separation of the oil and gas in a real system, while the methods to calculate the compositions and quantities of each phase during the separation process, or in other words, how to use the phase-state equations to calculate the gas–liquid equilibriums in oil reservoir engineering will be particularly introduced in the next lesson.

3.3.1 The Separation of Natural Gas from Cruel Oil

The phase changes of the reservoir hydrocarbons result from the convictions and equilibriums between the liquid and gas phases, which can be reflected by the solution and separation of the two phases.

The separation of oil and gas is the phenomenon of degasification which happens to the cruel oil when it is subjected to the dropping fluid pressure during production. It occurs in the reservoir, in the process of wellbore flowing or the conveyance of the produced oil and gas on the surface. According to different patterns in which the pressure is reduced, the oil and gas can be separated in three ways: flash liberation, discrepant liberation, and differential liberation.

1. Flash Liberation (Single-stage Degasification)

Flash liberation is also called "contact liberation," "contact degasification," or "single-stage degasification." The liberated gas, during the separation, maintains constant contact with the oil, and the composition of the system is not changed. As shown in Fig. 3.24 which illustrates a flash liberation experiment with a PVT cylinder, the system undergoes a pressure reduction as the piston goes down, and vice versa.

When a reduced-pressure experiment is carried out for the degasification of a gas–oil system, the pressure reduces gradually from the reservoir pressure until it

Fig. 3.24 Process of contact degasification

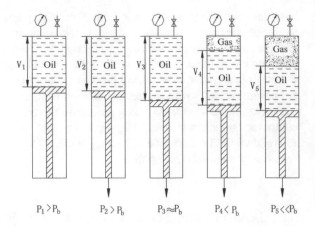

drops down to the atmosphere pressure and the system reaches equilibrium. And during the process, related parameters should be measured and a p–V (pressure–volume) curve as illustrated in Fig. 3.25 should be drawn.

During the whole pressure reduction, from p_1 to p_5, the oil and gas keep in contact, the gas is not discharged, and the total composition of the system is held constant. This is the characteristic of flash liberation which is the reason why it is also called contact degasification and single-stage degasification.

The p–V curve indicates that before the pressure is reduced to P_b what happens to the system is just the volume expansion of the liquid phase, which causes a small $\Delta V/\Delta P$ ratio, or a small slope of the line; but when the pressure drops lower than P_b, there are gases that are liberated out of the liquid, which causes a much greater $\Delta V/\Delta P$ ratio (i.e., the slope of the line). Therefore, the crossing point of the two lines that differ significantly in slope represents the point at which the degasification

Fig. 3.25 p–V relationship during the process of contact degasification

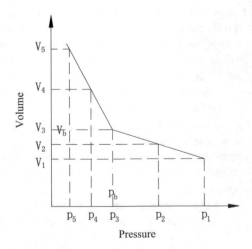

begins to occur. What's more, the pressure at this crossing point is rightly the saturation pressure P_b of the studied oil–gas system.

It should be noted that on a p–V curve the linear relation like this only appears nearby the point of saturation pressure, that is to say, when the pressure is remote from this point, the linear relationship does not necessarily exist.

As long as the total volume of the liberated gas at atmosphere pressure, symbolized by V_g (m³), and the on-surface volume of the oil, symbolized by V_o (m³) are measured, the gas–oil ratio of flash liberation can be obtained: $R_s = V_g/V_o$.

Under field situation, this single-stage pattern degasification corresponds to the process of feeding the well-mouth effluents directly into the separator or storage tank in which the oil and gas phases reach equilibrium instantaneously. The quantity of gas liberated through this way is comparatively large, or in other words, the gas–oil ratio tends to be high. Furthermore, gas from this process is quite heavy and contains comparatively much light oil. To reduce the wastage of light oil and gain more crude oil on the surface, a multistage degasification pattern should be adopted.

2. Discrepant Liberation (Multistage Degasification)

In a discrepant liberation, the liberated gas at each stage of the separation process is discharged out of the system and only the liquid is allowed to enter the next stage. In such a manner, the system undergoes the degasification in companion with a series of changes in its composition. Figure 3.26 shows the discrepant liberation process, and Fig. 3.27 is a schematic drawing that illustrates the processing steps under field situation. Figure 3.27a illustrates a two-stage degasification in which the equilibrium pressure on the second stage is the atmosphere pressure while the

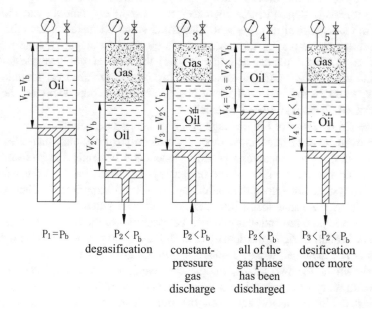

Fig. 3.26 Process of multistage degasification

Fig. 3.27 Flowchart of multistage degasification in oil-field practice

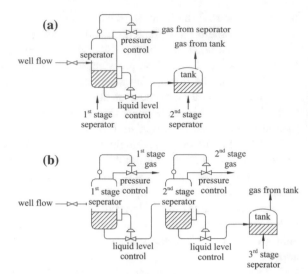

equilibrium pressure on the first stage is kept higher than the second one. Figure 3.27b illustrates a three-stage degasification in which the equilibrium pressure on the third stage is the atmosphere pressure while the equilibrium pressure on the second and first stages is gradually increased by sequence.

The calculation of the compositions and quantities of the two phases in flash and discrepant liberation processes also obeys the principles introduced in the lesson 2 of this chapter, and related examples will be presented in lesson 4.

The multistage degasification increases the cost in equipments and the complexity in technological management. Therefore, some oil fields retrieve the heavy hydrocarbons by delivering the natural gas after the first stage to some recovery devices or treatment plants, and then output the left natural gas. Besides, the determination of the optimum conditions for separation process, including the pressure, temperature, and how many stages needed, is dependent on the prosperities of crude oil and related calculations.

3. The Comparison of Single-stage and Multistage Degasification

As proven by experiments and theories, the compositions and quantities of the liberated oil from the separation process depend on not only the pressure, temperature, and the composition of the oil–gas system itself, but also the pattern in which the system undergoes degasification. In related experiment, when two oil samples from the same formation undergo different patterns of degasifications under the same pressure and temperature, the quantities of the liberated gas from the two are not equal. Besides, the density of both the oil and gas obtained from the two patterns is not identical, either. Tables 3.7 and 3.8 give more details.

As shown in the two tables, on the one hand, a single-stage degasification produces more gas but less oil than a multistage degasification, that is, the former one gives a higher gas–oil ratio; on the other hand, the gas liberated from a

Table 3.7 The influence of degasification patterns on the quantity of liberated gas

Oil field (oil samples form well bottom)	Gas–oil ratio, m^3/m^3		Relative error of the gas quantities from the two patterns (%)
	Single-stage degasification	Single-stage degasification	
Romashki, Russian	59.6	41.4	30.6
Cibuersic, Russian	83.0	71.5	13.9
Gimiteliyesic, Russian	175.4	144.5	17.6

Table 3.8 Comparison of single and multi degasification on same oil sample

Degasification pattern	Temperature (°C)	Equilibrium pressure (MPa)	Gas–oil ratio (m^3/m^3)	Specific gravity of tank-oil (15.5 °C)	Specific gravity of gas (15.5 °C)
Single-stage	25	0.1	273	0.7711	0.9737
Two-stage	25	The 1st stage 0.5	229	0.7807	0.8559
		The 2nd stage 0.1	10.3		

The saturation pressure of this crude oil is 12.6 MPa, and the reservoir temperature is 82°C

single-stage pattern has a comparatively higher specific gravity, which indicates that there is more light oil contained in it.

The reason for this phenomenon: During a single-stage degasification, the oil and gas phases keep in contact all along, and all the components, no matter the heavy ones or the light ones, will enter the gas phase according to their own equilibrium ratios $K_j = y_j/x_j$. As a result, the heavy hydrocarbons, due to thermo-dynamic equilibrium, also have a chance to be vaporized and become part of the gas phase, which leads to a high quantity and a high specific gravity of the liberated gas. In contrast to the situation discussed above, when a system undergoes a multistage degasification, what happens to every drop of pressure is that the light hydrocarbon components which always tend to be liberated more easily are firstly discharged out of the system while the heavy ones are left. The consequence is that the light components keep on escaping and the proportion of the heavy components in the left liquid phase keeps on increasing. Therefore, in a multistage degasification, the gas is liberated out of the system with a small quantity and low specific gravity. This paragraph provides a qualitative or descriptive explanation, and examples about related quantitative calculations will be presented in the lesson 4 of this chapter.

4. Differential Liberation

The differential liberation, also called differential degasification, refers to the pattern of degasification in which the liberated gas is discharged immediately after every minimum pressure drop and forced away from the contact with the liquid phase. That is to say, this pattern, as illustrated in Fig. 3.28, is a process of

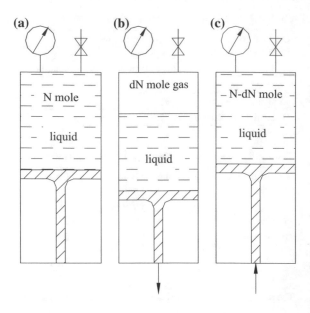

Fig. 3.28 Process of differential liberation. **a** The liquid is in a state of saturation, **b** pressure decreases and the two phases coexist, **c** gas is discharged and the remaining liquid is in a new state of saturation

continuous drop of pressure, continuous discharge of gas, and continuous change of system composition.

What distinguishes the differential liberation from the discrepant liberation is that it involves many more stages of degasification while the pressure drop in each stage is quite tiny. Different from the numerable several stages involved in a multistage degasification, in a differential liberation the pressure is reduced by a large number, even countless, times of pressure drops, until it falls to a designated value, for example, the atmosphere pressure.

During a differential liberation in the laboratory, the gases liberated from every drop of pressure will be discharged in time; at the same time, the volumes of them will be measured and accumulated to a summation V_g, which is the total volume of the gas discharged in the whole differential liberation process. Then, measure the volume of the left liquid phase of crude oil under the standard conditions on the land surface, V_o, and finally, we can figure out the gas–oil ratio of a differential liberation: $R = V_g/V_o$.

5. The Process of Degasification in Reservoir Development and Production

(1) In Formation

When a reservoir is developed at a pressure lower than its saturation pressure, the process of degasification is rather complicated. Ordinary reservoirs usually undergo degasification in a pattern which falls between the contact liberation and the differential liberation. Once the gas is liberated from the liquid, on the one hand, since the velocity difference existing between the two phases causes them to loose contact with each other, the underground degasification is quite close to the process of differential liberation; on the other hand, the gas still keeps in touch with the oil that is held in the rock porosities, which make the underground degasification similar to the situation of contact liberation.

For the reservoirs with comparatively great thickness and good vertical permeability, secondary gas-caps are formed as a result of the gravity segregation of the oil and gas, and the formation of them is close to contact liberation.

At any rate, it is probable for the degasification process to fall behind the pressure dropping in an oil reservoir (usually referred to as a hysteresis phenomenon), and the degasification process in a porous media is much more complicated than that in the laboratory. Therefore, there is some discrepancy between the laboratory measure and the real degasification process in the underground reservoirs.

In a real reservoir, there are pressure gradients resulting from the seepage flowing, or in other words, the pressures at different parts of a reservoir are not the same. For the regions whose pressure has already fallen lower than the saturation pressure, there will be gas liberated from the liquid when a minimum pressure drop happens. And at this moment, this local reservoir pressure is rightly the current saturation pressure. In the numerical simulation of oil reservoirs, the saturation pressure is treated as a parameter that changes with the change of time and location.

(2) In Oil Well

When the discrepancy between the flow velocities of the gas and liquid phases is not great enough, the degasification process in an oil well is close to contact liberation, too. The reason is that, as the reservoir oil flows from the well bottom to the well mouth on the surface, the gradually decreasing pressure makes the gas liberated from the liquid continuously. This process of degasification keeps on occurring in the well, and simultaneously, the long time of contact between the two phases enables a thermodynamic equilibrium to establish.

In case that the relative movement between the oil and gas is obvious and the gas phase flows ahead of the oil quite well, which is known as the slippage phenomenon, the degasification process in the oil well occurs in a pattern between the contact liberation and the differential liberation. Hence, it is said, in oil-field engineering practice, the classification of the contact liberation and the differential liberation does not exist in the strict sense.

(3) In the On-Surface Storage and Delivery

To some extent, the oil–gas separation on surface is an ideal process of separation. The patterns most commonly used include the two-stage and three-stage degasifications. Note that the single-stage degasification is a contact liberation.

3.3.2 The Solution of Natural Gas in Crude Oil

Compared with the tank-oil, one of the important characteristics of the reservoir oil is that there is natural gas dissolved in it. The solution and separation of natural gas in oil are essentially the conversions between the oil and gas, which occurs in the hydrocarbon systems that are subjected to changes in pressure. They are two facets of the phase conversion—what happens to the tank-oil is the separation of the natural gas from it, and what happens to the reservoir oil is the solution of the natural gas into it. The discussion of the solution process can help understand the separation process better.

1. The Solubility and Solubility Coefficient of Natural Gas in Crude Oil

As known from the physical chemistry, the solution of a single-component gas in a liquid obeys the Henry's Law which states that the solubility is proportional to the pressure when the temperature is held a constant. Expressed mathematically:

$$R_s = \alpha P \tag{3.31}$$

where

P the pressure of the system when solution occurs, MPa;
α solubility coefficient, indicating the volume of a gas that can be dissolved by a unit volume of solvent at the standard state of pressure and temperature when the gas is subjected to a unit of pressure increase, (standard) $m^3/(m^3 \text{ MPa})$;
R_s solubility, (standard) m^3/m^3.

The solubility is defined as the volume of a gas that can be dissolved by a unit volume of solvent at the standard state of pressure and temperature. It reflects the quantity of the gas that is dissolved in the liquid, while the solubility coefficient reflects the ability of the liquid to dissolve the gas. If the solubility coefficient is a constant, the relationship between the solubility and the pressure is linear.

Although applicable to the gas–oil systems in which a mutual solution does not happen easily, Eq. (3.31) produces considerable deviation when applied to a system in which the hydrocarbon gases dissolved in the crude oil have similar chemical structures. In such a case, the solubility coefficient α is no longer a constant, rendering the solubility-pressure curve a nonlinear one, or in other words, on the solution curve for such a system, different pressures correspond to different values of solubility coefficients. As shown in Fig. 3.29, when the pressure $P < P_A$, the solubility coefficient is comparatively great, caused by the readily soluble hydrocarbons such as ethane and propane which takes the initiative to dissolve into the oil; when the pressure $P > P_A$, the remnant gas phase which is dominantly comprised of methane will continue to dissolve into the crude oil, leading the solubility coefficient to nearly a constant. In pace with the increasing pressure, the solubility of gas in the crude oil keeps on growing, too.

The solution of natural gas in crude oil is quite different from the situation of a single-component gas in a liquid. The solubility of a natural gas depends on the

Fig. 3.29 Solution curve of
natural gas in crude oil

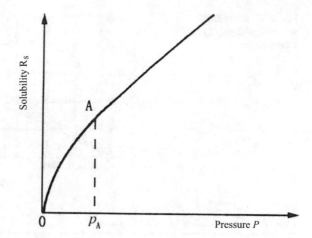

following factors: pressure, temperature, the compositions of the natural gas and the
crude oil.

With regard to given oil, if the temperature is held a constant the quantity of the
dissolved natural gas will increase with an increase in pressure, and the relationship
between its solubility and the pressure is a curve with a variable slope (Fig. 3.29),
which implies that at different pressures the solubility of a natural gas in crude oil is
not a constant. Under this circumstance, we should divide the curve into a certain
number of subsections of different pressures and determine the average solubility
coefficient for each of the subsections. The equation is:

$$\overline{\alpha}_i = \frac{\Delta R_{si}}{\Delta P_i} \tag{3.32}$$

where

ΔP_i pressure difference, MPa;
$\overline{\alpha}_i$ the average solubility coefficient when the pressure difference is ΔP_i,
       (standard) m^3/m^3 MPa;
ΔR_{si} the change of solubility corresponding to the pressure difference ΔP_i,
       (standard) m^3/m^3.

2. The Relationship between Solubility and Temperature
The solubility of gas in oil is in relation to temperature as shown in Fig. 3.30.
The quantity of dissolved gas decreases with the increase in temperature, and under
high pressure this decrease tends to happen by a larger margin, the reason of which
lies in the regularity that the vapor pressure of hydrocarbon gases increases when
the temperature increases.

3. The Relationship between Solubility and the Properties of Oil and Gas
The solubility values of gases, hydrocarbon or not hydrocarbon, are quite different in the same crude oil. Table 3.9 assumes that the solubility of methane is 1

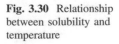

Fig. 3.30 Relationship
between solubility and
temperature

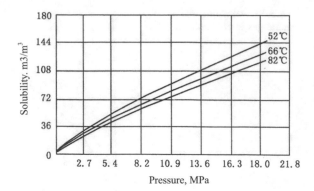

Table 3.9 The solubility of different gases

Gases	Methane	Ethane	Propane	CO_2	N_2
Solubility (multiple of the solubility to methane)	1	5	20	3.5	0.25

and lists the evaluated solubility values of C_2H_6, C_3H_8, CO_2, and N_2, which are
treated approximately as multiples of that of methane. Apart from this, the conclusion can also be drawn from an analysis of Fig. 3.31.

As shown in Table 3.9, by sequence, the solubility values of the mentioned
gases take the turns: $C_3H_8 > C_2H_6 > CO_2 > CH_4 > N_2$.

Subjected to the same pressure and temperature, natural gases with the same
composition have greater solubility in light crude oil than in heavy oil (i.e.,
high-density oil). Figure 3.32 illustrates this regularity.

Fig. 3.31 Solution curves of
different gases in crude oil

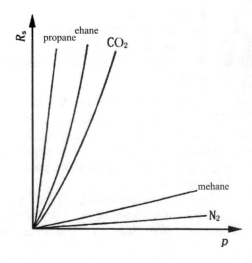

Fig. 3.32 Solubility of natural gas in crude oils of different density (from Beal 1946)

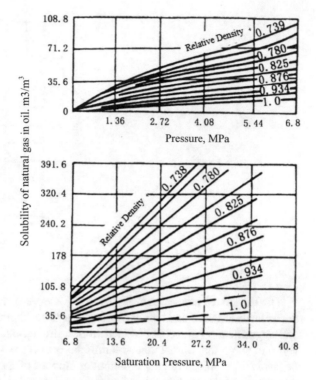

To conclude, the greater the density of natural gas, the greater solubility of it in oil; the smaller the density of oil, the easier for it to dissolve more natural gas. This is because small oil density and great gas density, when subjected to the same temperature and pressure, lead to a similarity between the oil and gas components, which help the crude oil to dissolve natural gas more easily.

The solubility of natural gas in crude oil is dependent on the properties of the natural gas and the oil.

In the case of a non-balanced solution process, the time of contact and the size of the contact area also have effect on the solubility of the studied gas.

4. The Relationship between Solution and Separation

As a unity of opposites, the two processes, solution and the separation, have close relationship when the given system with invariable composition is subjected to a constant temperature.

Holding the temperature a constant, if we dissolve all of the vaporized gases back into the reservoir oil sample where it comes from, will the solution curve seems the same as the degasification curve? Theoretically, for the flash liberation, the contact degasification curve and the solution curve should coincide with each other because that there is no change in the composition of the system and the two phases have reached equilibrium. But for the differential liberation, the two curves do not coincide since the change in the composition of the system during the process.

Fig. 3.33 Degasification
curve and solution curve

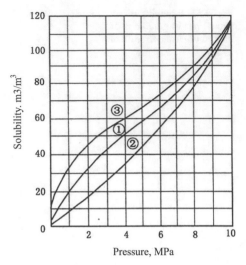

The difference can be seen in Fig. 3.33. Curve ① is the completely coincided degasification curve and solution curve in a flash liberation. The solution curve ② and degasification curve ③ obtained from a differential liberation do not coincide. By the way, the differential solution is a reverse process of the differential degasification, injecting only a minimum amount of gas into the liquid every time and repeat injections like this when the last amount is fully dissolved.

The difference between the solution curve and degasification curve also has something to do with the properties of the gas components. Methane, whose vapor pressure is comparatively high, has a set of solution curve and degasification curve that are quite close to each other (Fig. 3.34). However, ethane, whose vapor pressure is comparatively low, has a set of solution curve and degasification curve that are quite apart from each other (Fig. 3.35). The discussion above can well

Fig. 3.34 Degasification
curve and solution curve of
methane. ① Discrepant
solution, ② discrepant
degasification

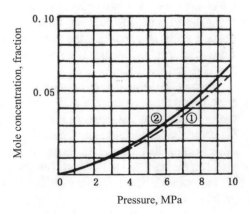

Fig. 3.35 Degasification curve and solution curve of petane. ① Discrepant solution, ② discrepant degasification

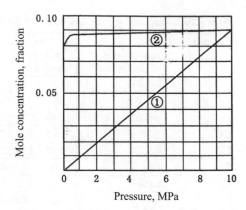

reflect the features shown in Fig. 3.33: At low pressure, the difference between the two curves is great, while under high pressure, the difference tends to be small.

Explanation for this difference: In natural gas, the heavy hydrocarbons (e.g., pentane) that have comparatively low vapor pressure can be dissolved easily when pressure increases; but for them it is hard to turn back into gas phase when the pressure decreases, which causes the disagreement between the two curves.

As talked above, there are many factors that influence the solution and separation of natural gas in crude oil. Among these factors, the most important ones are the properties of the gas and oil, the pressure and temperature at which the solution or degasification happens, and the pattern of degasification.

3.4   Calculation of Oil–Gas Separation Problems Using Phase-State Equations

Using the phase-state equations, from Eq. (3.18–3.21), as well as the bubble-point Eq. (3.22) and the dew-point Eq. (3.23), we can calculate the phase equilibriums that occur in oil–gas separation. The following part will present some examples.

3.4.1   Trial Calculation of Bubble-Point Pressure on Basis of Given Composition

[Example 3.7] According to the bubble-point pressure equation $\sum Z_j K_j = 1$, calculate the bubble-point pressure of the mixture whose composition Z_j is given in the second column of Table 3.10. It is already known that the reservoir temperature is 103°.

Table 3.10 Calculation process and results of the bubble-point pressure

Component	Mole concentration of the oil-well product (z_j)	Trial pressure $P = 21.0$ MPa (absolute)		Trial pressure $P = 22.45$ MPa (absolute)		Trial pressure $P = 23.2$ MPa (absolute)	
		K_j	$z_j K_j$	K_j	$z_j K_j$	K_j	$z_j K_j$
C_1	0.4404	2.150	0.9469	2.06	0.9072	2.020	0.8896
C_2	0.0432	1.030	0.0445	1.025	0.0443	1.020	0.044 1
C_3	0.0405	0.672	0.0272	0.678	0.0274	0.680	0.0275
C_4	0.0284	0.440	0.0125	0.448	0.0127	0.450	0.0128
C_5	0.0174	0.300	0.0052	0.316	0.0055	0.323	0.0056
C_6	0.0290	0.215	0.0062	0.230	0.0067	0.239	0.0069
C_{7+}	0.4011	0.024	0.0096	0.025	0.0100	0.026	0.0104
	1.0000		1.0521		1.0138		0.9969

Table 3.11 Conditions of single-stage and three-stage degasification

Temperature of degasification		49 °C
Single-stage degasification	Absolute pressure of degasification	0.1 MPa
Three-stage degasification	Absolute pressure of first-stage degasification	3.5 MPa
	Absolute pressure of second-stage degasification	0.46 MPa
	Absolute pressure of third-stage degasification	0.1 MPa

Solution: The calculation adopts a trial-and-error method. Table 3.10 gives the calculation process and results.

By interpolation method, the obtained bubble-point pressure (absolute) is 23.06 MPa.

3.4.2   Comparison Between Single-Stage Degasification and Multistage Degasification

[Example 3.8] The first and second columns of Table 3.12 show the composition of an oil-well product. According to the two patterns of degasification shown in Table 3.11, calculate the compositions and quantities of the vaporized gas, respectively, by a single-stage and multistage processes.

Solution:

(1)   Refer to related charts to obtain the value of K_j.

Since the pressure of degasification is not high (3.5, 0.46, 0.1 MPa), the K-chart with a convergence pressure of 35.0 MPa can be used. The equilibrium ratios of the components are shown in Table 3.12.

Table 3.12 Composition of an oil-well product and the equilibrium ratio for each component

Component	Mole concentration of the oil-well product (z_j)	Equilibrium ratios at 49 °C and different pressures (K_j)		
		P = 3.5 MPa	P = 0.46 MPa	P = 0.1 MPa
C_1	0.4404	8.1	60.0	265
C_2	0.0432	1.65	10.6	46.5
C_3	0.0405	0.59	3.4	14.3
C_4	0.0284	0.23	1.25	5.35
C_5	0.0174	0.088	0.415	1.72
C_6	0.0290	0.039	0.17	0.70
C_7^+	0.4011	0.003	0.013	0.052
Sum	1.000			

(2) Calculation of the Single-Stage Degasification

By resorting to Eq. (2.20) and adopting the trial-and-error method, carry out a trial calculation of $\bar{n}_L$ for three times. Table 3.13 shows the results.

As seen from Table 3.13, after the third trial we can get: for the single-stage degasification, from every mole oil-well product there are $\bar{n}_L = 0.3973$ mole tank-oil and $\bar{n}_g = 0.6027$ mole gas can be obtained by separation. Then, calculate the composition of the vaporized gas, y_j, according to the equation $y_j = K_j x_j$, and list them in the last column of the table.

(3) Calculation of the Multistage Degasification

① The conditions for the first stage of separation are: 3.5 MPa (absolute) & 49 °C. Assuming that the mole concentration of the gas phase is $\bar{n}_L$, as the value of x_j can be figured out according to Eq. (3.20), the result of the trial calculation shows that when $\bar{n}_g = 0.418$ the summation of x_j satisfies $\sum x_j = 1$; then, calculate the value of y_j according to Eq. (3.21). Table 3.14 shows the results.

② The conditions for the second stage of separation are: 4.6 MPa (absolute) & 49 °C. Now, the liquid that remains through the first stage will enter the second-stage separator to undergo further separation. The calculation equations are the same as that in the calculation of the first stage, except the different degasification conditions of pressure, which lead to different K_j values obtained from charts. The calculation results are shown in Table 3.15.

After the second-stage separator, the mole fraction of the crude oil that we can get from every mole oil-well product is:

$$\bar{n}_L = 0.582 \times 0.844 = 0.491$$

After the second-stage separator, the mole fraction of the gas that we can get from every mole oil-well product is:

Table 3.13 Results of the single-stage degasification calculation

Component	Mole concentration of the oil-well product (z_j)	Equilibrium ratio at 0.1 MPa &49 °C (K_j)	$\bar{n}_g = 0.602$ x_j	$\bar{n}_g = 0.603$	$\bar{n}_g = 0.6027$	$\bar{n}_g = 0.6027, \bar{n}_L = 0.3973$ y_j
C_1	0.4404	265	0.0028	0.0028	0.0028	0.7287
C_2	0.0432	46.5	0.0015	0.0015	0.0015	0.0707
C_3	0.0405	14.3	0.0045	0.0045	0.0045	0.0642
C_4	0.0284	5.35	0.0079	0.0079	0.0079	0.0420
C_5	0.0174	1.72	0.0122	0.0122	0.0122	0.0209
C_6	0.0290	0.70	0.0354	0.0354	0.0354	0.0248
C_{7+}	0.4011	0.052	0.9350	0.9371	0.9357	0.0487
Sum	1.000		0.9993	1.0014	1.0000	1.00

Table 3.14 Process and results of the first-stage calculation in a multistage degasification

Component	Mole concentration of the oil-well product (z_j)	Equilibrium ratio at 3.5 MPa (absolute) & 49 °C (K_j)	$\bar{n}_g = 0.418$	$\bar{n}_L$ = 0.582
			x_j	y_j
C1	0.4404	8.1	0.1110	0.8991
C2	0.0432	1.65	0.0339	0.0560
C3	0.0405	0.59	0.0489	0.0288
C4	0.0284	0.23	0.0419	0.0096
C5	0.0174	0.088	0.0281	0.0025
C6	0.0290	0.039	0.0485	0.0019
C7+	0.4011	0.003	0.6877	0.0021
Sum	1.0000		1.0000	1.0000

Table 3.15 Process and results of the second-stage calculation in a multistage degasification

Component	Mole concentration of the liquid from the first stage (z_j)	Equilibrium ratio at 4.6 MPa (absolute) & 49°C (K_j)	$\bar{n}_g = 0.156$	$\bar{n}_L$ = 0.844
			x_j	$y_j = K_j x_j$
C_1	0.1110	60.0	0.0109	0.6530
C_2	0.0339	10.6	0.0135	0.1430
C_3	0.0489	3.4	0.0355	0.1207
C_4	0.0419	1.25	0.0403	0.0504
C_5	0.0281	0.415	0.0309	0.0128
C_6	0.0485	0.170	0.0557	0.0095
C_{7+}	0.6877	0.013	0.8132	0.0106
Sum	1.0000		1.0000	1.0000

$$\bar{n}_g = 0.582 \times 0.156 = 0.091$$

③ The conditions for the second stage of separation are: 0.1 MPa (absolute) & 49 °C (K_j), and this separation happens in oil-tanks. Table 3.16 shows the calculation results.

After the last-stage oil-tank separation, the mole fraction of the crude oil that we can get from every mole oil-well product is:

$$\bar{n}_L = 0.491 \times 0.9486 = 0.466$$

After the last-stage oil-tank separation, the mole fraction of the gas that we can get from every mole oil-well product is:

$$\bar{n}_g = 0.491 \times 0.0514 = 0.025$$

Table 3.16 Process and results of the third-stage calculation in a multistage degasification

Component	Mole concentration of the liquid from the first stage (z_j)	Equilibrium ratio at 0.1 MPa (absolute) & 49 °C (K_j)	$\bar{n}_g = 0.0514$ x_j	$\bar{n}_L = 0.9486$ $y_j = K_j x_j$
C_1	0.0109	265	0.0007	0.1965
C_2	0.0135	46.5	0.0038	0.1885
C_3	0.0355	14.5	0.0210	0.3022
C_4	0.0403	5.35	0.0330	0.1775
C_5	0.0309	1.72	0.0299	0.0513
C_6	0.0557	0.70	0.0566	0.0396
C_{7+}	0.8132	0.052	0.8550	0.0444
Sum	1.0000		1.0000	1.0000

Compare the single-stage degasification and the multistage degasification:

After the single-stage degasification, from every mole oil-well product we can get 0.397 mol crude oil and 0.603 mol gas;

After the multistage degasification, from every mole oil-well product we can get 0.466 mol crude oil and 0.534 mol gas.

Obviously, compared with the single-stage degasification, the multistage one can separate out 0.07 mol more oil (0.466–0.397 = 0.07 mol) from every 1 mol oil-well product, which means that an extra of 15 % oil can be retrieved (0.07/0.466 × 100 % = 15 %) through this method. In fact, it is an amount quite substantial when reflected in oil-field production, and this is rightly the reason why the multistage degasification pattern is preferred by engineers.

3.4.3 Differential Liberation Calculation

Now, derive the calculation formula for differential separation. Refer to Fig. 3.28, assuming that there is Nmol liquid mixture in the constant pressure vessel (PVT cylinder), the pressure of the system in the original condition is adjusted at the saturation pressure (Fig. 3.28a)).

The pressure is slightly reduced when piston withdraws; then there is dNmol gas evaporated from the liquid (Fig. 3.28b). At the constant pressure, put off gas and at the same time push piston up until the liquid occupies the vessel, and all gases are released (Fig. 3.28c). Then, the saturation pressure of the system equals to the pressure in the container. Because dNmol gases are released, there is $(N - dN)$ mol liquid in the system. A degassing level is completed.

Repeat the above steps, until the saturation pressure of the system equals to the atmospheric pressure, and the differential separation process ends. As a certain amount of gas is exhausted each level, the composition of the liquid mixture changed constantly. In order to describe the change of material and components in the process, introduce the following parameters:

N—the total number of moles of liquid in the cell at the beginning of a differential liberation;

dN—the number of moles of gas formed during the differential liberation;

$N - dN$—the number of moles of liquid remaining at the end of the increment;

x_i—the mole fraction of the component i in the liquid at the beginning of the differential liberation;

dx_i—the change in the mole fraction of the component i caused by the loss of an increment of vapor, i.e., dN;

$(x_i - dx_i)$—the mole fraction of the component i in the liquid following the differential liberation, or in other words, the liquid of next stage of liberation;

x_iN—the number of moles of the component i in the liquid at the beginning of a differential liberation;

$(x_i - dx_i)(N - dN)$—the number of moles of the component i in the liquid at the end of the increment;

y_idN—the number of moles of component i in the gas removed.

Firstly, the material balance equation for the component i is: the number of moles of the component i in the gas is equal to the number of moles of component i lost by the liquid.

$$y_idN = x_iN - (x_i - dx_i)(N - dN) \tag{3.33}$$

This equation may be expanded to $y_idN = x_iN - x_iN + x_idN + Ndx_i - dx_idN$

Since dx_i is small, the term dx_idN may be neglected. Also, K_ix_i can be substituted for y_i to give:

$$K_ix_idN = Ndx_i + x_idN$$

or

$$\frac{dN}{N} = \frac{1}{K_i - 1}\left(\frac{dx_i}{x_i}\right) \tag{3.34}$$

If we assume that the values of K_i remain constant within a reasonable range of pressure change and use the subscripts b and f to indicate initial and final conditions of the differential liberation, Eq. (3.34) can be integrated as follows:

$$\int_{N_b}^{N_f} \frac{dN}{N} = \frac{1}{K_i - 1}\int_{x_{ib}}^{x_{i_f}} \frac{dx_i}{x_i} \tag{3.35}$$

$$\ln\frac{N_f}{N_b} = \frac{1}{K_i - 1}\ln\frac{x_{if}}{x_{ib}} \tag{3.36}$$

or

$$\frac{x_{if}}{x_{ib}} = \left(\frac{N_f}{N_b}\right)^{Ki-1} \tag{3.37}$$

Besides, another equation is required to complete the calculation:

$$\sum x_{if} = 1 \tag{3.38}$$

Equation (3.37) is called differential liberation equation. Combining Eqs. (3.37) and (3.38), the two unknown parameters N_f and x_{if} can be obtained, which are, respectively, the number of moles of the liquid that remains finally and the final composition of the liquid phase.

Differential liberation calculations of two types interest petroleum engineers most: In the first case, system compositions and final pressure are given and the number of moles to be vaporized is calculated; in the second case, the system compositions and the number of mole to be vaporized are given and the final pressure is calculated. Either case requires a trial-and-error solution. The following part will give related examples.

[Example 3.9] The following hydrocarbon mixture described in Table 3.17 has a bubble-point pressure of 4.1 MPa at 26.7 °C. Determine the number of moles of the gas that will be vaporized when this liquid is differentially vaporized to 2.7 MPa at constant temperature.

Solution: The value of K_i can be obtained from the charts with a convergence pressure of 35 MPa. For simplicity, we choose the average pressure 3.5 MPa to get the value of K_i. Then, select different values of N_f/N_b to undergo trial calculations, according to the equation $x_{if} = x_{ib}\left(\frac{N_f}{N_b}\right)^{Ki-1}$, and get the values of x_{if}. Table 3.17 gives the calculation process.

As shown in the last column of Table 3.17, when $N_f/N_b = 0.91$, the summation satisfies $\sum x_{if} = 1$. As a result, through a differential liberation, 0.91 mol oil and 0.09 mol gas can be obtained from every 1 mol of the given liquid at 26.7 °C.

[Example 3.10] Determine the final pressure required to differentially vaporize 0.1 mol natural gas from the mixture given in Example 3.9 at a constant temperature

Table 3.17 Calculation of the mole number of the vaporized gas

Component	x_{ib}	K_i (26.7°, 3.5 MPa)	Trial calculation to obtain x_{if}		
			First trial $N_f/N_b = 0.9$	Second trial $N_f/N_b = 0.95$	Third trial $N_f/N_b = 0.91$
C_1	0.20	5.1	0.130	0.162	0.136
C_3	0.30	0.40	0.320	0.309	0.317
n-C_5	0.50	0.05	0.553	0.525	0.547
Sum	1.00		1.003	0.996	1.000

The composition of the hydrocarbon mixture and the final pressure are already known

Table 3.18 Calculation of the final degasification pressure

Component	Mole fraction (x_{ib})	First trial (assume the final pressure is 2.7 MPa)		Second trial (assume the final pressure is 2.45 MPa)	
		Equilibrium ratio at 26.7 °C & 3.4 MPa	x_{if} from calculation	Equilibrium ratio at 26.7 °C & 3.27 MPa	x_{if} from calculation
C_1	0.20	5.1	0.130	0.52	0.128
C_3	0.30	0.40	0.320	0.41	0.319
n—C_5	0.50	0.050	0.553	0.0.0508	0.553
sum	1.00		1.003		1.000

The composition of the hydrocarbon mixture and the mole number of the vaporized gas are already known

of 26.7 °C. Equivalently, calculate the final pressure when $N_f/N_b = 0.9$. (The initial pressure is the bubble-point pressure 4.1 MPa).

Solution: Assume a final pressure and refer to related charts to obtain the value of K_i. Make a trial-and-error calculation using the Eq. (3.37), and then examine whether it satisfies the equation $\sum x_{if} = 1$ or not. Table 3.18 gives the calculation process.

As shown in Table 3.18, the final pressure required to differentially vaporize 0.1 mol natural gas is 2.45 MPa.

Chapter 4
Physical Properties of Reservoir Fluids Under Reservoir Conditions

Through the study of the preceding chapters, now we have a rudimentary recognition that the reservoir oil, which is subjected to the underground conditions, usually has a certain amount of natural gas dissolved in it. It differs greatly from the on-surface environment. Thus, reservoir oils exhibit such properties as volume, compressibility, viscosity, etc., quite different from those of the oil on surface. Moreover, during the extraction of oil from underground to surface, the oil undergoes a degasification, which leads to shrinkage in volume and increase in viscosity. To describe the relationship between underground oil and on-surface oil, a set of high-pressure physical parameters are introduced into the petroleum engineering, serving as a powerful tool in the reserve calculation, the evaluation of formation, the development design for oil fields, the dynamic analysis, and so on. In light of engineering application, this chapter will discuss the physical properties and related parameters of the reservoir fluids (reservoir crude oil and formation water) subjected to underground conditions of high temperature and high pressure.

4.1 High-Pressure Physical Properties of Reservoir Oil

4.1.1 Density and Specific Gravity of Reservoir Oil

1. Density of Reservoir oil

With huge amounts of natural gas dissolved in it, reservoir oil generally has density lower than degasified tank-oil. The density of reservoir oil decreases with the increase in temperature, but the variation of the density of oil with pressure is much more complicated. As shown in Fig. 4.1, when the pressure is lower than the

© Petroleum Industry Press and Springer-Verlag GmbH Germany 2017
S. Yang, *Fundamentals of Petrophysics*, Springer Geophysics,
DOI 10.1007/978-3-662-55029-8_4

Fig. 4.1 Variation of
reservoir oil density with
pressure

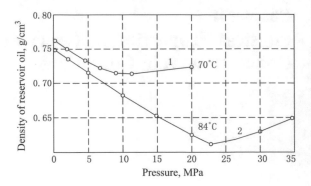

saturation pressure (the inflection point on the curve), with further solution of more natural gas, the density becomes lower as the temperature gets higher; when the pressure is higher than the saturation pressure, an increase in pressure will cause the density of the compressed oil to get higher because all of the natural gas has been dissolved in the oil already.

The density of reservoir oil can be measured through a high-pressure PVT experiment. Sometimes, related analytical data and charts can be used for calculation. (e.g., Standing and Katz Charts).

2. Specific gravity of reservoir oil

The specific gravity of reservoir oil can be expressed in the same way as the tank-oil, and among the expressions, the API specific gravity has unique advantages. According to the equation written below:

$$\text{API} = \frac{141.5}{\gamma_o} - 131.5 \tag{4.1}$$

where

γ_o the specific gravity of reservoir oil.

When subjected to definite temperature and pressure, the API specific gravity increases with the increase in the quantity of the dissolved natural gas in oil, or in other words, the API specific gravity is in direct proportion to the solubility. This is where the advantage of the API special lies.

Due to different temperature conditions, it should be noted that the specific gravity γ_o used in Western countries does not have the same meaning as the d_4^{20} used in China, and they have different values, either. For example, the specific gravity in the Lesson 3 of this chapter is γ_o rather than d_4^{20}.

4.1.2   Solution Gas–Oil Ratio of Reservoir Oil

The solution gas–oil ratio of reservoir oil R_s is referred to as the volume of the natural gas, measured at standard conditions, that can be dissolved by per unit volume or unit mass tank-oil (m^3/m^3 or m^3/t) when the oil is put under reservoir conditions of temperature and pressure. Usually, by the degasification process in a separator or a laboratory treatment, the solution gas–oil ratio can be obtained as long as the volume of the degasified oil V_{os} (m^3) and the volume of the liberated gas on the surface V_g ((standard) m^3) are known:

$$R_{si} = V_g/V_{os} \qquad (4.2)$$

where

V_g volume of the liberated gas on the surface (under standard conditions);
V_{os} volume of the degasified oil on the surface, or volume of tank-oil, m^3;
R_{si} solution gas–oil ratio of the oil at a pressure P_i and a temperature T_i, m^3/m^3.

Since the degasification experiment mentioned above begins its pressure reduction with the original condition (including original pressure, original temperature, and original saturation state), the R_{si}, representing the quantity of the gas dissolved in the oil under original conditions, is also called original solution gas–oil ratio.

Obviously, in case that the starting point (p, V) in a degasification is not the original condition, a part of dissolved gas has already escaped from the oil sample before it undergoes the degasification, and the obtained R_{si} is the solution gas–oil ratio at p and V.

When comparing two oil samples, the one with a greater R_{si} value contains more dissolved gas.

The solution gas–oil ratio and the solubility are uniform essentially. By degasification experiments on reservoir oil sample, the solution gas–oil ratios of reservoir oil upon various pressures can be obtained. When the pressure is lower than the saturation pressure, the solution gas–oil ratio has the same significance with the solubility, while the original solution gas–oil ratio is a fixed value because that the composition, temperature, and pressure are definite upon the original condition.

Besides, the quantity of the liberated gas V_g varied as different methods of processing the fluids are chosen. In order to make easier comparisons, we take the solution gas–oil ratios obtained from single-stage degasifications as the studied objects in our discussion.

Figure 4.2 shows the relationship between the solution gas–oil ratio (from a single-stage degasification) and the pressure at 71 °C, and this presented curve is a typical solution curve of unsaturated oil reservoir. As shown in the figure, when the reservoir pressure is higher than the saturation pressure, the solution gas–oil ratio is

Fig. 4.2 Typical solution
gas–oil ratio curve of
reservoir oil in a contact
degasification

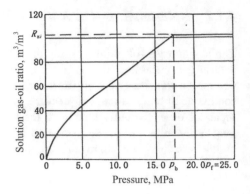

equal to the original solution gas–oil ratio and held a constant before the pressure
decrease to the inflection point. When the pressure has fallen lower than the saturation pressure, a quantity of gas will be liberated from the reservoir oil at every
drop of pressure, and thus, the dissolved gas that remains in oil becomes less and
less, leading to the decreasing value of the solution gas–oil ratio R_{si}. When the
pressure equals 1 atm, the value of R_{si} equals zero. Conversely, as the pressure
increases, the solution gas–oil ratio will grow greater and greater until $P = P_b$ (the
saturation pressure), after the solution gas–oil ratio is equal to R_{si} and all the gas has
already dissolved into the oil. Then, the solution gas–oil ratio will be held a constant until the pressure is raised up to the original reservoir pressure.

　　Table 4.1 shows the solution gas–oil ratios in some oil fields both at home and
abroad. From this table, we can see that there are great differences among the
solution gas–oil ratios from different oil fields. The solution gas–oil ratio of the oil
from Huabei Oil Field is $R_{si} = 70$ m^3/m^3, meaning that the oil has a low content of
gas, while the solution gas–oil ratio of the oil from the North Sea Oil Field of
Norway is as great as $R_{si} = 580$ m^3/m^3. A greater original solution gas–oil ratio
indicates a larger quantity of dissolved gas in oil, which makes it easier for the oil to
gush up to the surface by itself.

Table 4.1 High-pressure physical-property data of some oil fields at home and abroad

Oil fields in China	R_{si} (Sm^3/ m^3)	B_o	C_o ($\times10^{-4}$ MPa^{-1})	Oil field abroad	R_{si} (Sm^3/ m^3)	B_o
Daqing	48.2	1.13	7.7	Roumania	1.1	1.05
Dagang	37.3	1.09	7.3	America	11.0	1.07
Shengli	70.1	1.22	–	Venezuela	85.1	1.26
Gudao, Shengli	27.5	1.10	7.3	Canada	89.0	1.25
Renqiu, Huabei	7.0	1.10	10.35	Iran	190.0	1.42
Yumen	65.8	1.16	9.6	Norway	580.0	1.78

4.1.3 Formation Volume Factor of Oil

1. Oil Formation Volume Factor

The oil formation volume factor, also called the "oil volume factor" for short, is the ratio of the crude oil volume at reservoir conditions to the tank-oil volume under standard conditions, or in another way, the volume of reservoir oil required to produce one unit volume of oil in the stock tank on the surface. Represented by the symbol B_o, it can be expressed mathematically as:

$$B_o = \frac{V_f|P,T}{V_{os}|P,T} = B_o(P,T) \tag{4.3}$$

where

V_f the oil volume at the pressure P and the temperature T, m^3;
V_{os} the tank-oil volume on the surface at 0.1 MPa and 20 °C, m^3.

The oil volume factor at original reservoir conditions (p, T) is called "oil formation volume factor," symbolized as B_{oi}.

Comparing the volume of the reservoir oil with the oil that enters the stock tank on the surface, there are three points of distinctions: the underground oil has dissolved gas, expands in volume at high temperature, and undergoes a contraction due to high pressure. The change in volume resulting from these three factors is expressed in terms of the formation volume factor of oil. Owing to the evolution of the dissolved gas, the gas-saturated underground oil shrinks in volume when lifted to the surface from the reservoir. In ordinary circumstances, when an oil–gas mixture is put underground, its volume changes due to the dissolved gas and the heat expansion goes far beyond its contraction under high pressure, the reason why the volume of oil that enters the stock tank at the surface is less than the volume of the oil underground and the reason why the volume factor B_o of oil usually has a value greater than 1. Table 4.1 shows that the oil volume factor B_o is proportional to the gas–oil ratio.

Possessing a characteristic of high volatility, the volatile oil has very great volume factor and high shrinkage rate. For instance, as a reservoir oil with a volume of 5 m^3 flows from the formation to the well tubing, and then to the separator on the surface, the decreasing pressure applies a multistage degasification to the fluid, resulting from which the tank-oil finally obtained on the surface is only 1 m^3. So, the volume factor B_o for this oil is not the conventional 1.5, but 5. This is mainly because the volatile light hydrocarbons that are abundantly contained in this type of oil are converted from a liquid state to a gas state during the reduction of pressure. Besides, the reduction of temperature is another factor that causes the oil's shrinkage in volume.

2. Shrinkage Factor of Reservoir oil

When reservoir oil flows up to the surface, its volume inevitably becomes smaller, and this phenomenon is called the shrinkage of reservoir oil. The degree to which the reservoir shrinks is expressed by the "shrinkage factor of reservoir oil" or the "shrinkage rate of reservoir oil":

(1) Define the shrinkage factor as the reciprocal of the volume factor of crude oil, i.e., $\delta_o = 1/B_o = V_{os}/V_f$. The volume of the tank-oil on the surface can be obtained through multiplying the shrinkage factor by the volume of the crude oil under reservoir conditions. Conversely, the volume of the reservoir oil can be obtained through multiplying the volume factor by the volume of the tank-oil. In this manner, the conversion between the two volumes can be easily performed.

(2) Define the shrinkage rate as $\beta = (V_f - V_{os})/V_f = (B_o - 1)/B_o$. From the perspective of physical significance, $V_f - V_s$ reflects the shrinkage volume of the reservoir oil when lifted up to the surface.

Owing to the difference in definition, the shrinkage factor and the shrinkage rate have different values, requiring special attention for their physical significances when cited and used.

3. Relationship between the Volume Factor of Reservoir oil and Pressure

The formation volume factor of oil varies with the change in reservoir pressure. The relationship between the formation volume factor of oil and the reservoir pressure is given in Fig. 4.3 (the solid line). As shown by this line, we can know that:

(1) When $P < P_b$, a reduction in reservoir pressure results in the liberation of gas. Consequently, as the reservoir pressure drops lower, the liquid remaining in the reservoir has less gas in solution, leading to a smaller volume V_f and thence a smaller volume factor B_o.

Fig. 4.3 Relationship between P and B_o, B_t

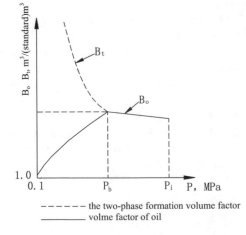

----- the two-phase formation volume factor
———— volme factor of oil

(2) When $P > P_b$, the volume factor decreases with the increase in reservoir pressure. This is because that the volume of oil V_f shrinks due to the increasing pressure, leading to a smaller volume factor B_o.

(3) When $P = P_b$, both the solution gas–oil ratio R_s and the volume factor B_o reach their maximum values.

4. Two-Phase Formation Volume Factor

The term "volume factor of oil" takes only the volume change of oil in consideration. However, when the reservoir oil is reduced to be lower than the saturation pressure, some gas will be released out of the oil, leading to a two-phase situation. Therefore, a concept of the two-phase formation volume factor of oil is introduced to describe the relationship between the total volume of the two phases in reservoir and the volume of the corresponding tank-oil on the surface.

The two-phase formation volume factor of oil is defined as the ratio of the volume occupied by the underground oil plus the gas evolved from it in reservoir to the volume of the corresponding tank-oil, symbolized by B_t.

Assuming that the original GOR of a reservoir oil is R_{si}, and its GOR at the pressure P is R_s, when $P < P_b$, the volume of the evolved gas converted to the conditions on the surface is $(R_{si} - R_s)V_{os}$, in which V_{os} is the volume of the tank-oil.

Assuming B_g be the volume factor of the evolved gas, in reservoir, the volume of the gas evolved from the oil at the pressure P is:

$$(R_{si} - R_s)V_{os}B_g$$

With V_f and V_{os} to, respectively, represent the volume of a reservoir oil at the pressure P and the volume of its corresponding tank-oil on the surface, we can get B_t by resorting to the definition of the two-phase formation volume factor of oil mentioned above.

$$B_t = \frac{V_f + (R_{si} - R_s)V_{os} \cdot B_g}{V_{os}} = \frac{V_f}{V_{os}} + (R_{si} - R_s)B_g$$
$$= B_o + (R_{si} - R_s)B_g \tag{4.4}$$

The variation of B_t with the change in pressure is also given in Fig. 4.3 (the clashed curve). We can draw the following conclusions from this figure and Eq. (4.4):

(1) When the reservoir pressure is higher than or equal to the saturation pressure (i.e., $P \geq P_b$), $R_s = R_{si}$, equivalently $R_{si} - R_s = 0$, and consequently $B_t = B_o$, which means that the two formation volume factors are identical.

(2) When the reservoir pressure is reduced to the atmosphere pressure, as no more dissolved gas can be liberated out of the oil, $R_s = 0$; in this situation, because $B_g = 0$, and $B_o = 1$, $B_t = 1 + R_{si}$, which is the maximum value of B_t.

(3) Since B_o, B_g, and R_s are all functions of the pressure P, B_t, the two-phase formation volume factor of oil is a function of the pressure, too. The dashed curve in Fig. 4.3 gives the B_t–P relationship.

(4) Since when $P > P_b$, only a single-phase oil is present, and the $B_f–P$ curve exists exclusively in case of $P < P_b$.

4.1.4   Compressibility Coefficient of Reservoir Oil

As shown in Fig. 4.3, when $P > P_b$, it is only an elastic expansion or contraction of the oil that responds to a change in the reservoir pressure. Usually, the elasticity of the reservoir oil is expressed by the compressibility coefficient, or the elastic volume factor, Co. The so-called compressibility coefficient refers to the fractional change of the reservoir oil volume with the change in pressure. Under isothermal conditions, the compressibility coefficient of oil is defined as:

$$C_o = -\frac{1}{V_f}\left(\frac{\partial V_f}{\partial P}\right)_T \approx -\frac{1}{V_f}\cdot\frac{\Delta V_f}{\Delta P} = -\frac{1}{V_f}\frac{V_b - V_f}{P_b - P} \tag{4.5}$$

where

C_o isothermal compressibility coefficient of oil, MPa^{-1};
V_b, V_f the volumes of the reservoir oil, respectively, at the pressure P_b and P, m^3;
P_b, P   the saturation pressure of oil and the reservoir pressure, respectively, MPa.

Note that when $P > P_b$, the reservoir oil volume decreases as the pressure increases. In order to make sure Co possesses a positive value, a minus should be added to the right side of the equation. Divide both the numerator and the denominator in Eq. (4.5) by the volume of the tank-oil V_{os}:

$$C_o = -\frac{1}{B_o}\frac{B_{ob} - B_o}{P_b - P} \tag{4.6}$$

In the equation, B_{ob} and B_o are the formation volume factors, respectively, at the pressure P_b and P.

To determine the value of C_o by experiment, what are needed to be measured are merely the corresponding volumes of crude oil, V_b and V_f, respectively, at the pressure P_b and P. Then, the value of C_o can be calculated by resorting to Eq. (4.5). In the case that the values of B_{ob} and B_o, the formation volume factors of oil, are already known, Eq. (4.6) can be used to calculate the value of C_o, the compressibility coefficient of oil.

The compressibility coefficient of reservoir oil is dependent on its solution gas–oil ratio and the conditions of temperature and pressure to which the reservoir oil is subjected. Firstly, a high solution GOR, which means that the reservoir oil has a great mount of gas dissolved in it, implies a low density and a great elasticity and therefore a high compressibility coefficient of the studied oil. Secondly, a higher the reservoir temperature implies that the oil owns smaller density, greater elasticity, and, therefore, higher compressibility coefficient. Thirdly, an increase in pressure

Table 4.2 Average compressibility factors within different pressure intervals

Pressure interval (MPa)	Average compressibility factor, C_o (MPa^{-1})
$P_b = 19.0$	
19.0–19.4	38.9×10^{-4}
19.4–24.2	36.0×10^{-4}
24.2–29.2	30.2×10^{-4}
29.2–34.4	24.7×10^{-4}

leads to a greater density and therefore a lower compressibility coefficient of the oil. Table 4.2 shows that the compressibility coefficient of reservoir oil varies with the change in pressure, and that's to say, within different pressure intervals, the values of the coefficient are different. Note that the compressibility coefficient has comparatively higher value when the pressure is close to the saturation pressure.

The values of the compressibility factor within different pressure intervals approximately vary within such a range:

For the tank-oil on the surface $(4–7) \times 10^{-4}$ MPa^{-1};

For the reservoir oil $(10–140) \times 10^4$ MPa^{-1}.

In reservoir engineering, the compressibility factor of oil is usually used to calculate the elastic reserve when reservoir pressure decreases.

4.1.5 Viscosity of Reservoir Oil

As a parameter commonly encountered in petroleum engineering calculations, the viscosity of reservoir oil is an important factor influencing the oil production. Sometimes, a high viscosity of oil may cause the production unable to continue.

The viscosity of oil around the word varies dramatically: For underground oil, it ranges from 1 mPa s to several thousand mPa s; for tank-oil on the surface, it ranges from 1 mPa s to hundreds of thousand mPa s. From a visual point of view, some oil is thinner than water, but some others can be as thick as a semisolid-state plastic micelle.

1. Internal Factor Determining the Viscosity of Oil

The chemical composition of crude oil, as the internal cause, is an important influencing factor for the viscosity of it. Figure 4.4 shows that crude oil exhibits an increase in viscosity when the size of the hydrocarbon molecules increases. The content of the non-hydrocarbons (colloid and asphalt) in crude oil affects its viscosity significantly. Asphalt is an oxygen–sulfur compound, or polycyclic aromatic hydrocarbons, with short side chains. It has a carbon–hydrocarbon ratio of about 10 and a molecular weight ranging from 10^3 to 10^5. The colloid possesses similar composition with asphalt but smaller molecular weight. Both of them are viscous, black, and semisolid amorphous substances.

Fig. 4.4 Relationship
between viscosity and
molecular weight

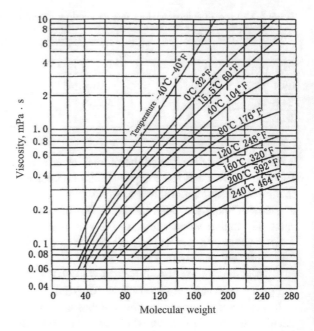

Causing increased internal friction between the liquid sheets, the existence of macromolecular compounds (colloid and asphalt) results in higher viscosity of oil.

In the case of high content of colloid and asphalt or low pressure, a colloidal structure can be formed, which results in some non-Newton characteristics of oil. For example, highly viscous oil exhibits thixotropy, possesses static shear stress (τ_o), and suffers a decrease in viscosity with shear rate increasing.

Table 4.3 shows the relationship between the colloid and asphalt content and the viscosity of the tank-oil from Shengli Oil Field. Seen from the table, it is obvious that the higher content of the colloid and asphalt crude oil has the higher viscosity and the higher density of it.

The content of heavy hydrocarbons contained in a crude oil has effect on the viscosity, too. Generally, the crude oil with a high density also exhibits a high viscosity.

The quantity of the gas in solution is another dominant influencing factor for the viscosity of oil. As shown in Table 4.4, the higher the solution GOR, the lower the

Table 4.3 Relationship between viscosity and colloid and asphalt content

Well number	Xin-1, Shengli	Tuo-2, Shengli	Guan-12, Shengli	Xin-2, Shengli	Xin-6, Shengli
Colloid and asphalt content (%)	27.6	36.7	41.9	49.5	64.5
Specific gravity	0.8876	0.9337	0.9414	0.9521	0.9534
Viscosity (mPa s)	58.5	490	885	5120	6400

Table 4.4 Relationship between viscosity and solution GOR

Oil sample	Original GOR (R_{si}) (standard) (m³/m³)	Viscosity of reservoir oil (μ_o) (mPa s)
Gudao, Shengli	27.5	14.2
Dagang	37.3	13.3
Daqing	48.2	9.30
Yumen	68.5	3.20
Shengli	70.1	1.88

viscosity. This regularity can be explained in such a way: As some gas is dissolved in oil, the liquid–liquid attraction force that originally exists in the oil is partly replaced by the much weaker liquid–gas attraction force, leading to lower inner friction resistance within the gas-contained reservoir oil and thence lower viscosity of it. When a reservoir oil is depleted up to the surface, its viscosity gets lower owing to, on the one hand, the evolution of the gas once dissolved in the liquid, and, on the other hand, the decreasing temperature from the underground to the surface.

2. External Factors Influencing the Viscosity of Oil

The viscosity of oil is quite sensitive to the change in temperature. Crude oil tends to exhibit a lower viscosity when the temperature is increased. Different oils possess different levels of sensibility for temperature. Figure 4.5 shows that, for some oils, a by-half decrease in oil viscosity may result from an increase of 10 °C.

With regard to the effect of pressure shown in Fig. 4.6, within different pressure intervals, above or below the saturation pressure P_b, the pressure affects the viscosity of oil differently.

In the case that the pressure is higher than the saturation pressure ($P > P_b$), an increase in pressure, which is followed by an elastic compression of the reservoir oil, leads to an increase in viscosity because of the increased density and the consequently occurring stronger friction force between the liquid sheets.

In the case that the pressure is lower than the saturation pressure ($P < P_b$), a decrease in reservoir pressure causes unremitting liberation of the gas form the oil, and, as a result, the sharp increase in the viscosity of oil.

Fig. 4.5 Relationship between temperature and oil

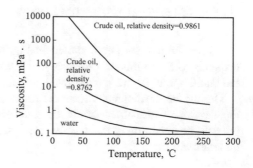

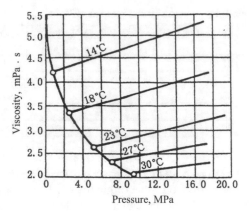

Fig. 4.6 Relationship between pressure, temperature, and reservoir oil viscosity

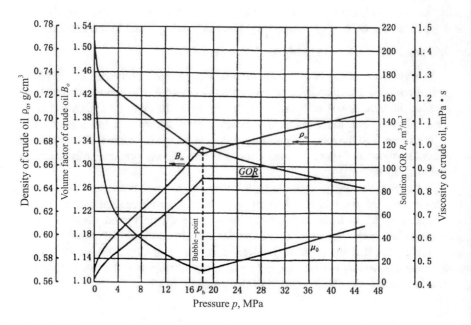

Fig. 4.7 High-pressure physical curves of typical unsaturated reservoir

As for the relation curve shown in Fig. 4.7, the lowest point of viscosity corresponds to the saturation pressure, and the inflection point of the curve is rightly the bubble-point of the studied reservoir oil.

During the development of an oil reservoir, as analyzed above, the pressure in the developed formation should be maintained above the saturation pressure to prevent the liberation of the dissolved gas, so as not to increase the viscosity of the underground oil.

With the oil from a typical unsaturated oil reservoir as the studied object, the changing regularities of these mentioned high-pressure parameters are gathered up in Fig. 4.7

4.1.6 Freezing Point of Oil

The freezing point of oil depends on not only the composition of it, such as the paraffin content, the colloid and asphalt content, and the light oil content, but also the quantity of its dissolved gas and the conditions of temperature and pressure to which the oil is subjected to.

In crude oil, usually, under original reservoir conditions, there exist dissolved paraffins which precipitate in a crystalline state due to the pressure reduction and the oil degasification during the depletion of oil. The precipitation of them does cause higher resistance of the flowing oil and, consequently, lower production capacity of the oil well.

As mixtures of various macromolecule alkenes, paraffins have not only different compositions, but also different chemical structures, crystalline temperatures, and crystalline states. Therefore, for the paraffins in oil, their crystalline experience, from the state of solution to the state of "starting to crystallize" and then to the state of "complete crystallization," lasts somewhat a temperature range rater than a temperature point.

The primary crystalline temperature for the paraffins in crude oil decreases with the increase of the quantity of the dissolved gas, and, in a similar pattern, decreases with the increase of the content of the light alkenes contained in oil. However, the effect of pressure on the initial crystalline temperature of paraffins is comparatively small.

The colloid and asphalt contained in crude oil are polar materials and can be easily attached to the paraffin crystal nucleus, helping the gathering and growing of the paraffin crystal nucleus.

In the production of high-freezing point crude oil, great difficulty is encountered because of the crystallization of the paraffins. Therefore, the problem of paraffin inhibition deserves special attention during the production.

4.2 Physical Properties of Formation Water Under High Pressure

Formation water is a general name for the edge water and bottom water on the edge and bottom of a reservoir formation, the interlayer water, and the connate water that exist in the same layer with crude oil. In particular, the connate water, coexisting with the oil and gas and not moving to flow, is the residual water that remained in the porosities when the reservoir started to form.

The formation water is in close contact with reservoir oil and gas. Among them, the edge water and bottom water are often utilized as forces to drive oil, and the connate water, whose distribution feature in the microscopic rock pores directly influences the oil saturation of the formation, has important significance for the reservoir exploration, development, and EOR methods.

The research on the composition and properties of formation water can be used (1) to judge the direction to which the edge-water flows and the connectivity among fault blocks and to analyze the causes for water exit in oil well; (2) to study the compatibility of the injected water and to analyze the cause and level of reservoir damage; (3) to offer basis for the oil-field sewage treatment and the pollution discharge design; and (4) to determine the depositional environment according to the water type.

4.2.1   Composition and Chemical Classification of Formation Water

Refer to Lesson 1.5 in Chap. 1 for details.

4.2.2   High-Pressure Physical Properties of Formation Water

There exist amounts of salts but just a spot of natural gas dissolved in the formation water under reservoir conditions of high pressure and high temperature. For example, the quantity of dissolved natural gas in formation water is not more than $1-2$ m^3/m^3 at 10.0 MPa. This is regarded as one of the characteristics of formation water.

After all, the high-temperature and high-pressure environment to which the formation water is subjected is different from the conditions on the surface, resulting in many differences in more than one aspects between the underground and surface water. For this reason, some high-pressure physical properties of formation water such as the volume factor, the solubility of natural gas, the compressibility coefficient, and the viscosity should be studied and described in order to provide necessary information for the reservoir engineering calculation and numerical simulation.

1. Gas Solubility in Formation Water

The solubility of natural gas in formation water can be defined as the volume of gas that can be dissolved in per unit volume of surface water when it is put to

Fig. 4.8 Solubility of natural gas in formation water (Dodson and Standing 1944)

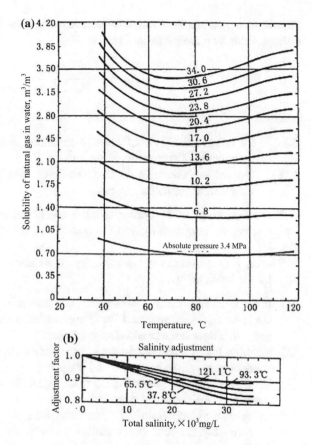

reservoir conditions. The solubility of natural gas in pure water has been found to be dependent upon the temperature and pressure of the water as shown in Fig. 4.8a: The solubility increases with the increase in pressure. With regard to the temperature, below 70–80 °C, the solubility of natural gas in pure water decreases with the increase in temperature, while above 70–80 °C, it increases with the increase in temperature. From the discussion above, we can know that in the case of high temperature and pressure, a large volume of formation water have huge amounts of gas in solution, which means a considerable reserve of natural gas.

The solubility of natural gas is also in connection with the water salinity. As shown in Fig. 4.8b, in the water with higher salinity, a lower solubility of natural gas would be obtained. Thence, a salinity correction must be done before reading the solubility values of natural gas from related references.

2. Compressibility of Formation Water

The elastic energy of reservoir water is a very important part of the total elastic energy of formation. The compressibility of formation water is defined as the rate of volume change when the studied formation water is subjected to a unit of pressure change. Expressed mathematically:

$$C_{\mathrm{w}} = -\frac{1}{V_{\mathrm{w}}}\left(\frac{\partial V_{\mathrm{w}}}{\partial P}\right)_T \tag{4.7}$$

where

C_{w} the compressibility coefficient of formation water, MPa^{-1};
V_{w} the volume of formation water, m^3;
$\left(\frac{\partial V_{\mathrm{w}}}{\partial P}\right)_T$ the volume change rate of formation water with pressure at constant temperature, m^3/MPa.

The relationships of the compressibility coefficient of formation water C_{w} with the pressure, the temperature, and the quantity of the dissolved gas are shown in Fig. 4.9.

The steps to obtain the compressibility coefficient of formation water by using Fig. 4.9 are listed below:

(1) In accordance with the already known reservoir pressure and temperature, resort to Fig. 4.9a and read the compressibility coefficient of the formation water that does not contain dissolved gas;
(2) Resort to Fig. 4.8 and obtain the quantity of the dissolved gas in the formation water at reservoir temperature and pressure;
(3) Read the correction factor for C_{w} from Fig. 4.9b on the basis of the obtained quantity of the dissolved gas; and
(4) Calculate the compressibility coefficient of the gas-containing formation water by correcting the read compressibility coefficient of the gas-free formation water.

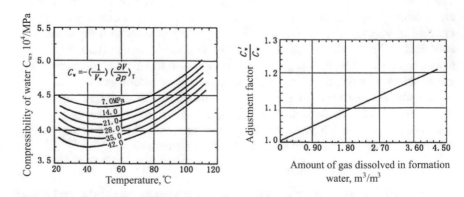

Fig. 4.9 Compressibility coefficient of formation water (Dodson and Standing 1944). **a** Compressibility coefficient of gas-free formation water and **b** correction for the compressibility coefficient of gas-containing formation water

Approximately, the compressibility coefficient of formation water which ranges between $(3.4–5.0) \times 10^{-4}$ MPa^{-1} is lower than that of crude oil. Besides, within different intervals of pressure and temperature, the values of C_w are different, either.

[Example 4-1] Calculate the compressibility coefficient of the formation water with a salinity of 30,000 mg/L under reservoir conditions of 20.4 MPa and 93 °C.

(1) In accordance with the reservoir conditions of 20.4 MPa and 93 °C, resort to Fig. 4.8a to obtain the value of the solubility of natural gas in pure water by adopting the linear interpolation method: 2.7 m^3/m^3;
(2) In accordance with the known salinity, 30,000 mg/L, read the correction factor 0.88 from Fig. 4.8b; then, calculate the solubility of natural gas in formation water:

$$R_w = 2.7 \times 0.88 = 2.4\,\text{m}^3/\text{m}^3$$

(3) According to the pressure and temperature that are already known, the compressibility coefficient of pure water under the given conditions can be obtained by referring to Fig. 4.9a: 4.53×10^{-4} MPa^{-1}.
(4) As the quantity of the dissolved gas figured out in (2) is 2.4 m^3/m^3, the correction factor can be read from Fig. 4.9b, which turns out to be 1.12. Finally, the compressibility coefficient of the given reservoir water can be calculated:

$$C_w = 4.53 \times 10^{-4} \times 1.12 = 5.07 \times 10^{-4}\,\text{MPa}^{-1}$$

3. Volume Factor of Formation Water

The volume factor of formation water is defined as the ratio of the volume of formation water under formation conditions to the volume of equivalent water under surface conditions:

$$B_w = \frac{V_w}{V_{ws}} \tag{4.8}$$

where

B_w volume factor of formation water, fraction;
V_w volume of formation water under reservoir conditions, m^3;
V_{ws} volume of this formation water under surface conditions, m^3.

The relationship of the volume factor of formation water with its pressure, temperature, and quantity of dissolved gas is shown in Fig. 4.10. It is obvious that an increase in pressure produces a decrease in the volume factor, whereas, at constant pressure, an increase in temperature produces an increase in the volume factor.

Naturally, at a given pressure and temperature, gas-saturated water has a higher volume factor than pure water. However, since the quantity of the dissolved natural

Fig. 4.10 Volume factor of formation water (Dodson and Standing 1944) (*solid line* for water with dissolved natural gas, *dashed line* for pure water)

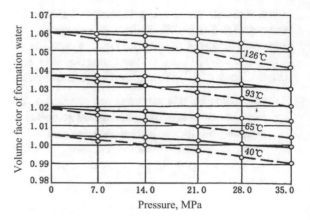

gas in formation water is rather limited, the volume factor of formation water, ranging between 1.01 and 1.06, is comparatively quite small.

4. Viscosity of Formation Water

The viscosity of formation water is dependent upon its pressure, temperature, and salinity. The relationships are shown in Figs. 4.11 and 4.12. It is seen that the temperature has significant effect on the viscosity of formation water—the viscosity decreases dramatically with the increase in temperature, while the pressure has indistinct effect on the viscosity.

The viscosity of formation water also varies indistinctly with salinity. For example, the viscosity–temperature curve of the brine water with a salinity of 60,000 mg/L is quite closed to that of pure water. The effect of dissolved gas on water viscosity is little due to the small quantity of it in solution, so no correction for it is needed.

Fig. 4.11 Relationship between pressure and the viscosity–temperature curve (1956)

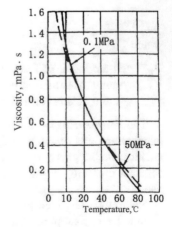

Fig. 4.12 Relationship between salinity and the viscosity–temperature curve (1956)

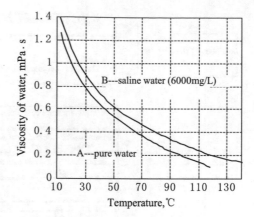

4.3 Measurement and Calculation of High-Pressure Physical Parameters of Reservoir Oil and Gas

The calculation of reserves and the dynamical analysis of reservoirs are heavily reliant on the high-pressure physical parameters of reservoir fluids, such as B_o, B_g, Z, C_o, and C_g. Particularly in the engineering calculation of volatile oil reservoir and gas-condensate reservoirs, quite high accuracy is required for these parameters. In circumstances like these, the way to read the commonly used charts, for example, the Z-charts, tends to produce unallowable error, and to alleviate the problem, an experimental measurement or calculations on the basis of the compositions should be used to obtain high-accuracy data. In the measurement of the reservoir fluid parameters, accordingly, high technical requirements should be met in the sample preparation, the choice of the experimental equipments, and the manipulation of the conditions of high temperature and high pressure.

This lesson focuses on the ways to obtain the high-pressure physical parameters, for example, the saturation pressure, the solution gas–oil ratio, the volume factor, the compressibility coefficient, and the viscosity, mainly through actual measurement, the charts, and the empirical equations. Actual measurement is the most direct way to obtain the parameters, and the other two methods are based on the actual measured data and the in situ information.

4.3.1 Method to Measure the High-Pressure Properties of Oil and Gas

In the oil-field practice, in-house experiments are most commonly used to measure the high-pressure properties of oil and gas. The process of analysis is shown as a flowchart in Fig. 4.13.

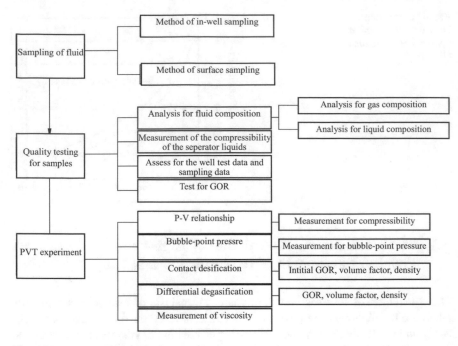

Fig. 4.13 Analysis flowchart of the high-pressure physical properties of crude oil

1. PVT Experimental Facility

A PVT (pressure, volume, and temperature) instrument is an apparatus used to measure the high-pressure physical properties. It comprises several main parts given below, and a flowchart of its using is shown in Fig. 4.14.

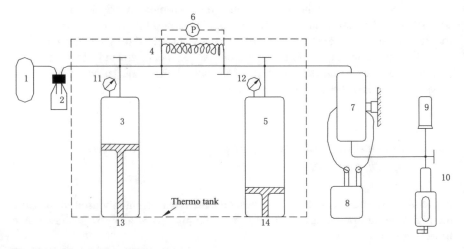

Fig. 4.14 Mercury-free PVT instrument

(1) PVT Cylinder (3, 5): As the core of the whole facility, they are high-pressure containers where in the studied oil sample or gas sample reaches their equilibrium. Usually, a PVT cylinder is equipped with observation windows, and the one above the cylinder is used to watch the bubble-point of black oil, while the one below the cylinder is used to watch the dew-point of condensate gas. The material observation windows are sapphire or high-strength glass. The instrument shown in the figure consists of two PVT cylinders—a main vessel and an auxiliary vessel, between which there are pipelines for connection as well as a bypass capillary viscometer.

(2) High-Pressure Metering Pump (13, 14): It applies pressure to the liquid within the PVT cylinder and measure the volume of it. In the figure, the one connected with the piston is rightly the screw of the high-pressure metering pump. In the work of measurement, a transducer is used to detect the position of the piston and related computer programs are resorted to for calculation.

(3) Precision Pressure Gauge or Pressure Transducer (11, 12): They are used to measure the pressure of the liquid within the high-pressure container.

(4) Oil and Gas Separator (2) and Gas Meter (1): In discrepancy liberation experiments, they are used to separate the oil and gas and measure the volume of the liberated gas.

(5) High-Pressure Viscometer: It is comprised of the capillary tube (4) and the differential pressure transducer (6), used to conduct an online measurement of the viscosity of the high-pressure and gas-containing oil.

(6) Thermo-Tank: All the devices, such as the PVT cylinder and the viscometer, should be put in the thermo-tank to keep an environment in reservoir temperature.

(7) Outside Sample-Transfer Devices: This part includes a sampler (7), a water bath for sample transfer (8), a sample-transfer pump, (10) and a storage tank of the pump's working fluid (9).

Characteristics of this facility:

(1) High pressure and high temperature: The whole pipeline system withstands a high temperature up to 150 °C and a high pressure up to 70 MPa.

(2) High Measurement Accuracy: The temperature error produced by the thermo-tank should be controlled within ±1 °C, and the measurements of the pressure, the volume of oil, and the volume of liberated gas must satisfy high-accuracy requirement.

(3) High requirement for the technological level of the operating personnel: Because of the complexity of the experimental facility, the control system, and the flowing paths of the pipelines, plus the large number of valves, it is required for the experiment operators to be familiar with the principles, the measuring content, and the measuring procedure. Otherwise, the experiment may suffer failure in the accuracy in the obtained results or, more seriously, damages of the facility.

Mercury is once used as pressure-transducer medium for old-fashioned PVT instruments, but it is poisonous and dangerous especially when it overflows in experiments. Therefore, a kind of mercury-free PVT instrument, which uses screw piston pump to provide pressure, is commonly adopted in recent years.

2. Preparation of Oil or Gas Samples

(1) Sampling

Sampling is the process of gaining samples that represent the reservoir fluids, including the on-surface sampling and the underground sampling. The key to catch qualified samples is to keep the obtained ones remaining in their reservoir states (e.g., phase state, composition). The specific procedure of sampling should be in accord with petrochemical industry standards.

(2) Sample Transfer

Sample transfer is the process of transferring the samples obtained by the subsurface samplers into the PVT cylinders on the promise of maintaining the reservoir temperature and pressure conditions, so that the subsequent experiments can be carried on. According to different internal structures, the samplers can be divided into two types: mid-piston samplers and mid-piston-free samplers. The former one makes use of hydraulic oil or water as pressure-transducer medium to push the piston, whose movement impels the oil and gas to move; the latter one makes use of mercury as pressure-transfer medium to directly push the oil and gas to move. The procedure to transfer samples into the PVT experimental facility is as follows:

(1) Place the sampler which gets the samples in a thermo-tank or a heat-preservation water jacket and connect the sampler with the PVT facility.
(2) Heat the sampler to the reservoir temperature and, at the same time, employ the high-pressure metering pump to hold the pressure at the reservoir pressure;
(3) Advance the high-pressure metering pump (10) that connects with the sampler from which the gas and oil sample is impelled into the PVT cylinder, while at the same time withdraw the high-pressure metering pump (3) and (5); besides, it is required in the two processes to keep the pressure constant;
(4) Read the volume of the transferred oil and gas, V_f, on the computer screen or the scale of the pump (10); after the transferring work, close the upper valve of the PVT cylinder to make sure the system contained in the cylinder would not change.

Thus far, we have finished the preparation work.

3. Measurement of the Phase State and Physical Properties of Black Oil

Adopting the devices mentioned above, the parameters of crude oil such as the saturation pressure, the oil–gas ratio, the volume factor, the compressibility coefficient, and the viscosity can all be measured.

(1)   Measurement of Saturation Pressure and Compressibility Coefficient

To measure the saturation pressure and compressibility of crude oil, it will be sufficient to make use of only one of the PVT cylinders shown in Fig. 4.14. In the measurement, with all of the valves closed, the oil sample must be confined in the cylinder at constant temperature and pressure that is equal to the reservoir conditions. Then, withdraw the metering pump connected to the cylinder to reduce the pressure of the oil sample, and read the gauge pressure P_1 as well as the volume of the holding capacity V_1 (i.e., the volume of the oil sample) of the PVT cylinder. Repeat the pressure reduction every time the system fully reaches its equilibrium, and obtain the volumes of the oil sample $(V_1, V_2, ..., V_5)$ at all the stages of pressure $(P_1, P_2, ..., P_5)$. This measurement follows a pattern in which the composition of the system does not change, which means that there is no gas liberated from the liquid during the process.

Correlating the corresponding values of P and V, a P–V relationship curve like the one shown in the Fig. 3-25 can be made. The pressure exactly at the inflection point of the curve, or in other words, the crossing point of the two straight lines, represents the saturation pressure of the studied reservoir oil.

On the basis of the P_i and V_i above the saturation pressure, the compressibility coefficient C_o of crude oil under reservoir conditions can be calculated through the equation written below:

$$C_o = -\frac{1}{V_f}\frac{V - V_f}{P - P_f} = -\frac{1}{V_f}\frac{V_b - V_f}{P_b - P_f} = -\frac{1}{V_f}\frac{\Delta V}{\Delta P} \tag{4.9}$$

where

P_f, P_b, P the original reservoir pressure, the saturation pressure, and an arbitrary pressure higher than P_b $(P > P_b)$, respectively, MPa;

V_f, V_b, V respectively the volumes of the oil sample at the original reservoir pressure, the saturation pressure, and the arbitrary pressure, cm³;

C_o the compressibility coefficient of the oil sample under reservoir conditions, MPa⁻¹.

In the case of calculating the compressibility coefficients within different pressure intervals, another equation should be adopted:

$$C_j = -\frac{1}{V_j}\frac{\Delta V}{\Delta P} \tag{4.10}$$

where

ΔP the pressure differential within the pressure interval between the jth stage and the $j + 1$th stage, MPa;

ΔV the volume differential within the pressure interval between the jth stage and the $j + 1$th stage, cm³;

V_j the volume at the jth stage of pressure, cm^3;
C_j compressibility coefficient within the pressure interval between the jth stage
          and the $j + 1$th stage, MPa^{-1}.

(2) Measurement of Solution GOR and Volume Factor

Repressurize the oil sample within the PVT cylinder to the reservoir pressure, and during the pressurization, keep agitating the cylinder or initiate the electromagnetic agitator, which moves up and down, to make the gas dissolve into the oil completely and evenly. Then, wait for a period of time until the pressure and temperature reach equilibrium.

Holding the conditions of reservoir temperature and pressure, open slightly the valve of the PVT cylinder and advance the pump slowly, letting out a small quantity of oil sample from the cylinder and into the separator. In addition, the volume difference of the oil sample within the cylinder before and after the discharge, ΔV_f, is rightly the underground volume of the discharged oil. Then, the discharged oil undergoes a separation in the separator where the volume of the liberated gas V_g can be measured by a gas meter; besides, the volume of the degasified oil is calculated in the manner of dividing the mass of the degasified oil (W_o) by the density of it (ρ_o).

The volume factor and original solution GOR of oil can be obtained through the equations below:

$$B_{oi} = \frac{\Delta V_f \rho_o}{W_o} \tag{4.11}$$

$$R_{si} = \frac{V_g \rho_o}{W_o} \tag{4.12}$$

where

B_{oi}, R_{si} the volume factor of oil under original conditions and the original solution
                   GOR, m^3/m^3;
ΔV_f the underground volume of the discharged oil, m^3;
W_o the mass of the degasified oil in the separator, kg;
ρ_o the density of the degasified oil in the separator at 20 °C, kg/m^3;
V_g the volume (under standard conditions) of the gas liberated from the
                   discharged oil, m^3.

(3) Measurement of Viscosity

The high-pressure viscometer matching the PVT facility has two types—the rolling ball viscometer and the capillary viscosity, which can measure the viscosity of the oil with gas in solution under the conditions of reservoir temperature and pressure.

The falling ball viscometer is a circular tube of diameter D which is filled with high-pressure gas-bearing oil and has a steel ball of diameter d in it. The measuring mechanism: When the stell ball falls downward in the oil, the falling velocity or falling time is proportional to the viscosity of the oil, or in other words, the time that the steel ball falls in the oil is proportional to the viscosity of the oil. Expressed mathematically:

$$\mu_o = k(\rho_b - \rho_o)t \tag{4.13}$$

where

μ_o the viscosity of reservoir oil, mPa s;
ρ_b the density of steel ball, g/cm^3;
ρ_o the density of oil, g/cm^3;
t the falling time of steel ball, s;
k a constant of the viscometer which depends on the inner diameter and the inclination angle of the circular tube, as well as the diameter of the steel ball.

The value of k is calibrated by experimenting with standard oil with known density.

By substituting the measured falling time t to the equation above, the viscosity of oil μ_o can be obtained, or read the value of μ_o from the standard curves (μ_o–t curve) which are already made for reference.

The capillary viscometer works in another mechanism: When rendering the liquid flowing through a slender and long capillary tube with a definite quantity, a higher viscosity of the liquid would cause larger pressure difference between the two ends of the tube. So, the viscosity of the liquid can be obtained through a measurement of the pressure difference. The calculation equation is based on the Poiseuille's law. In the case that the crude oil is a Newton fluid:

$$\mu = \frac{\pi R^4 \Delta P}{8QL} \tag{4.14}$$

where

M the viscosity of reservoir oil, Pa s;
ΔP the pressure difference between the two ends of the capillary tube, Pa;
R the diameter of the capillary tube, m;
L the length of the capillary tube, m;
Q the flow rate through the capillary tube, m^3/s.

4. Measurement of the Phase Diagram and Physical Properties of Condensate Oil

The key to the measurement of the phase diagram of condensate oil is determining the dew-points at different temperatures based on which the envelope line can be drawn. Since only a single or a batch of liquid droplets are formed at the

dew-point pressure, the volume of the liquid phase is too small to be reflected on the
p–V curve, and therefore, the dew-point can only been observed and measured in
experiments. The quantity of the droplets is so small that it is hard for human to
observe them directly with naked eyes. In order to solve this problem, microscopes
and a camera system should be used to measure and record the volume of the
minute amount of condensate liquid formed in gas. The experimental procedure
goes as follows:

(1) Beginning with the room temperature, carry out an expansion experiment
     every other 10 °C. In this series of expansion experiments in which each one
     holds their temperature and composition constant and begins the pressure with
     a value higher than the original gas reservoir pressure, the bubble-point at each
     temperature should be measured. Finally, join together the bubble-points at all
     of the experimented temperatures with a line to form the envelope of the phase
     diagram.
(2) PVT measurement of the physical properties of the condensate gas samples:
     Pressurize the condensate sample to above the reservoir pressure, and under
     the conditions of constant temperature and constant composition, cause it to
     expand by reducing the pressure at the reservoir temperature; during the
     experiments, measure the relationship between the pressure and the volume of
     the gas sample, and calculate the $Z(P)$, $B_g(P)$, $C_g(P)$, and $\rho(P)$ relationships at
     this temperature; then, draw the experimental variation curves of Z, B_g, C_g,
     and ρ at changing pressure.

The specific experimental procedure should be in accord with the Standard SY/T
5543-1992 and the Standard SY/T 6101-1994.

To ensure the success ratio and the accuracy of the experiment, before the
experiments, an auxiliary computation based on PVT simulation software should be
carried out to predict the experimental procedure, the stage grading for pressure, the
quantities of the discharged oil and the liberated gas, the saturation pressure, the
dew-point pressure, and so on, so that the right moment for observation would not
be missed.

5. Analysis of the Physical Properties of Dry Gas

On the whole, the PVT method to measure the high-pressure physical properties
of dry gas is the same as what is presented previously. By pressurizing the gas
sample above the reservoir pressure and reducing the pressure at reservoir temperature to let the gas sample undergo expansion under constant temperature and
constant composition conditions. The relationship between the volume of the gas
sample and the pressure can be measured, based on which the relationships at the
given temperature such as $Z(P)$, $B_g(P)$, $C_g(P)$, and $\rho(P)$ are calculated.

6. Adjustment of Experimental Data

Various errors inevitably occur in companion with the experiment. The accuracy
of the data obtained from measurement depends on the systematic error of the

whole set of experimental facility, the personal error, and the technical level of the operating personnel. Therefore, before citing the experimental data for use, an adjustment of the data should be conducted to delete accidental errors.

The significant digit of the parameters and experimental data also play an important part in the problem of accuracy. For example, when a volumetric method (equation: $N_o = Ah\varphi S_o/B_o$) is adopted to do reserve calculations, it is accurate enough to employ the value of B_o that has three significant digits. In light of the precision of PVT instruments commonly used and the current technical level of the operating personnel, it is not difficult to make sure the value of B_o has an accuracy of three significant figures. However, in the case that a material balance method is adopted to calculate the reserve, the employed value of the volume factor B_o must have an accuracy of five significant figures, which means that the data obtained from experiments could not meet the accuracy requirement any more. Under such a situation, an adjustment of the data should be conducted to produce an available B_o that satisfies the accuracy requirement in engineering calculations.

(1) Adjustment of the Volume Factor of Oil above the Saturation Pressure

When the reservoir pressure is above the saturation pressure, the formation volume factor of oil, which is also dependent upon the temperature and the composition of the studied fluid, builds up a nonlinear relationship with the pressure.

In order to adjust the data, replace the volume factor B_o with the relative volume factor which is equal to the ratio of B_o to the volume factor of oil at the saturation pressure (B_{ob})—expressed mathematically, $\overline{B} = B_o/B_{ob}$. Because that $B_o \leq B_{ob}$, $\overline{B} \leq 1$.

For example, the $\overline{B}$–P relationship of a reservoir oil is shown in Fig. 4.15. It is already known that the original reservoir pressure is 34.0 MPa, the saturation

Fig. 4.15 Adjustment of the volume factor of a reservoir oil. ① *Straight line*, ② measured data, and ③ departure value of between the measured data and the *straight line*

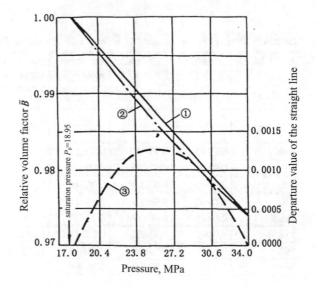

pressure is 18.33 MPa, and the volume factor of oil at the saturation pressure is $B_{ob} = 1.391$. The method to adjust B_o is as follows:

(1) Connect the two end points of the measured curve with a straight line, and at the same time, list the values of P, $\overline{B}$ in the first and second columns of Table 4.5.
(2) As the slope of the straight line is easy to obtain, we can now write the straight-line equation:

$$\overline{B}' = 1.0000 - \frac{1.0000 - 0.9739}{340.14 - 183.33}(P - 183.33)$$
$$= 1.0000 - 1.66443 \times 10^{-4}(P - 183.33) \qquad (4.15)$$

(3) Figure out the values of $\overline{B}'$ at different pressures by substituting the pressures into the equation written above, and list them in the third column of Table 4.5.
(4) Subtract the measured $\overline{B}$ from the calculated $\overline{B}'$ to obtain the departure values, and Table 4.5 shows them in the forth column.
(5) Plot the departure values on the plate of Fig. 4.15 and connect them with a line which turns out to be a smooth parabolic departure curve.
(6) Dealing with the drawn departure curve, read the values corresponding to different pressures, and Table 4.5 shows them in the fifth column.
(7) By subtracting the adjusted departure values from the values of $\overline{B}'$ obtained through Eq. (4.15), we can get the adjusted values of the relative volume factor $\overline{B}'$, which are shown in the sixth column of Table 4.5.
(8) Finally, according to the equation $B_o = \overline{B}' \times B_{ob}$, calculate the adjusted value of the formation volume factor of oil at any pressure, and Table 4.5 shows them in the seventh column.

Table 4.5 Data for the Adjustment of $\overline{B}$ (above the Saturation Pressure)

P (0.1 MPa)	Measured $\overline{B}$	Calculated $\overline{B}$	Departure value	Departure value after adjustment	$\overline{B}'$ after adjustment	Volume factor B_o after adjustment
1	2	3	4	5	6	7
			3–2		3–5	6 × B_{ob}
340	0.9739	0.9739	0.0000	0.0000	0.97390	1.354695
320	0.9768	0.97730	0.00050	0.00050	0.97680	1.358729
300	0.9790	0.98069	0.00079	0.00090	0.97979	1.362888
280	0.9829	0.98409	0.00119	0.00115	0.98294	1.36727
258.5	0.9862	0.98749	0.00129	0.00125	0.98624	1.37186
231	0.9909	0.99202	0.00112	0.00112	0.99090	1.378342
204	0.9960	0.99655	0.00088	0.00090	0.99594	1.385353
183	1.0000	1.00000	0.00000	0.00000	1.00000	1.391

2) Adjustment of the Two-Phase Volume Factor of Oil below the Saturation Pressure

Since the relationship between the two-phase volume factor B_t and the pressure P follows a hyperbolic type, the adjustment of B_t can not be conducted directly in a linear manner. Acting upon this problem, a new function, named Y-Function, is adopted to satisfy the adjustment requirement. In this function, the two-phase volume factor B_t is replaced by the two-phase relative volume factor, the ratio of the two-phase volume factor to the single-phase volume factor of oil at the saturation pressure $(\overline{B_t} = B_t/B_{ob})$. Obviously, the value of $\overline{B_t}$ is greater than 1. The Y-Function is expressed in such equation:

$$Y = \frac{P_b - P}{P(\overline{B_t} - 1)} \tag{4.16}$$

Seen from the equation, the value of Y is linear with the pressure P, which favors the adjustment work. By the way, in the equation, P_b is the saturation pressure.

Now, let us finish the adjustment work for the below saturation pressure part of the example mentioned in 1. It can be seen that the saturation pressure of the given oil sample is 183.3 atm, at which the volume factor is $B_{ob} = 1.391$. The other relevant data are all listed in Table 4.6.

Procedure of the Y-Function Adjustment:

(1) On the basis of the measured values of the two-phase volume factor B_t at different pressures, calculate the values of $\overline{B_t}$, $\overline{B_t} - 1$, $P_b - P$, and $(P_b - P)/P$, and Table 4.6 shows them in the corresponding columns.
(2) According to Eq. (4.16), figure out the values of Y at different pressures, and Table 4.6 shows them in the sixth column, and Fig. 4.16 shows the relation between Y-Function and pressure, which usually turns out to be a straight line;
(3) Get the slope of the straight line (8.575×10^{-3}) and its ordinate at the origin (1.830), and then, write the straight-line equation:

$$Y = 8.575 \times 10^{-3}P + 1.830 \tag{4.17}$$

Table 4.6 Data for the adjustment of two-phase volume factor

1	2	3	4	5	6	7	8
P (atm)	Measured $\overline{B_t}$	$\overline{B_t} - 1$	$P_b - P$	$\frac{P_b - P}{P}$	Y	$\overline{B_t}$ after adjustment	B_t after adjustment
183.33	1.0000	0.0000	0	–	–	1.0000	1.391
170.27	1.0233	0.0233	13.7	0.07671	3.2923	1.0233	1.42341
136.05	1.1160	0.1160	48.9	0.34750	2.9957	1.1159	1.552217
105.44	1.2691	0.2691	80.3	0.73871	2.7451	1.2700	1.76657
67.48	1.7108	0.1708	119.8	1.71673	2.4152	1.7124	2.381948
36.74	2.8606	1.8606	152.6	3.99074	2.1449	2.8600	3.97826

Fig. 4.16 Relationship between Y-Function and pressure

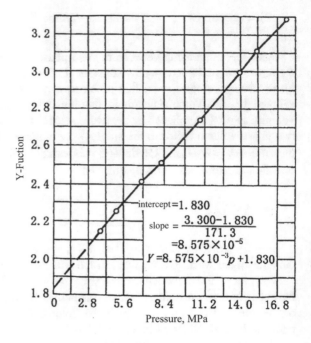

(4) Substituting Eq. (4.16) into Eq. (4.17), we can get:

$$\frac{P_b - P}{P(\overline{B_t} - 1)} = 8.575 \times 10^{-3}P + 1.830$$

(5) By substituting the known $P_b = 183.3$ atm into the equation above, we can get the $\overline{B_t}$–P relationship:

$$\overline{B_t} = 1.0000 + \frac{183.3 - P}{8.575 \times 10^{-3}P^2 + 1.830P} \tag{4.18}$$

(6) Resorting to Eq. (4.18), calculate the adjusted values of $\overline{B_t}$ at different pressures and list them in the seventh column of the table.

(7) Substitute the adjusted values of $\overline{B_t}$ into the equation $B_t = \overline{B_t} \times B_{ob}$, and list the obtained values of the adjusted two-phase volume factor B_t at different pressures, in the eighth column.

3) Adjustment of the Volume Factor of Natural Gas (B_g)

Figure 2-7 show that the B_g–P relation curve can be expressed as a quadratic polynomial within certain pressure interval. Since the value of B_g is far less than 1, what is usually used for the work of curve fitting is B_g', the reciprocal of B_g

(mathematically, $B'_g = 1/B_g$). The quadratic polynomial can be written as:

$$B'_g = a + bP + cP^2 \tag{4.19}$$

Taking a, b, and c as the unknown coefficients to be determined, we can build up a three-element linear equations set. The values of the three coefficients can be obtained by solving the set of equations with the least square method.

For example, from experiments, n groups of B'_{gi} at different pressures P_i are obtained:

$$\begin{cases} P_1, P_2, \ldots, P_n \\ B'_{g1}, B'_{g2}, \ldots, B'_{gn} \end{cases}$$

Utilizing the principle of minimum square method, three equations can be listed:

$$\begin{cases} \sum B'_g = na + b\sum P + c\sum P^2 \\ \sum PB'_g = a\sum P + b\sum P^2 + c\sum P^3 \\ \sum P^2 B'_g = a\sum P^2 + b\sum P^3 + c\sum P^4 \end{cases} \tag{4.20}$$

where

$$\sum B'_g = B'_{g1} + B'_{g2} + \cdots + B'_{gn}$$

$$\sum P = P_1 + P_2 + \cdots + P_n$$

$$\sum PB'_g = P_1 B'_{g1} + P_2 B'_{g2} + \cdots + P_2 B'_{gn},$$

and the others can be reasoned out by analogy.

Actually, the values of $\sum B'_g, \sum P, \sum P^2, \sum P^3, \sum P^4, \sum PB'_g, \sum P^2 B'_g$, n, etc., are already known, in light of which the three unknown coefficients, a, b, and c, can be figured out through Eq. (4.20).

The B_g'–P equation can be obtained by substituting the values of a, b, and c back into Eq. (4.19), upon which the values of B_g' at different pressures are gained.

[Example 4-3] Table 4.7 shows a list of gas pressures P_i as well as the reciprocal, B_g', corresponding to each of them. Adjust the values of B_g' by utilizing the minimum square method.

Substitute relevant data given in Table 4.7 into Eq. (4.20):

$$\begin{cases} 750.75 = 7a + 98.74b + 1472.24c \\ 11189.53 = 98.74a + 1472.24b + 23023.38c \\ 174909.85 = 1472.24a + 23023.38b + 374487.57c \end{cases}$$

Table 4.7 Adjustment of the values of $B_g{}'$ through minimum square method

P (MPa)	B_g	B_g' (standard) (m³/m³)	PB_g'	P^2B_g'	P^2	P^3	P^4	After adjustment	
								$1/B_g$	B_g
19.4	0.00681	146.75	2847.38	55,247.10	376.47	7304.55	141,728.64	146.38	0.006832
17.3	0.00759	131.82	2286.68	39,666.19	300.90	5219.67	90,543.54	131.80	0.007587
15.3	0.00860	116.35	1774.97	27,078.82	232.74	3550.70	54,169.06	116.51	0.008583
14.5	0.00906	110.44	1600.98	23,209.23	210.16	3046.67	44,167.27	110.83	0.009022
12.4	0.01055	94.75	1178.13	14,648.68	154.60	1922.26	23,900.97	95.11	0.010515
10.7	0.01228	81.47	871.83	9330.26	114.53	1225.70	13,117.27	81.54	0.012263
9.1	0.01446	69.17	629.55	5729.57	82.83	753.84	6860.81	68.72	0.014552
$\Sigma = 98.74$		$\Sigma = 750.75$	$\Sigma = 11,189.53$	$\Sigma = 174,909.8$	$\Sigma = 1472.24$	$\Sigma = 23,023.38$	$\Sigma = 374,487.6$		

By solving the equations, we can get:

$$a = -9.48823$$
$$b = 9.087975$$
$$c = -0.05435$$

Then, substitute the obtained coefficients into Eq. (4.19) to get the $B_g'-P$ equation:

$$B_g' = -9.48823 + 9.087975P - 0.05435P^2 \qquad (4.21)$$

Calculate the values of B_g' through the equation above and list B_g' and B_g in the last two columns of Table 4.7. Seen from the data, it is obvious that the accuracy of the values of B_g after adjustment is improved and high enough to satisfy the requirement in reserve calculation.

4.3.2   To Obtain the High-Pressure Physical-Property Parameters of Reservoir Oil by Charts

Bridged by relevant charts, the high-pressure physical-property parameters (e.g., the saturation pressure, the volume factor, and the viscosity of reservoir oil) can be inferred from the physical-property parameters of the surface tank-oil (e.g., the relative density of crude oil or natural gas, the solution GOR, the viscosity of degasified oil, and the reservoir temperature).

As the physical-property parameters of the surface tank-oil are easy to measure, the method by charts is simple and convenient to implement. As long as the properties of oil or gas are not too exceptional to apply to the charts, this method is feasible.

The mentioned charts are drawn on the basis of actual statistical data from oil fields. The following part will introduce how to utilize these charts.

1. To Obtain the Saturation Pressure of Reservoir oil by Charts

Figure 4.17 shows the chart commonly used in oil fields for the determination of the saturation pressure of reservoir oil. This chart, which takes many factors relevant to the saturation pressure into consideration, has a comparatively high accuracy—there is only an error of 7 % apart from the measured data. Usage of the chart: Start with the solution GOR and take account of the density of the surface oil, the reservoir temperature, and the density of the natural gas, use turning and folding straight lines to find the value of the next corresponding parameter one by one, and finally obtain the value of the saturation pressure.

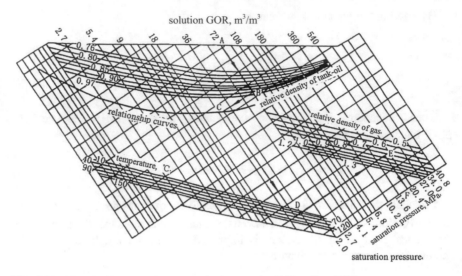

Fig. 4.17 Chart for the saturation pressure of reservoir oil (Standing 1947)

Reverse seeking from the present saturation pressure to the solution GOR can also be done with this chart. Note that the employed starting pressure must not be higher than the original saturation pressure of the studied oil reservoir, because in real oil reservoirs, the solution GOR is constant in value at pressures greater than or equal to saturation pressure.

[Example 4-3] It is already known that the present solution GOR is $R_s = 72$ m³/m³, the relative density of tank-oil is $\gamma_o = 0.88$, the reservoir temperature is $t = 93.3$ °C, and the relative density of natural gas is $\gamma_g = 0.8$. (1) Determine the present saturation reservoir pressure and (2) draw the relation curve of the saturation pressure vs the solution GOR.

Solution: (1) Follow the arrows marked on the chart and go ahead along with the folding line $A \rightarrow B \rightarrow C \rightarrow D \rightarrow E \rightarrow F$, which stops at the present saturation reservoir pressure, 16.6 MPa.

(2) Select 13.6, 10.2, 6.8, and 3.4 MPa as the starting values of the saturation pressure and do reverse seeking for each of them to get their corresponding solution GOR, respectively, 60, 44, 28, and 12 m³/m³. Then, on the basis of the obtained data, the relation curve of the saturation pressure vs the solution GOR can be drawn.

1. To Obtain the Formation volume factor of oil by Charts

The chart shown in Fig. 4.18 presents the relationship among the formation volume factor of oil, the solution GOR, the relative density of natural gas, the relative density of surface tank-oil, and the reservoir temperature.

[Example 4-4] It is already known that the present solution GOR is $R_s = 72$ m³/m³, the relative density of natural gas is $\gamma_g = 0.8$, the relative density of the surface degasified oil is $\gamma_o = 0.88$, and the reservoir temperature is $t = 93.3$ °C.

[Solution] (1) Start with the solution GOR, follow the marked arrows, and go ahead along with the dashed folding line: $A \rightarrow B \rightarrow C \rightarrow D \rightarrow E$, which ends at the volume factor of the given reservoir oil, $B_o = 1.245$.

In addition to the given chart for the volume factor of single-phase oil, there are also similar charts for the two-phase (oil and gas) volume factor. These charts can be found in references about the fluid prosperities of oil (Fig. 4.18).

3. To Obtain the Viscosity of Reservoir oil by Charts

As the composition of reservoir oil is quite complex, the viscosity of it varied within a wide range. Therefore, laboratory measurements, which have advantage in accuracy, are preferred to the charts.

To obtain the viscosity of reservoir oil, firstly resort to Fig. 4.19 to read the viscosity of the surface tank-oil corresponding to the known relative density of the surface tank-oil and the reservoir temperature, and then, resort to Fig. 4.21 and read the viscosity of the reservoir oil corresponding to the solution GOR and the obtained viscosity of the surface tank-oil.

For example, If it is known that the relative density of a surface tank-oil is $\gamma_o = 0.850$ and the reservoir temperature is $T = 87.8$ °C, the viscosity of the surface tank-oil can be read from Fig. 4.19 (1 mPa s). In accordance with the solution GOR $R_s = 72$ m³/m³, the viscosity of reservoir oil can be read from Fig. 4.21 (0.5 mPa s).

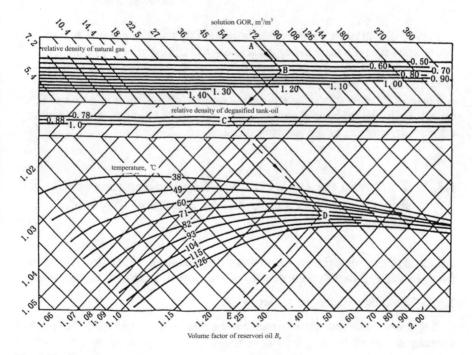

Fig. 4.18 Chart for the formation volume factor of oil (Standing 1947)

Fig. 4.19 Relationship of viscosity of degasified oil with temperature and oil density (Beal 1946)

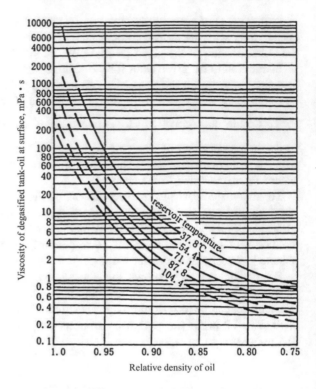

Relative density of oil

Figure 4.20 shows the relationship between the reservoir pressure and the viscosity when the pressure is above the bubble-point. If the viscosity of the crude oil under the original reservoir conditions is known, the viscosity at the bubble-point pressure can be obtained, and vice versa (Fig. 4.21).

Figure 4.22 shows the viscosity–temperature curves for crude oils with different density, drawn on the basis of statistical data. The curves serve as reference sources for the determination of the viscosity of oil.

4.3.3 To Obtain the High-Pressure Physical-Property Parameters of Reservoir Oil by Empirical Relation Equations

Empirical relation equations correlating all the parameters can be generalized from the abundant statistical data combined with the further processing of the charts.

1. To Obtain the viscosity of Reservoir oil

Beggs and Robinson once presented the equation to determine the original viscosity of the oil under reservoir conditions according to the viscosity of the degassed oil at the reservoir temperature:

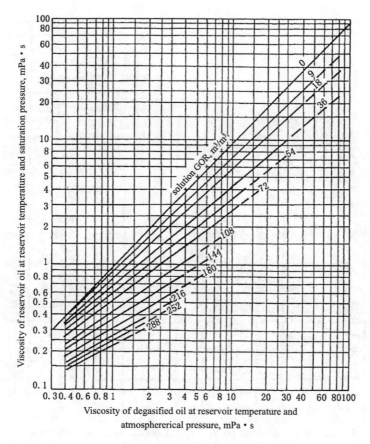

Fig. 4.20 Viscosity of reservoir oil at reservoir temperature and saturation pressure

$$\mu_o = A\mu_{od}^B \qquad (4.22)$$

$$\left. \begin{array}{l} A = 4.4044(\rho_o R_s + 17.7935) \\ B = 3.0352(\rho_o R_s + 26.6904) \end{array} \right\} \qquad (4.23)$$

where

$\mu_o\mu$ viscosity of the oil under reservoir conditions, mPas;
μ_{od} viscosity of the degasified oil at reservoir temperature, mPa s;
ρ_o density of the surface tank-oil, g/cm^3;
R_s Solution GOR, m^3/t.

2. To Obtain the Volume Factor of Oil at the Saturation Pressure

   Utilizing the standing empirical relation, volume factor of oil at the saturation pressure can be calculated through the equation written below:

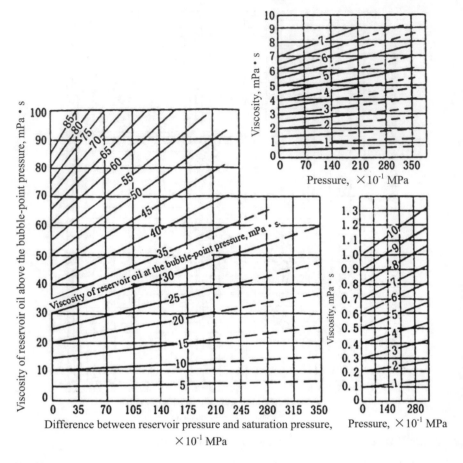

Fig. 4.21 Relationship between pressure and the viscoisty of reservoir oil above saturation pressure (Beal 1946)

Fig. 4.22 Viscosity–temperature curves of oils with different density

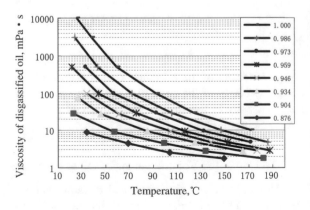

$$B_{ob} = 0.972 + 1.1175 \times 10^{-3}[7.1174 + R_s(\gamma_o\gamma_g)^{0.5} + 0.4003t]^{1.175} \quad (4.24)$$

where

B_{ob} volume factor of oil at the saturation pressure, m^3/m^3;
R_s Solution GOR, m^3/t;
γ_o relative density of the surface tank-oil, dimensionless;
γ_g relative density of the surface natural gas, dimensionless;
T reservoir temperature, °C.

The computed results have an arithmetical mean error of 1.17 %, which is within a reasonable error range allowed in engineering.

[Example 4-6] It is known that the solution GOR of a reservoir oil is $R_s = 77$ m^3/t, the relative density of the surface tank-oil is $Y_o = 0.9213$, the relative density of the natural gas is $Y_g = 0.7592$, and the reservoir temperature is $t = 54$ °C. Determine the volume factor of the oil in this reservoir at its saturation pressure.

Solution: By substituting the given parameters into Eq. (4.24), we can get:

$$B_{ob} = 0.972 + 1.1175 \times 10^{-3}\left[7.1174 + 77(0.9213 \times 0.7592)^{0.5} + 0.4003 \times 54\right]^{1.175} = 1.20$$

4.4 Application of the Fluid High-Pressure Property Parameters: Material Balance Equation of Hydrocarbons in Reservoirs

In this lesson, what will be introduced is the material balance equation of hydrocarbons in underground reservoirs. On the one hand, you are going to see the application of the high-pressure physical-property parameters of crude oil and natural gas presented in this chapter; on the other hand, your comprehension of these parameters will be assisted. Of course, the application of the high-pressure physical-property parameters does not go only so far.

4.4.1 Derivation of Material Balance Equation

The material balance equation of hydrocarbon in reservoir is based on the material balance principle. For a reservoir with a certain capacity, the material balance equations, respectively, for the oil and gas phases can be built up after a period of development, during which a part of each of the two phases is removed and depleted out. Take the material balance of the gas phase as example: initial volume of the natural gas existing = the volume removed + the volume remaining.

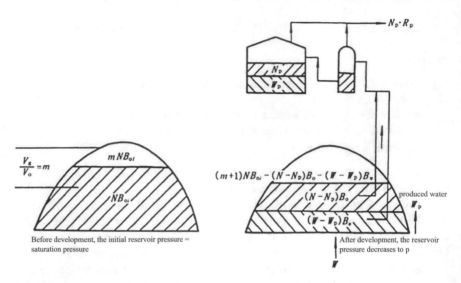

Fig. 4.23 Fluid volumetric change before and after reservoir development

There are various types of oil and gas reservoirs, but they all obey the principle of material balance. Here, let us take a reservoir with gas-cap and suffering water invasion during development for example, and examine the volumetric and pressure changes occurring the productive life. The changes of the oil, gas, and water are shown in Fig. 4.23. The symbols listed below are used to denote certain terms for brevity.

N initial oil reserve in place (volume under surface standard conditions), m^3;

G initial gas reserve in place, m^3;

R_{si} initial solution GOR, m^3/m^3;

N_p cumulative oil produced, m^3;

R_p cumulative mean producing GOR, m^3/m^3;

R_s solution GOR when the reservoir pressure drops at P, m^3/m^3;

W cumulative water influx when the reservoir pressure drops at P, m^3;

W_p cumulative water produced, m^3;

$m = V_g/V_o$ ratio of initial gas-cap volume to initial reservoir oil volume;

B_{oi} formation volume factor of oil at initial reservoir pressure;

B_o formation volume factor of oil at the pressure P;

B_{gi} volume factor of gas at initial reservoir pressure;

B_g volume factor of gas at the pressure P;

B_{ti} two-phase volume factor at initial reservoir pressure;

B_t two-phase volume factor at the pressure P;

B_w volume of formation water at the pressure P.

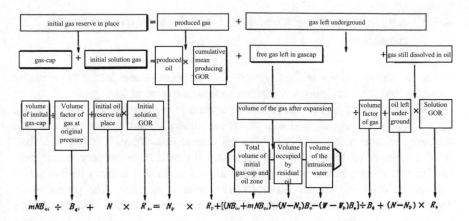

Fig. 4.24 Material balance relationship in reservoir and derivation of material balance equation

The derivation process is illustrated as a flowchart shown in Fig. 4.24, upon which we can obtain the material balance equation of the hydrocarbons in such a reservoir:

$$\frac{mNB_{oi}}{B_{gi}} + NR_{si} = N_pR_p + \frac{(m+1)NB_{oi} - (N - N_p)B_o - (W - W_p)B_w}{B_g}$$
$$+ (N - N_p)R_s$$

By substituting the relation equations about the two-phase volume factor, $B_t = B_o + (R_{si} - R_s) B_g$ and $B_{ti} = B_{oi}$, into the equation above, we can get a simplified form:

$$N = \frac{N_p[B_t + (R_p - R_{si})B_g] - (W - W_p)B_w}{(B_t - B_{ti}) + mB_{ti}\left(\frac{B_g - B_{gi}}{B_{gi}}\right)} \tag{4.26}$$

Equation (4.26) is rightly the general form of the material balance in reservoir. In dealing with different types of reservoirs, this equation can be simplified in light of the specific conditions. For example:

(1) If a reservoir is free from gas-cap, we can know that $m = V_g/V_o = 0$, in light of which the denominator can be simplified.
(2) If a reservoir is free from water invasion (including edge water, bottom water, and injected water), we can know that both W and W_p equal zero, in light of which the numerator can be simplified.

4.4.2   Analysis of the Parameters Used in Material Balance Equation

The numerous parameters and data involved in the material balance equation can be classified into three categories according to their different obtaining accesses.

The first category is the high-pressure physical-property parameters such as R_{si}, R_s, B_g, B_g, B_{oi}, B_o, B_{ti}, B_t, and B_w, which are usually obtained from experimental measurements. In order to obtain these parameters accurately, it is required that: (1) The value of the present reservoir pressure P should be accurately determined. (2) The phase equilibrium in experiments should be in agreement with the phase equilibrium occurring in the underground reservoir. However, the processes of degasification that occurs in experimental measurements and that naturally occurs in underground reservoirs are not all the same, which somewhat leads to some deviations in the values of the volume factor and the GOR, as well as some errors in the reserve calculation. (3) The measured experimental data should be adjusted before being adopted (See the lessen 3 of this chapter).

The second category is the statistical data from production such as N_p, W_p, and R_p. Generally speaking, in the information from field, the statistical data about the quantity of the produced oil is comparatively accurate. However, the accuracy of the produced gas and the produced water is rather low owing to the lack of attention for them. So, in the data processing, it is important to take stringent precautions against the errors about them, which can be as high as 10 % plus. In consideration of this problem, the statistical work of all the produced materials should be done carefully, so that the production performance of the reservoirs can be correctly estimated, based upon in which exact and reliable data can be provided for the material balance equation.

The third category is the unknown data such as N, m, and W. In the material balance equation, there are three unknown parameters: N (the initial oil reserve in place), m the ratio of initial gas-cap volume to initial reservoir oil volume), and W (the cumulative water influx).

The value of m can be obtained from the geologic data and well logging data. But in real reservoirs, neither the oil–water interface nor the oil–gas interface is clear, whereupon there exists actually an oil–water or oil–gas transition zone. Because of this phenomenon, it is very difficult to demarcate the oil band and the gas-cap. However, in this manner, even if the value of m is already known, there are still two parameters whose values are not easy to determine. In fact, the three unknown parameters, m, N, and W can be calculated out simultaneously by listing and solving a set of material balance equations that describe different developmental periods of the studied reservoir.

The solution and application of equations set mentioned above are given in the course of reservoir engineering.

In reservoir engineering calculations, the several commonly employed methods such as the seepage-flow calculation method, the volumetric method, and the material balance method are interpenetrating and complementary to each other. For example, the volumetric method is used at the initial stage of reservoir development to calculate the reserve, the seepage-flow calculation is used to deploy development program, and the material balance method is used during the reservoir development to conduct the calibration and dynamic forecast work.

Part II
Physical Properties of Reservoir Rocks

As oil and gas reservoirs are porous rock stratifications deep in the earth, the characteristics of the underground storage space, including the structure and properties of the reservoir porous medium, decide the occurrence behavior, the storage abundance, and the reserve of the hydrocarbons existing in the reservoirs. They also govern the production capacity of the oil wells as well as the level of difficulty and the final effect of the development.

The deposits of oil and gas are mainly sedimentary rock strata. Table 1 shows that sedimentary rocks can be classified into two types—clasolite (clastic rock) and carbonatite (carbonate rock). The deposits of the major oil and gas regions in the world are mainly clastic reservoir rocks, which consist of various types of sandstone, conglomerate, pebbly sandstone, and mudstone. As the predominant type among the reservoir rocks, the clasitc rock is widely spreading and has better physical properties than the others.

Carbonate rock constitutes another important type of oil & gas deposits. According to worldwide statistics, the oil & gas reserve stored in the carbonate rock deposits accounts for half of the total reserve, and the oil production of them also takes up 50 % of the total production. The Persian Gulf Basin concentrates numerous carbonate reservoirs, and in China a crop of reservoirs of this type has also been discovered. Practice exhibits that the search for oil & gas resources from carbonate reservoirs has broad prospects. This part will take the clastic rock (sandstone) as the studied object.

Sandstone deposits are porous medium resulting from the sedimentation of sand grains cemented together by cementing materials. The solid granular materials form the framework of rock and the intergranular space within rock exists as voids or pores. In this part, what will be introduced include the framework properties such as the granulometric composition, also called the particle-size composition, and the specific surface (the surface area to volume ratio), as well as some other properties

Table II.1 Classification and Instances of Reservoir Rock

Sedimentation Type	Lithology	Classification	Typical Reservoir Example
Clastic rock	Sandstone	Loose sandstone	Sa'ertu, Shengtuo
		Silt sandstone	Wendong
		Compacted sandstone	Zuoyuan
		Fractured sandstone	Yanchang
	Conglomerate	Conglomerate	Kelamyi
	Pebbly sandstone	Pebbly sandstone	Shuguang
		Fractured pebbly sandstone	Menggulin
	Mudstone	Porous fractured & vuggy muddy limestone	Nanyish
Carbonate rock	Dolomite	Fractured & cavity dolomite	Renqiu
		Fractured porous argillaceous dolomite	Fengchengch
	Limestone	Fractured & cavity limestone	Suqian
		Bioaccumulated limestone	Zhangxi
		Porous fractured algal limestone	Yidong
	Igneous rock	Fractured porous andesite	Fenghuadian
		Fractured tuff	Hadatu
		Volcanic rock	Chepaizi
		Basalt, andesite	417 Block, Kelamyi
	Metamorphic rock	Fractured metamorphic rock	Yaerxiao
		Fractured granite	Jinganbo

such as the permeability, the saturation, the compressibility, the thermal property, the electrical property, the radioactivity, the acoustic property.

These properties or parameters are not always constant, but vary more or less under the influence of the operations during well drilling and oil production. The sensitivities of reservoir (the velocity sensitivity, the water sensitivity, and the acid sensitivity) and the problem of their evaluation will be presented as another contributing section of Part II.

Chapter 5
Porosity of Reservoir Porous Medium

5.1 Constitution of Sandstone

Sandstone is composed of sand grains of different properties, shapes, and sizes which are cemented up by cementing materials. The property of a reservoir is significantly affected by the size, shape, and arrangement pattern of the grains as well as the constituents, quantity, property, and cementation pattern of the cementing materials. The parameters and curves about the size distribution can be used by the geologists to judge the sentimental environments, and by the reservoir engineers to evaluate the reservoirs. Here, the relationship between the particle size of sand grains and the physical properties of reservoirs will be discussed.

5.1.1 Granulometric Composition of Sandstone

1. Concept of Granulometric Composition

By smashing a sandstone rock into individual grains with a rubber hammer, we can find that the sand grains vary in size. Particle size, which is the diameter of a particle, is a term referred to as the measurement of the size of the rock grains (unit: mm or μm). Ranked groups according to different particle-size ranges are called "grain grades." The grain gradation can be given in many ways, and the one shown in Table 5.1 is the most commonly encountered one.

The granulometric composition of sandstone rocks is the percentages of the sand grains within the different particle-size ranges in the total quantity of the sandstone grains. The particle sizes and granulometric composition of sandstone rock are usually measured by adopting the sieve analysis and the sedimentation method.

© Petroleum Industry Press and Springer-Verlag GmbH Germany 2017
S. Yang, *Fundamentals of Petrophysics*, Springer Geophysics,
DOI 10.1007/978-3-662-55029-8_5

Table 5.1 Grain gradation

Grain gradation	Clay (mud)	Silt		Sand				Gravel		
		Fine silt	Coarse silt	Fine sand	Medium sand	Coarse sand		Fine gravel	Medium gravel	Coarse gravel	Boulder gravel
Diameter of grain (mm)	<0.01	0.01–0.05	0.05–0.1	0.1–0.25	0.25–0.5	0.5–1	1–10	10–100	100–1000	>1000

1) Sieve Analysis

A sieve analysis is performed on column of sieves fixed on a mechanical shaker. The top sieve has the largest screen openings, and each lower sieve has smaller screen openings than the one above. After the shaking is completed, the sand grains remaining on each sieve should be weighed by a balance, by analyzing, which the mass distribution of the sand grains can be known.

There are two ways to express the screen openings of sieve: (1) the number of the openings within every inch of length, named mesh or order. For example, a sieve of 200 meshes has 200 openings within every inch length of the screen. (2) The size of each opening is directly expressed as their diameter. For example, the grading differential between two adjacent sieves is $\sqrt{42}$ or $\sqrt{2}$. Obviously, the gradation test of particle size is dependent upon the size of the screen openings of the employed sieve column.

Generally speaking, the range of sand-grain diameter of reservoir sandstones is 1–0.01 mm. Table 3-3 shows the sieve analysis results of the rock sample from some oil field.

2) Sedimentation Analysis

The material captured by the narrowest meshed sieve is the extremely fine powder. Further gradation of the powder will be accomplished through a sedimentation analysis which is usually employed to determine the content of the grains smaller than 72–53 μm. As grains with different-sized sink at different rates in liquid, they require different length of time to complete their travel and reach the container bottom. Hypothesis:

(1) The sand grains are hard, ball-typed particles;
(2) The fine particles move quite slowly in viscous, non-compressible liquid and exist in an infinite distance away from the container walls and bottom;
(3) The fine particles sink with constant velocity, and the individual fine particles keep being in a dispersed state during their sinking.
(4) There is no relative movement on the interface between the moving fine particles and the dispersion medium.

Based upon the assumptions listed above, the sinking rate of a particle can be obtained according to the Stokes equations in fluid mechanics:

$$V = \frac{gd^2}{18v} \left(\frac{\rho_s}{\rho_L} - 1 \right) \tag{5.1}$$

Then, the calculation equation for particle size can be derived out:

$$d = \sqrt{\frac{18vV}{g\left(\frac{\rho_s}{\rho_L} - 1\right)}} \tag{5.2}$$

where

d Particle diameter, cm;
V Sinking rate in liquid of the grain with a diameter d, cm/s;
ρ_s Density of particle, g/cm³;
ρ_L Density of liquid, g/cm³;
v Kinematical viscosity of liquid, cm²/s;
g Acceleration of gravity, $g = 981$ cm/s².

As shown in the equation above, since the density of particle ρ_s can be obtained through a weighting bottle or some other methods, it is obvious that the last parameter needed for the employment of this Eq. (5.2) is the in-liquid sinking rate V of the grains. As long as the suspension liquid has been selected (in fact, clear water is usually used), in other words, the values of ρ_L and v are already known. Besides, through a counting process, the number percentage of the grains with a diameter d can be measured.

Experience has shown that when the particle diameter ranges between 50 and 100 μm, this method possesses enough accuracy. In addition, because the concentration of the grains has great effect on their sinking rate in dispersion liquid, the weight concentration of the rock grains in suspension liquid should be controlled below 1 %.

The particle diameter d measured through either the sieve analysis or the sedimentation analysis is merely a range, the values within this range are smaller than the opening diameter of the sieve above (d_1') and larger than the opening diameter of the sieve below (d_i''). In view of what is mentioned previously, the average value of the particle size should be calculated:

$$\frac{1}{\overline{d_i}} = \frac{1}{2}\left(\frac{1}{d_i'} + \frac{1}{d_i''}\right) \tag{5.3}$$

where

$\overline{d_i}$ Average particle diameter and
d_i'', d_i' the opening diameters of two adjacent sieves, respectively.

In dealing with comparatively compacted fine-grained rocks, another way to measure their granulometric compositions is to make them into petrographic thin sections for microscope inspection or the image instrument analysis. In recent years, various methods and instruments based on the optics are researched and used at home and abroad for the measurement of particle size.

2. Illustration of Granulometric Composition

The granulometric composition shown in Table 5.2 can also be illustrated by different graphs such as histrogram, cumulative-curve graph, and frequency-curve graph, among which the most commonly used ones in oil field practice are the particle-size distribution curve (Fig. 5.1) and the cumulative particle-size distribution curve (Fig. 5.2).

Table 5.2 Granulometric composition of a rock sample

Particle size (diameter of screen openings)		Mass remaining on the sieve g	Average particle size $\overline{d}$ mm	Mass percentage (%)	Accumulative percentage (%)
Mesh	Mm				
15	1.332	0.243			
20	0.900	0.068	1.074	0.080	100.000
40	0.450	6.161	0.600	7.198	99.920
60	0.280	36.504	0.345	42.649	92.722
80	0.180	30.291	0.219	35.390	50.073
100	0.154	4.697	0.166	5.488	14.683
120	0.125	3.057	0.138	3.572	9.195
140	0.105	0.120	0.114	0.140	5.623
160	0.098	1.616	0.101	1.888	5.483
180	0.090	0.58	0.094	0.685	3.595
200	0.074	2.491	0.081	2.910	2.910
Sum		85.591			

Fig. 5.1 Particle-size distribution curve

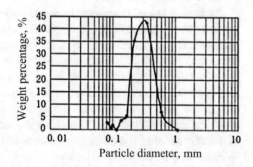

Fig. 5.2 Cumulative particle-size distribution curve

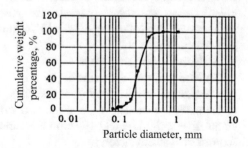

The percentages of differently sized grains in rock are illustrated in the particle-size distribution curve. The peak of the curve indicates the studied rock is primarily composed of grains with a certain size. Firstly, the higher the peak, the

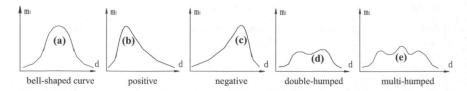

Fig. 5.3 Different patterns of commonly encountered particle-size distribution curve

Table 5.3 Particle-size composition attributes and particle-size parameters

Attribute	Technical measurement
Dispersibility	Median particle diameter, average particle size, specific surface
Sorting	Sorting coefficient, non-uniformity coefficient
Symmetry	Skewness
Sharpness	Kurtosis
Concentration tendency	Median, average modulus

more uniform the grains are in size; secondly, as the peak is more to the right, the grains are coarser, and vice versa.

As for the cumulative particle-size distribution curve, the steeper the ascending section, the more uniform the grains is in size. By utilizing this curve, the particle-size parameters can be obtained according to some characteristic points, based upon which a quantitative analysis of the homogeneity (uniformity) of the granulometric composition of rock can be performed.

Figure 5.3 shows several patterns of commonly encountered particle-size distribution curve (m_i is the mass percentage and d is the particle diameter). As the graphical notation is clear-cut and straight-forward, these graphs distinctly show the uniformity level of the distribution characteristic of the particle size of rock grains. In order to quantitatively calculate the uniformity level and characteristic of granulometric composition, the introduction of the particle-size parameters should be given. Table 5.3 shows particle-size distribution attributes and particle-size parameters. The following part will introduce several commonly used particle-size parameters to you.

3. Particle-Size Parameters

(1) Non-uniformity Coefficient α

The non-uniformity coefficient is defined as the ratio of the particle diameter d_{60} to the particle diameter d_{10}, and the two symbols, respectively, correspond to the cumulative weight 60 and 10 % on the cumulative particle-size distribution curve. Expressed mathematically:

$$\alpha = d_{60}/d_{10} \tag{5.4}$$

Generally speaking, the non-uniformity coefficient of reservoir rock ranges is 1–20. It is obvious that the more uniform the particle size is distributed, the closer to 1 the non-uniformity coefficient is.

(2) Sorting Coefficient S

By dividing the cumulative curve into four sections with three characteristic points, the cumulative weights 25, 50, and 75 %, Trask selected two of them to define the sorting coefficient:

$$S = \sqrt{\frac{d_{75}}{d_{25}}} \tag{5.5}$$

where

d_{25} The particle diameter corresponding to the cumulative weight 25 % on the cumulative particle-size distribution curve, mm;

d_{75} The particle diameter corresponding to the cumulative weight 75 % on the cumulative particle-size distribution curve, mm.

According to the appointment given by Trask, $S = 1$–2.5 suggests good sorting, $S = 2.5$–4.5 suggests moderate sorting, and $S > 4.5$ suggests bad sorting.

(3) Standard Deviation σ

Ford and Warburg presented a way to rank the sorting property of particles by employing the normal distribution standard deviation σ. The equation of this method is:

$$\sigma = \frac{(\varphi_{84} - \varphi_{16})}{4} + \frac{(\varphi_{95} - \varphi_{5})}{6.6} \tag{5.6}$$

$$\phi_i = -\log_2 d_i = \log_2 \frac{1}{d_i}$$

where

φ_i The base-2 logarithm of the ith particle diameter;

d_i The diameter of the ith particle diameter (i.e., the diameters, respectively, corresponding to the positions of 95, 84, 16, and 5 % on the cumulative curve) mm.

As shown in Eq. (5.6), this method embraces not only the main interval of the cumulative particle-size distribution curve, but also the characteristics of the head point (φ_5) and the tail point (φ_{95}) of the curve. Therefore, this parameter, as one of the most commonly used particle-size parameters, offers an ideal path to evaluate the sorting property of rock grains. Table 5.4 shows that the smaller the standard deviation σ, the better the sorting property of the rock.

Table 5.4 Ranking of sorting according to standard deviation (by Fork and Ward 1957)

Ford and Warburg standard deviation (σ)	Sorting rank
<0.35	Extremely good
0.35–0.50	Very good
0.50–0.71	Good
0.71–1.00	Moderate
1.00–2.00	Bad
2.00–4.00	Very bad
>4.00	Extremely bad

4. Average Size of Particle

1. Median Grain Diameter d_{50}: the particle diameter corresponding to the mass 50 % on the cumulative particle-size distribution curve, mm.
2. Average Particle Size d_m: the arithmetic average of the particle diameters corresponding to several characteristic points. Three averaging methods are presented below:

$$d_m = \frac{d_5 + d_{15} + d_{25} + \cdots\cdots + d_{85} + d_{95}}{10} \tag{5.7}$$

or

$$d_m = \frac{d_{16} + d_{50} + d_{84}}{3} \tag{5.7'}$$

or

$$d_m = \frac{d_{25} + d_{50} + d_{75}}{3} \tag{5.7''}$$

The carbonate rocks (e.g., limestone and dolomite) do not have the problem of particle size. This is because a carbonate rock, where the framework, the cementing materials, and the filling materials in voids are almost of the same substance, is impossible to be divided into individual grains.

5.1.2 Specific Surface of Rock

1. Concepts and Significance of the Research

The dispersivity of rock framework can also be described by the term special surface area, which is also called special surface. It is defined as the total internal surface area of the pores per unit volume of rock, or the total superficial area of the rock framework per unit volume of rock:

$$S = \frac{A}{V} \qquad\qquad (5.8)$$

where

S Specific surface of rock, cm^2/cm^3;
A Total internal surface area of the pores within rock, cm^2;
V Bulk volume of rock, cm^3.

In the case that the grains are at point-contact with each other, the total internal surface area of the pores is rightly equal to the total superficial area of all the grains. For example, if the radius of the spheres that comprise a porous medium is R, the specific surface of the medium turns out to be $S = 8 \times 4\pi R^2/(4R)^3 = \pi/2R$. As shown in this equation, it is obvious that the smaller the R, the greater the specific surface of the porous medium.

In a similar manner, the sandstone with finer grains possesses greater specific surface. Since the particle size of sandstone is quite small, this type of rock has very great specific surface. Examples are listed in Table 5.5.

As the surface of the pores within rock provides the fluid flowing boundaries, the specific surface decides many properties of rock, exerting comparatively great influence on the fluid flow in deposits. For example, the specific surface affects the surface phenomena at the contact of the fluid and the rock, the flow resistance that drags the fluids in rock, the permeability of rock, and the absorption of the fluids.

Apart from the particle size, it is found that many other elements, such as the shape and arrangement of the grains, the content of the cementing materials also considerably affects the value of specific surface. For example, the specific surface of a packing of spherical grains is greater than that of a packing of oval grains; and, the less the cementing materials among grains, the greater the specific surface is.

The specific surface can also be defined as the total internal surface area of the pores per unit mass of rock. According to this mass-oriented definition, the specific surface of sandstone is about 500–5000 cm^2/g, while the specific of shale rock is about $10^6 cm^2/g$ (i.e., 100 m^2/g).

The framework volume V_s and the pore volume V_p are also used to define the specific surface in practical application and formula derivation:

$$S_s = \frac{A}{V_s}; \quad S_p = \frac{A}{V_p} \qquad\qquad (5.9)$$

Table 5.5 Sandstone particle size and specific surface

Type of sandstone	Particle diameter, mm	Specific surface, cm^2/cm^3
Ordinary sandstone	1–0.25	<950
Fine sandstone	0.25–0.1	950–2300
Muddy sandstone	0.1–0.01	>2300

where

S_s Specific surface taking the skeleton volume of rock as denominator, cm^2/cm^3;
S_p Specific surface taking the pore volume of rock denominator, cm^2/cm^3.

The relationship among the different specific surfaces is:

$$S = \phi \cdot S_p = (1 - \phi)S_S \tag{5.10}$$

where

φ The porosity of the rock core, fraction.

In normal times, the specific surface of core is the one that takes the bulk volume as the denominator.

2. Measurement and Calculation of Specific Surface
(1) Penetrant Method
The penetrant method is performed on the basis of the perviousness of a fluid to rocks or grained layers.

Figure 5.4 shows the experimental procedure. This porosimeter consists of Mariotte vessel (1), core holding unit (2), and water manometer (3). A rock pore is

Fig. 5.4 Apparatus for the measurement of specific surface

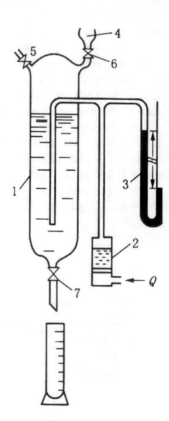

put into the core holding unit (2). Negative pressure is formed above the core, owing to the discharge of water from the Mariotte vessel, and the gas flows through the core. The differential pressure between the two ends of the core is measured by the water manometer (3), and the volume of the discharged water is rightly equal to the gas volume that flows through the core.

The switch (5) should be turned on to let the water flow through the funnel (4), fall into the Mariotte vessel before starting the experiment. Once the water level in the vessel mounts up to a specified height, the switch (5) and (6) should be turned off. During the experiment, the discharge flow rate is controlled by the switch (7). Calculate the flow rate after the differential pressure exhibited by the water manometer reaches its stability. Then, the specific surface can be obtained by through the Kozeny–Carman equation (1927):

$$S = 14\sqrt{\phi^3}\sqrt{\frac{AH}{Q_0\mu L}} \tag{5.11}$$

where

Q_0 Gas flow rate through the core, cm^3/s;
S The specific surface taking the bulk volume of rock as denominator, cm^2/cm^3;
φ The porosity of core, fraction;
A The cross-sectional area of core, cm^2;
L The length of core, cm;
μ The viscosity of air at room temperature, mPa s;
H The stabilized pressure differential as the gas passing the core, cm water column.

Although the air penetrant method, which takes air as the performing fluid, is most commonly used, water can also be employed as the fluid for the sandstones whose grains are comparatively coarser and do not disperse or expand in water.

(2) Estimation of Specific Surface by Granulometric Composition of Rock

Assuming that the sandstones are arrangements of uniform spheres, we can easily know that, accordingly, the superficial area of each individual sphere is $S_i = \pi d^2$, and the volume of each individual sphere is: $V_i = \frac{1}{6}\pi d^3$.

If we use the symbol φ as the notation of the total volume of the pores in per unit volume of rock, it is obvious that the volume occupied by the grains in per unit volume of rock is $V_s = 1 - \varphi$. Consequently, the number of the grains in per unit volume of rock is:

$$N = \frac{1-\phi}{V_i} = \frac{6(1-\phi)}{\pi d^3} \tag{5.12}$$

Then, we can obtain the specific surface of per unit volume of rock:

$$S = N \cdot S_i = N \times \pi d^2 = \frac{6(1 - \phi)}{d} \tag{5.13}$$

In fact, the grains involved in a sandstone rock are not uniform in size. According to relevant analysis of data:

The content of the grains with an average diameter of d_1 is G_1;
The content of the grains with an average diameter of d_2 is G_2;
....
The content of the grains with an average diameter of d_n is G_n.

So, in per unit volume of rock, the specific surfaces for these differently sized grains are:

$$S_1 = \frac{6(1-\phi)}{d_1} \cdot G_1\%$$
$$S_2 = \frac{6(1-\phi)}{d_2} \cdot G_2\%$$
$$\cdots\cdots\cdots\cdots$$
$$S_n = \frac{6(1-\phi)}{d_n} \cdot G_n\%$$

Thus far, we can get the specific surface of the studied rock:

$$S = \sum S_i = \frac{6(1 - \phi)}{100} \sum_{i=1}^{n} G_i/d_i \tag{5.14}$$

However, it is impossible for all of the grains in a real rock to be spherical. In order to fit reality better, a particle-shape correction coefficient C, which usually ranges from 1.2 to 1.4, is introduced into the equation. Now, the specific surface can be written as:

$$S = C \cdot \frac{6(1 - \phi)}{100} \sum_{i=1}^{n} G_i/d_i \tag{5.15}$$

When this method is adopted to estimate the specific surface of a rock, as regularity, weaker cementation among the grains and higher roundness degree of the grains leads to higher accuracy.

5.1.3 Cementing Material and Type of Cementation

As discussed previously, sandstones result from the cementation of the grains. After the introduction of the property of the sand grains, let us talk about of the cementing materials.

1. Mud (Clay) Cementing Material

The cementing materials of reservoir rocks refer to the precipitate materials apart from the detrital particles, usually existing as crystalline or amorphous authigenic minerals. It makes a share not more than 50 % and enables the grains to be hard solid stone by acting as a binder. The cementing materials always weaken the oil-bearing capacity and permeability of rock.

The composition, quantity, and cementation type of the cementing materials in rock have significant effect on the rock properties such as the compact level, the porosity, and the permeability. The most commonly encountered constituents of cementing materials are mud and lime, and sulfate and silica followed by.

(1) Mud Cementing Materials

Clay is the fine-grained materials normally considered to be less than 0.0l mm in size. It is naturally occurring earthly of the fine grains, which possesses plasticity when mixed with a little amount of water. The major chemical constituents contained in clay are silicon dioxide, alumina, water, and a small amount of iron, alkali metals, and alkaline earth metals.

Clay minerals refer to the minerals that comprise the main body of clay. According to the characteristics of them, the clay minerals can also be defined as the general name of the finely dispersed hydrated layer-silicate minerals and the hydrated non-crystalline silicate minerals. As seen from what is discussed above, the clay minerals contribute the main part of the clay in reservoir rock, and the mud is dominantly comprised of clay. The clay content in sandstone is usually higher than 10 % and sometimes can be up to 20 %. They are one of the main causes of the weak reservoir physical property and the reservoir damage during development.

The clay minerals commonly existing in oil & gas reservoirs are hydrated layer silicates such as kaolinite, montmorillonite, illite, chlorite, and their mixed layers. In some particular geologic environments, some chain silicates like sepiolite may also be encountered.

The water involved in clay minerals can be classified into three types according to their existing state: (1) absorbed water—water that is absorbed to the surface of the clay minerals, (2) interlayer water—water that exists between the unit crystalline layers of clay minerals, (3) constitution water—water that exists within the crystalline lattices in the form of hydroxide radicals. Among the three types mentioned above, usually the absorbed water and the interlayer water, which are comparatively loosely combined with the minerals, can be readily released at the temperature of 100–200 °C, while a higher temperature of about 400–800 °C is required to release the constitution water.

The water involved in clay minerals can be classified into two types according to the causes of formation: The first one is the terrigenous clay minerals, which usually suffer a lack of good crystalline forms due to the abrasive erosion during the process of transportation and sedimentation; the other one is the authigenic clay minerals that are formed in the sedimentary and diagenetic process and usually developed in porous reservoir sandstones with good sorting property, little terrigenous clay, and nice permeability. Besides, the authigenic clay minerals usually have good

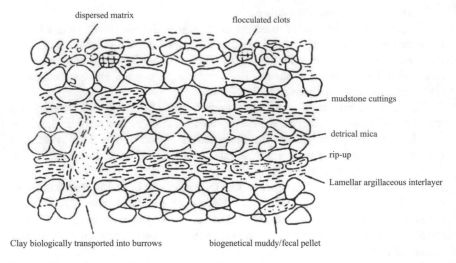

Fig. 5.5 Occurrence of terrigenous clay materials (By Qin jishun, 2001)

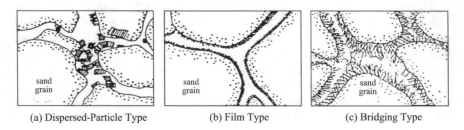

(a) Dispersed-Particle Type   (b) Film Type   (c) Bridging Type

Fig. 5.6 Occurrence of authigenic clay minerals

crystalline forms, and their degree of crystallization is associated with the development level of the pores in reservoir rocks.

Figure 5.5 shows the occurrence of the terrigenous clay materials including the dispersed matrix, the flocculated clots, the old-aged mudstones, the synchronous mud rock-masses, laminated mud interlayers, and the percolated residues. During the diagenetic compacting process, the clay grains are deformed and forced into the pores, leading to the decrease of the porosity of sandstone.

As shown in Fig. 5.6, the occurrence of the authigenic clay minerals in the sandstone pores can be classified into three types: dispersed-particle type, film type, and bridging type. They exert different influences on the permeability of the deposits.

  I. *Dispersed-Particle Type*: the clay materials packing the intergranular pores
     of sandstones and existing in the form of dispersed particles. Kaolinite is
     typical of this type. Kaolinite, as complete pseudo-hexagonal idiomorphic
     crystals or aggregates often arranged regularly in book-like or vermicular

stacks, fills the intergranular pores of sandstones. The influence of this dispersed-particle-typed clay on deposits twofold: on the one hand, occupying part of the intergranular space and partitioning the initial pores into fine and micropores, clay particles among the grains decreases not only the porosity but also the permeability of rock; on the other hand, since the clay particles are loosely packed in the pores and have rather weak adhesive force with the sandstone grains, they may move with the flowing fluids and cumulate at the pore throats, causing the problem of blockage.

II. *Film Type*: the clay minerals that perform in an oriented arrangement on the particles' surface to form continuous clay films attached on the pore walls. The montmorillonite, the chlorite, the illite, and the mixed layer clay minerals are typical of this type. Most of these mentioned clay materials are arrayed paralleled to the surface of the grains, or, in other words, the pore walls, and exist as a film thinner than 5 μm. The potential impact of this type of clay minerals on the deposits lies in two aspects: ① they greatly reduce the effective radius of pores and often cause blockage of the pore throats. It has been proven by large amounts of sample analysis that the deposits containing this kind of clay have poor permeability worse than that of the deposits containing dispersed-particle-typed clay. ② It is this type of clay materials that cause the reservoir damages during the developments, owing to the incompatibility of these minerals with the fluids that enter the formations and react with the clay films chemically and physically.

III. *Bridging Type*: the clay minerals that grow from one side of the pore wall and finally reach the other side, forming a clay crystalline bridge over the pore space. The most commonly encountered clay minerals of this type are the various streaked-schistic or fibrous autogenetic illites, whose distribution within the pores takes the form of networks. Besides, the montmorillonites and mixed layer clay minerals can form clay bridges at the pore throats, too.

Partitioned by the bridging-typed clays, the initial pores of sandstone become twisted and turned micropore spaces among the grains.

Because the fibrous illites have very great specific surface, a large area of bond water is formed within the sandstone pores, which reduces the oil-bearing capacity of the rocks.

To sum up, the terrigenous clay minerals and the authigenic clay minerals in reservoir rocks are different in occurrence, distribution pattern. Also they have different impact on the reservoir rock physical properties. The former one is the main cause of the heterogeneity of the reservoir, while the latter one is the main cause of the reservoir damage and the reduction of production capacity.

(2) Lime Cementing Materials

The lime cementing materials mainly consist of carbonate minerals, among which the most commonly encountered ones in sandstones are the calcite ($CaCO_3$) and the dolomite [$CaMg(CO_3)_2$]. These carbonate minerals in sedimentary rocks can be divided into two types: the primary and secondary.

Fig. 5.7 Relationship
between porosity and content
of carbonate minerals

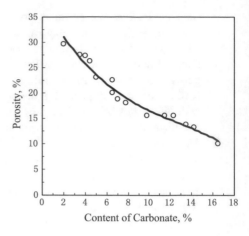

Through the analysis of the content of the carbonate minerals in rocks, especially the content of the secondary carbonate minerals, we can learn the law of the hydrodynamic-field activity, as well as the features of a geological period when a certain formation is formed. It is been proven by research that a lower content of the carbonate minerals in a reservoir rock leads to a greater rock porosity. Figure 5.7 shows the relationship between porosity and content of carbonate minerals. Explanation for this regularity is as follows: the underground acidulous water, being vigorously active deep in the earth, invades into the formations and erodes the carbonate minerals within them; obviously, the parts that confront larger amount of acidulous water are eroded more seriously—more carbonate minerals are eroded away and less carbonate minerals are left in formation. As a result, the rock finally has comparatively greater porosity. On the contrary, the more the left carbonate minerals in formation, the smaller the porosity of that part.

(3) Sulfate Cementing Materials

The sulfate cementing materials contained in reservoir rocks mainly consist of gypsum ($CaSO_4 \cdot nH2O$) and anhydrite ($CaSO_4$).

2. Type of Cementation

The distribution of the cementing materials in rocks and their contact relationships with the detrical particles are called the type of cementation. The type of cementation is usually dependent upon many a factor of the cementing materials, for example, the composition, the content, the sedimentary conditions, and the changes happened to the materials after their sedimentation. As shown in Fig. 5.8, there are three types of cementation:

(1) Basal Cementation

This type features the high content of the cementing materials—the total volume of the cementing materials is larger than that of the detrital particles and that the particles are distributed in the cementing materials and isolated from each other. Since the sedimentation of these cementing materials happens simultaneously with that of the detrital particles, this type is also called primary cementation, which is born with very

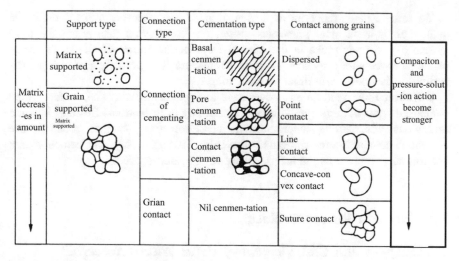

Fig. 5.8 Cementation type (By Liu Yujun, 1980)

high consolidation strength. Besides, because the pores existing in the cementing materials are quite micro, it has very poor capability of store oil and gas.

(2) Pore Cementation

The cementing materials, only filling the intergranular pores, do not exist so abundantly in a rock with this type of cementation, in which the grains shows a pattern of bracket contact. Here, most of the cementing materials, which often can only be found in big pores, are secondary and distributed unevenly in pores. Besides, the consolidation strength of this type ranks only next to the basal cementation.

(3) Contact Cementation

This type implies that very little cementing materials are contained in rock (usually less than 5 %). With the cementing materials playing as a binder at where the grains touch each other, the rock grains show a pattern of point-contact or line-contact. These cementing materials, usually taking the form of mud, are usually comprised of primary matter or debris resulting from weathering. Both the porosity and permeability of the rocks of this cementing type are good. In Daqing Oil Field, for example, the rocks that belong to this type of cementation possess a porosity larger than 25 % and a permeability ranging from dozens of 10^{-3} μm^2 to couples of μm^2 (Table 5.6).

Cementation type	Porosity, %	Permeability, 10^{-3} μm^2
Contact cementation	23–30	50–1000
Pore cementation	18–28	1–150
Basal cementation	8–17	<1

Table 5.6 Porosity and permeability of formations with different cementation types

To some extent, the type of cementation of the reservoir rocks decides the quality of their physical properties. Based on the relevant data about the third system of reservoir rocks in the Huabei Depression, Table 5.3 shows the comparison between the physical property parameters (the porosity and the permeability) of the rocks with different cementing types.

Reservoir rocks seldom and never have a single cementing, and, in fact, some composite types of cementation appear commonly in formations. Besides, in some formations with strong heterogeneity, there are also grumose rocks whose cementing materials are distributed unevenly. That is to say, the cementation's heterogeneity represents the heterogeneity of the reservoir rocks.

5.2   Pores in Reservoir Rocks

5.2.1   Types and Classification of Pores Within Reservoir Rock

The intergranular space within the rocks which is not filled with cementing materials or other solid materials is called voids. In fact, rocks without voids do not exist on the earth, and the various rocks are different from each other only in the size, shape, and development level of their voids: for a sandstone, there are naturally existing voids among its grains; for a carbonate rock, voids can occur when the dissoluble parts within it is chemically corroded by formation water; for an igneous rock, voids were formed as a result of the occupancy of gas masses during the diagenetic process. Besides, fracture and micro-fracture are another kind of void present within the various rocks, developed in virtue of the terrestrial stress, tectonic stress, the geological processes, etc.

According to the geometric measurement and shape of the voids, they can be classified into three types: pores (usually for sandstones), cavities (usually for carbonate rocks), and fractures. Since the pores are most commonly encountered in the rocks, all types of voids are usually referred to as the name "pores."

Since the pores provide the space for the storage and flowing of crude oil and natural gas, their properties such as size, configuration, interconnectivity, and development level directly affect the reserve volume of the formations and the production capacity of the oil and gas wells.

As the mentioned properties of reservoir pores (voids) are quite complex, classifications and descriptions of the pores are given from different perspectives.

1. Classification of Reservoir Pores—Meinzer Classification

Considering the composition and interconnectivity of the reservoir pores, Meinzer classified them into six types shown in Fig. 5.9: (a) deposition of materials with good sorting property and high porosity; (b) deposition of materials with bad sorting property and low porosity; (c) deposition of gravel materials with a high porosity, due to the fact that the constituent gravels are porous themselves;

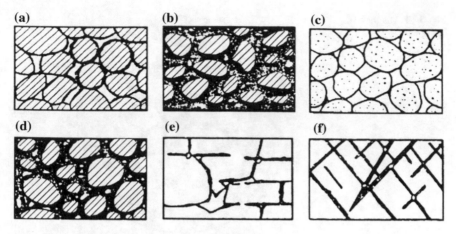

Fig. 5.9   Type of rock pores (By Meinzer 1942)

(d) deposition of materials with good sorting property but low porosity because of the sedimentation of the precipitated cementing materials among the rock particles; (e) porous rocks resulting from the chemical corrosion; and (f) cementing-material-containing porous rocks resulting from the faulting.

2. Classification of Reservoir Pores—according to the Genesis

(1) Intergranular Pores

In consideration of those grain-supported or miscellaneous-base-supported rocks that have a small amount of cementing materials, the intergranular spaces bounded by the particles are called intergranular pores, the predominant and most common pore type for sandstone.

The sizes and morphologies of the intergranular pores depend on the particle size, the sorting property, the sphericity, the packing pattern, and the degree of compaction. Moreover, the distribution of this kind of pores, which is directly related to the sedimentary environment, has been changing with the geological processes that occur after the diagenesis.

The sandstone reservoirs where intergranular pores predominate usually have large pores, thick throats, and good interconnectivity, possessing comparatively large porosity (usually larger than 20 %) and high permeability (usually higher than $100 \times 10^{-3} \ \mu m^2$). Figure 5.10 shows microscopically the typical morphology of intergranular pores.

(2) Microscopic Pores within Miscellaneous-Base

The microscopic pores within the miscellaneous-base mainly refer to the pores' voids that are formed as a result of the shrinkage of the miscellaneous-base-deposited matters during weathering and that are developed as intercrystalline pores during the recrystallization of clay minerals. This kind of pores can be found in miscellaneous-base mediums such as kaolinite, chlorite, mica, and carbonatite.

With width usually narrower than 0.2 μm, the microscopic pores within miscellaneous-base are so small that they can only be easily discernible under

Fig. 5.10 Diagram of
intergranular pores

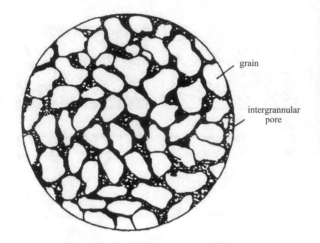

high-power microscope. Although this kind of pores, which sometimes can contribute more than 50 % of the total volume of all the pores in rock, is quite
considerable in quantity and volume, they actually entail tremendously low permeability of the rocks that bear them. In addition, as shown in Fig. 5.11(1), the
microscopic pores within the miscellaneous-base almost exist in all sandstone
rocks.

(3) Secondary Intercrystalline Pores

This kind of pore refers to the residual space present in rocks after the hys-
terocrystallization of quartz, which consequently causes an enlarged encroachment
of the crystallized products in the initial pore spaces. As a result of the secondary
growth of quartz, the pore volume is lessened and the throats are narrowed,
leading to a reduced porosity and lowered permeability of the rocks. Figure 5.11
(2) shows it.

Fig. 5.11 Diagram of
various pores under
microscope. 1-Microscopic
pores within
miscellaneous-base,
2-intercrystalline pores,
3-texture and bedding
fissures, 4-mica cleavage
fissure

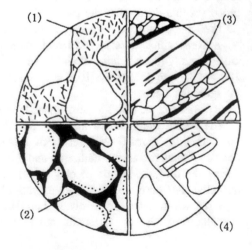

(4) Texture and Bedding Fissures

For those sandstones with texture or bedding structures, there are usually microfissures existing along with the textures or beddings due to the disparity of the different sand layer in lithology or grain-arrangement orientation. Figure 5.11(3) shows it. This kind of pore imparts to the rocks directional permeability.

(5) Fracture Pores

Microfissures are developed as a result of the crust stress actions. Passing around the grain outlines, the tiny-slender platelike microfissures are distributed and arranged under the influence of crust stress. Normally, the surface of fracture is perpendicular to the direction of the minimum crust stress. The crack with the fractures within sandstone reservoirs ranges from several tenths of a micron to dozens of microns. The fractures, although taking a quite small share (usually smaller than 5 %) in the total porosity of rock, significantly improve the permeability of rock. The characteristic of this kind of pore is that the fractures change with every remarkable change of the crust stress. For example, when the crust stress along the vertical direction of a fracture grows, this fracture tends to undergo a closure, leading to sharp reduction of the formation permeability. An illustration of a group of typical stress fracture pores is shown in Fig. 5.12(1). The cleavage fissure of mica, which is shown in Fig. 5.11(4), also belongs to the fracture pores.

(6) Solution Pores

Solution pores originate from the dissolution of the soluble substances in rocks, such as carbonate, feldspar, and sulfate. Table 5.7 gathers the genesis and characteristic of each of the pore types mentioned above.

3. Classification of Reservoir Pores—according to the Pore Size

According to the size, the pores in reservoir rocks can be classified into three types:

(1) Supercapillary Pores

Supercapillary pores refer to the pores with a diameter larger than 0.5 mm, as well as the fractures with a width wider than 0.25 mm. Most of the huge fractures, solution cavities, and the pores present in uncemented or loosely cemented sandstones belong to this kind of pore, within which the fluids enjoy free flowing with the aid of gravity.

Fig. 5.12 Diagram of various pores

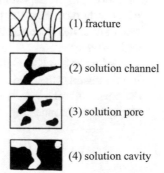

(1) fracture

(2) solution channel

(3) solution pore

(4) solution cavity

Table 5.7 Origin and characteristics of pore types

Rock type		Origin
Pores developed from primary sedimentation	Intergranular pore	Sedimentation
	Texture and bedding fissures	Sedimentation
Pores developed from secondary sedimentation	Vugular solution pores	Dissolution
	Secondary intercrystalline pores	Pressure solution
	Fractured pore	Effect of terrestrial stress
	Broken-grain pores	Fractures of rock
Mixed pores	Microscopic pores within miscellaneous-base	Compound origin

(2) Capillary Pores

Capillary pores refer to the pores with a diameter ranging 0.5–0.0002 mm, as well as the fractures with a width ranging 0.25–0.0001 mm. Most of the pores present in sandstones belong to this category. As the molecules of the solid pore walls exert quite great acting force on the liquid molecules of the fluids, the capillary force that occurs when there are two phases of fluids coexisting in the pores restrains the fluids from free flowing, unless an pressure differential great enough is.

(3) Microcapillary Pores

Microcapillary pores refer to the pores with a diameter smaller than 0.0002 mm, as well as the fractures with a width smaller than 0.0001 mm. Generally speaking, the pores present in shales and clays belong to this category. Within these pores, the reservoir pressure gradient is too weak to enable the fluids to flow, because the intermolecular attraction forces are rather powerful. Therefore, the channel diameter of 0.2 μm is usually regarded as a dividing line to judge whether a pore allows the fluids in it to flow. This type of pores is also called inactive pore.

4. Classification of Reservoir Pores—according to the other factors

1) According to the Generated Time

   The pores can be classified into original pores and induced pores. Original pore is that developed in the deposition of the materials, for example, the intergranular pores. Induced pore is that developed by various geologic processes subsequent to the deposition of rock, for example, the vugs and solution cavities which originate from the geological process of groundwater, and, another representative, the fractures that result from the rock breakings tunnel under tectonic stresses.

2) According to the Syntagmatic Relation

   In this manner, the voids can be divided into two types: pore and throat. Here, the pores refer the voids that are obviously big, while the throats are the slender channels interconnecting the pores.

3) According to the Interconnectivity
The pores can be subdivided into interconnected pores and disconnected pores. Most of the pores in rock belong to the first type of the mentioned, while the other type, the disconnected pores, also naturally exists.

5.2.2 Size and Sorting of Pores

The pores involved in a rock, big and small, are not uniform in size. For petroleum engineers, the size and sorting of rock pores are important basic data to help them know a reservoir. The pore radius can be obtained by using the corresponding capillary pressure curve (see Unit 9 for more details).

The term "pore-composition of rock" is used to describe the numerical percentage accounted by the pores of a specified diameter among all of the pores present in a rock. Two curves, the pore-size distribution curve and the pore-size cumulative distribution curve, can be drawn according to the pore-composition of rock. Figure 5.13 shows the former one of the two mentioned curves, in which the axis of ordinate indicates the volumetric percentage of the pores of a specified radius, while axis of abscissa gives the radius of pore (μm). Besides, sometimes $\log_2 d$ is also used as the horizontal axis where d represents the pore diameter.

Taking the cumulative volumetric percentage of pores as the axis of ordinate, Fig. 5.14 presents a pore-volume cumulative distribution curve.

Similar to the pore-size distribution curve, by analyzing certain character points the cumulative curve can also be used to quantitatively describe the pore-composition and pore-sorting property of rock.

Fig. 5.13 Pore-size distribution curve

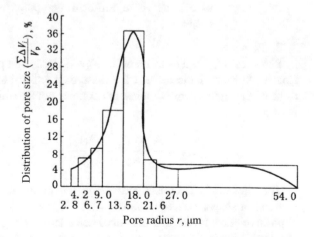

Fig. 5.14 Cumulative
pore-size distribution curve

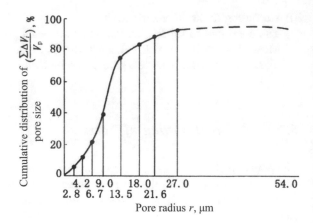

1. Sorting Coefficient
Sorting coefficient (S_p) is defined as:

$$S_p = \frac{(\phi_{84} - \phi_{16})}{4} + \frac{(\phi_{95} - \phi_5)}{6.6} \qquad (5.16)$$

$$\phi_i = -\log_2 d_i$$

where

d	Pore diameter, μm;
subscript of φ	Percentage of the character points on the cumulative distribution curve;
S_p	The uniform level of the distribution of pores; the smaller the value of S_p, the more uniform the pores and the better the sorting property.

2. Skewness
Skewness (S_{kp}) is a measure of the lopsidedness of a pore-size distribution curve. Coarse skewness indicates that the mass of the distribution is concentrated on the right of the figure, which represents coarser pores; while fine skewness indicates finer pores.

$$S_{kp} = \frac{\phi_{84} + \phi_{16} - 2\phi_{50}}{2(\phi_{84} - \phi_{16})} + \frac{\phi_{95} - \phi_5 - 2\phi_{50}}{2(\phi_{95} - \phi_5)} \qquad (5.17)$$

The skewness S_{kp} equals zero when the pore-size distribution curve is symmetrical in shape. Actually, the skewness of real rocks ranges between −1 and 1. A positive value means that the curve has a tail tapering into the coarser pores and, obviously, implies a coarse skewness, while a negative-valued S_{kp} implies a fine skewness.

3. Kurtosis

As a quantitative measure of the steepness or peakedness level exhibited by the pore-size distribution curve, the kurtosis is defined as:

$$K_p = \frac{\phi_{95} - \phi_5}{2.44(\phi_{75} - \phi_{25})} \tag{5.18}$$

When a curve is normally distributed, $K_p = 1$; when a curve is plat-topped, the value of K_p is smaller than 0.6; and when a curve has a peak, its K_p ranges between 1.5 and 3.

In the case that a pore system is composed of different types of pore, the curve of this system may exhibit a bimodal or multimodal distribution.

4. Average Diameter of Pore

(1) Median Pore Diameter D_{50}: the pore diameter corresponding to the 50 % on the cumulative pore-size distribution curve, μm.
(2) Average Pore Diameter D_m: the arithmetic average of the pore diameters corresponding to several characteristic points. Three averaging methods are presented below:

$$D_m = \frac{D_5 + D_{15} + D_{25} + \cdots\cdots + D_{85} + D_{95}}{10} \tag{5.19}$$

Or

$$D_m = \frac{D_{16} + D_{50} + D_{84}}{3} \tag{5.19'}$$

Or

$$D_m = \frac{D_{25} + D_{50} + D_{75}}{3} \tag{5.19''}$$

In a word, according to the parameters like the average pore diameter, the kurtosis K_p, the skewness S_{kp}, and the sorting coefficient S_p, the size and distribution of the pores present in a rock can be determined.

5.2.3 *Pore Structure*

1. Definition

Pore structure refers to all the characteristics and structure patterns of pores, including size, shape, interconnectivity, type, and pore-wall coarseness. Directly influencing the storage and conduction capacity of a porous rock, the pore structures play a role of basis in the study of a rock's porosity and permeability.

2. Pore-Structure Parameters

(1) Pore to Throat Ratio: the ratio of the pore diameter to the channel diameter.
(2) Connectioning Number: the number of the channels interconnected with a pore, normally ranging from 2 to 15 in sandstones.
(3) Tortuosity τ: a parameter used for pores to describe their level of being tortuous (twisted, having many turns). The tortuosity τ is defined as the ratio of the path length flowed by the fluid particles l to the apparent length of the rock L, and its value ranges within 1.2–2.5.

The three pore-structure parameters, pore-throat ratio, connection number, and tortuosity τ can be determined through an observation of the rock's cast-thin-section under high-power microscope. Besides, with a microscope we can also observe the inner pore-wall coarseness, the arrangement, combination of pores, etc.

Besides thin-section method, cast electron microscopic method that are commonly used in geology. In the study of the pore structures, some other methods are also employed to obtain the size and distribution of the pores present in a rock. For example, in petrol physics the capillary pressure curve and pore distribution curve of a rock can be determined by means like mercury-injection method, centrifuge method, or partition-board method.

3. Pore-Structure Types

The classification of the pore-structure types present in reservoir rocks depends on the aims of research and the requirements in application. Microscopically considering the oil and gas reservoir developments, the pore structures in rock can be divided into six pore-structure types existing in three levels of porous mediums. They will be introduced briefly in the following part of this lesson.

(1) Single Porous Medium

a. Intergranular Pore Structure

As the basic pore-structure type for clastic rocks and an important type existing in part of carbonate rocks, intergranular pore structure takes shape among the differently sized and shaped grains, where the space is also partly filled with cementing materials. In other words, the intergranular voids that are not occupied by cementing materials form the intergranular pores, which play in reservoirs as both the reserving space and the seepage-flowing channels for reservoir oil.

In the early study of this kind of structure, a single porous medium with intergranular pore structure is usually treated as a packing of uniform spheres. Later, the pore space was seen as a capillary tube model—a cluster of microcapillary tubes with uniform diameter or bending capillary tubes with changing area sections. Currently, another concept, network model, was introduced in reservoir developments (see Lesson 6.6 to get more details).

b. Unmixed Fracture Structure

Compacted carbonate reservoir rocks are almost impermeable. The microfractures that occur in such a carbonate rock are usually referred as "unmixed fracture structure," and they undertake the work of reserving space and seepage-flowing

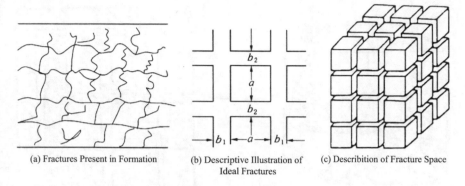

(a) Fractures Present in Formation   (b) Descriptive Illustration of   (c) Describition of Fracture Space
                                          Ideal Fractures

Fig. 5.15 Ideal model of unmixed fractures. **a** Fractures present in formation. **b** Descriptive illustration of ideal fractures. **c** Description of fracture space

channels. As the development and extending of these fractures are irregular more often than not, it is hard to quantitatively describe their conformation. Therefore, an ideal cubic-lattice model is sometimes employed for simplification, partitioning the whole rock into a number of square blocks with crisscrossed fractures. Figure 5.15 shows ideal model of unmixed fracture.

(2) Dual-Porosity Medium

a. Fracture-Pore Structure

Fracture-pore structure is developed in limestone and dolostone in particular. In rocks with this type of pore structure, which are actually intergranular porous matrix partitioned by fractures into lots of block units, the intergranular pores contained in the block units act as reserving space for oil and gas, while the fractures among the blocks are main channels for the seepage flow of these fluids. That is to say, in a dual-porosity medium with such a pore structure, the intergranular pores have large porosity but low permeability, while, conversely, the fractures have high permeability but small porosity. Because the physical parameters, the porosity and the permeability, involved in each of the two mentioned coexisting pore-space systems are entirely different from that involved in the other, a couple of hydrodynamic systems are simultaneously formed in the rock. In conclusion, the fracture-pore structure has characteristics such as dual porosity, dual permeability, and owning a couple of hydrodynamic systems of extraordinary disparity. Figure 5.16 shows the ideal model for this type.

b. Cavity-Pore Structure

Cavity-pore structure is also particularly developed in carbonate rocks. In rocks of this pore-structure type, big cavities with size beyond the magnitude of capillary tubes are distributed within the intergranular porous rock matrix. As a result, the flowing laws that govern the fluids in the two different kinds of pore spaces are different—the fluids contained in the intergranular pores obey the seepage flowing

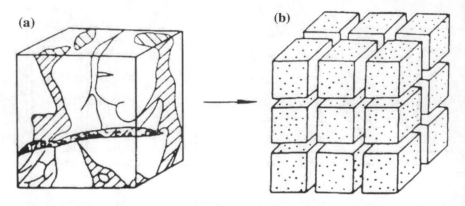

Fig. 5.16 Ideal model of pore-fracture double-porosity medium (By Ge Jiali 1982)

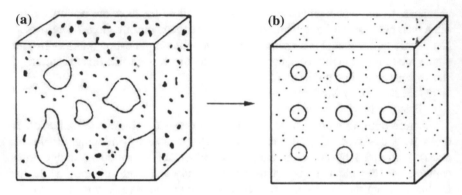

Fig. 5.17 Ideal model of pore-cavity double-porosity medium (By Ge Jiali 1982)

laws, while the fluids contained in the cavities obey the Navier–Stokes equation, which belong to the category of hydromechanics. Therefore, the cavity-pore structure should also be treated as a dual-porosity medium where two governing flowing mechanisms coexist. The ideal model for this type is shown in Fig. 5.17.

(3) Triple-Porosity Medium

    a. Pore-Microfracture-Macrocavity: A mixed structure composed of inter-granular pores, microfractures and macrocavities.

    b. Pore-Microfracture-Macrofracture: A mixed structure composed of inter-granular pores, microfractures and macrofractures. This type is particularly developed in carbonate rocks. The research of the seepage flowing laws obeyed by triple-porosity mediums is still in its infancy.

5.3   Porosity of Reservoir Rocks

5.3.1   Definition of Porosity

The total volume V_b of a rock, also called bulk volume or bulk volume, is composed of the pore volume V_p and the volume of the solid grains (the volume of the matrix) V_s:

$$V_b = V_p + V_s \qquad (5.20)$$

Porosity (φ) is defined as the ratio of the pore volume V_p in a rock to the bulk volume V_b of that rock in percentage. Expressed mathematically as:

$$\phi = \frac{V_p}{V_b} \times 100\% \qquad (5.21)$$

Substitute Eq. (5.20) into the equation above, we can get:

$$\phi = \frac{V_b - V_S}{V_b} \times 100\% = \left(1 - \frac{V_S}{V_b}\right) \times 100\% \qquad (5.21')$$

1. Absolute Porosity of Reservoir Rocks

The total pore volume (V_a) in rock can be divided into:

(1) Interconnected pore volume (also called effective pore volume): ① mobile pore volume; ② immobile pore volume.
(2) Unconnected pore volume.

Absolute porosity φ_a is the ratio of the total void space in the rock to the bulk volume of the rock, expressed in percentage:

$$\phi_a = \frac{V_a}{V_b} \times 100\% \qquad (5.22)$$

2. Effective Porosity of Reservoir Rocks

Effective porosity φ_e is the ratio of the interconnected void space in the rock to the bulk volume of the rock, expressed in percentage:

$$\phi_e = \frac{V_e}{V_b} \times 100\% \qquad (5.23)$$

The effective porosity is more frequently used in reserve calculation and reservoir evaluation.

3. Mobile Porosity of Reservoir Rocks

Not all the interconnected microcapillary pores let the fluid through. Normally, for the pores with throats of extremely small diameter, it is almost impossible for usual production pressure differential to enable the fluids to flow through them. Besides, this problem also exist in hydrophilic rocks where there are water films absorbed on the rock surfaces and consequently the channels present in the pores are reduced in volume. The so-called mobile porosity is the ratio of the volume of the space V_f that is occupied by mobile fluids to the bulk volume of the rock:

$$\phi_f = \frac{V_f}{V_b} \times 100\% \tag{5.24}$$

Since neither the volume of the dead pores nor that of the microcapillary pores are included in the calculation of mobile porosity, it is different from effective porosity. The former one, the mobile porosity, is not a fixed value but changing with the changes of the pressure gradient and physical or chemical properties of the liquids present in reservoirs. In reservoir developments, the parameter of mobile porosity is practically valuable to some extent.

Seeing from the definitions given above, we can know that: absolute porosity φ_a > effective porosity φ_e > mobile porosity φ_f.

5.3.2   Grade Reservoir Rocks According to Porosity

Porosity is an important index in reserve calculation and reservoir evaluation. Effective porosity is adopted by engineers to evaluate formations in the petroleum industry, so generally speaking, the mentioned "porosity" in relevant context specially refers to the "effective porosity."

The porosity of sandstone formations usually ranges between 5 and 25 %, while that of carbonate rock matrix is apt to be smaller than 5 %. Normally, a sandstone formation with porosity lower than 5 % is not valuable for development. The ranking of sandstone formations according to porosity is shown in Table 5.8.

5.3.3   Porosity of Dual-Porosity Medium Rock

Taking the fractured reservoir as an example, here we will introduce the two coexisting porosity systems in dual-porosity medium rocks. The porosity can be

Porosity %	Evaluation
25–20	Extremely good
20–15	Good
15–10	Moderate
10–5	Bad
5–0	No value

Table 5.8 Ranking of sandstone formations according to porosity

Fig. 5.18 Porosity of
dual-porosity medium rock

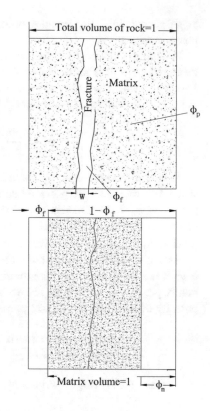

divided into two types for convenience: the initial porosity or matrix porosity which
reflects the intergranular pores present among rock grains and the induced porosity
(also referred to as cavity porosity and fracture porosity) which reflects the space
formed in the cavity and fracture systems. Therefore, the total porosity of this kind
of medium should be described with the term "dual-porosity." Figure 5.18 shows it.

For fractured reservoirs: bulk volume of rock = fracture volume + matrix volume, and total pore volume = fracture volume + initial pore volume.

The correlation among the total porosity φ_t, the fracture porosity φ_f, and the
initial porosity φ_p is written below:

$$\varphi_t = \varphi_p + \varphi_f \qquad (5.25)$$

where

φ_p Initial pore volume present in matrix/bulk volume of rock;
φ_f Fracture volume/bulk volume of rock.

Both φ_p and φ_f mentioned above take the bulk volume of the studied
dual-porosity medium rock as denominator. As proven by the amounts of

experimental data, the fracture porosity φ_f is obviously smaller than the initial porosity of rock φ_p.

In coring practice, it is really hard to obtain a core with fractures and the measured core in laboratory experiments are nothing but a part of matrix from a fractured rock. Therefore, the porosity measured in laboratory takes the matrix volume as denominator and referred to as matrix porosity φ_m:

$$\varphi_m = \frac{\text{pore volume in matrix}}{\text{matrix volume}} \tag{5.26}$$

Note that the parameter φ_p named initial porosity of rock is in light of the bulk volume of rock and the pores reflected by it are the intergranular ones, while the parameter φ_m named matrix porosity is in light of the matrix volume itself. Since the matrix volume of a rock is smaller than the rock's bulk volume $\phi_m > \phi_p$. Their relation equation will be derived hereinafter.

Assuming the bulk volume of a rock is 1 (a unit of volume), the fracture porosity is φ_f, it is easy to know that the matrix volume present in this rock is $1—\varphi_f$ and the matrix porosity is $(1—\varphi_f)\,\varphi_m$. Now, we can get the relationship between the initial porosity of rock and the matrix porosity:

$$\varphi_p = \frac{\text{pore volume in matrix}}{\text{bulk volume of rock}} = \frac{(1-\varphi_f)\varphi_m}{1} = (1-\varphi_f)\varphi_m \tag{5.27}$$

And Eq. (5.25) can be written as:

$$\phi_t = (1-\phi_f)\phi_m + \phi_f \tag{5.28}$$

Empirically, when $\varphi_t < 10~\%$, the maximum fracture porosity is related to the total porosity by $\varphi_{f\max} < 0.1\varphi_t$ and when $\varphi_t > 10~\%$, $\varphi_{f\max} < 0.04\varphi_t$. From the standpoint of reserving space present in rock, the value of the fracture porosity φ_f, which is much lower than the matrix porosity φ_m, can often be ignored. But for the compacted rocks (e.g., $\varphi_t < 5~\%$), the existence of fractures is quite meaningful to the reserving and, especially, the seepage flowing of oil and gas.

5.3.4 Measurement of Porosity

The determination of rock porosity can be accomplished via two approaches: (1) Direct measurement through experiments in laboratory; (2) Indirect measurement based on a variety of logging methods. Of the two methods, the indirect measurement is usually subjected to so many influencing factors that great error is produced during its usage. In contrast, the former one, the method on the basis of laboratory measurements, is able to accomplish the determination of porosity more accurately through routine analysis methods for rock samples, on which we will put emphasis in the following part.

The measurement of rock porosity in laboratory starts from the definition of porosity. With the values of V_b and V_p (or V_S) obtained from experiments, the porosity can be computed by definition:

$$\phi = \frac{V_p}{V_b} = \left(1 - \frac{V_S}{V_b}\right) \qquad (5.29)$$

1. Measurement of Bulk (Apparent) Volume of Rock
(1) Geometric Method
The bulk volume can be computed from measurements of the dimensions of the uniformly or regularly shaped samples. For example, with a vernier caliper the diameter d and length L of a cylinder sample can be directly measured for the determination of its bulk volume V_b.

This method is particularly desirable for the rocks that are good in cementation and kept completed and unbroken during drilling and coring.
(2) Wax-Sealing Method
The bulk volume of an irregularly shaped rock can be measured gravimetrically by coating it with wax. Procedure of this method: firstly, weigh the mass of the studied sample and designate it by ω_1; secondly, dunk it in molten wax for a while where the rock surface will be covered with a coat of wax film, and then weigh the mass of the wax-sealed sample, designated as ω_2; thirdly, weigh the wax-sealed sample in water and designate the obtained mass as ω_3; finally, calculate the value of V_b through the equation given below:

$$V_b = \frac{\omega_2 - \omega_3}{\rho_w} - \frac{\omega_2 - \omega_1}{\rho_p} \qquad (5.30)$$

where

V_b	Bulk volume of rock, cm^3;
ρ_w, ρ_p	Water density and wax density, usually $\rho_w = 0.918$ g/cm^3
ω_1, ω_2, ω_3	Mass of sample under different conditions, g

This method is applicable to those loosely cemented rocks that are ready to collapse or shatter to pieces, and the served rocks can be arbitrarily irregularly shaped.
(3) Kerosene-Saturating Method
Firstly, vacuumize a sample and then immerse it in kerosene, where the rock pores will be saturated with kerosene; then, weigh the kerosene-saturated sample, respectively, in air and in kerosene and get the mass of this sample under the two conditions, designated as ω_1 and ω_2. According to the Archimedean principle of buoyancy, the net upward buoyancy force is equal to the magnitude of the weight of fluid displaced by the body. Then, we can get the value of V_b:

$$V_b = \frac{\omega_1 - \omega_2}{\rho_o} \tag{5.31}$$

where

ρ_o Kerosene density, g/cm^3.

This method is applicable to irregularly shaped rocks.

(4) Mercury Method

In the case that the rock needed to measure has an irregular shape or that the mercury intrusion method has already been chosen to draw the capillary pressure curve, it is suitable to employ the "mercury method" to determine the bulk volume of rock, in which a mercury volumetric metering pump is utilized to undertake volume measurements. Brief introduction of the procedure: with the volume V_1 and V_2, respectively, read from the scale of the mercury volumetric metering pump before and after putting the sample into the sample chamber, the bulk volume of the sample which is rightly equal to the value of $(V_1 - V_1)$ can be easily obtained. Although being convenient and swift, this method is allowed to be employed on the premise that the mercury never enters the pores present in the studied sample.

2. Measurement of Pore Volume V_p

By introducing a fluid medium to fill the pore space of the sample and measuring the volume of the introduced fluid, the pore volume V_p can be determined. Considering that different fluid mediums used in experiments fill the pores to different levels, they usually yield different values of pore volume, which represent porosities of different physical meaning. For example, gas can enter the tiny and slender pores that give no admittance to water. In the following part, two commonly adopted methods for pore-volume determination will be presented.

(1) Gas Porosimeter

The instruments and flowchart to measure the pore volume with a gas porosimeter is shown in Fig. 5.19.

According to Boyle's law, the pore volume of rock can be computed:

$$V_k P_k = P(V + V_k)$$

Then, we can get:

$$V = \frac{V_k(P_k - P)}{P} \tag{5.32}$$

Fig. 5.19 Measurement of pore volume of core

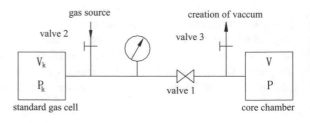

(2) Liquid-Saturating Method

Weigh the cleaned and dried sample in air and symbolize its mass as w_1; then, vacuumize the sample before saturating it with kerosene, and weigh the kerosene-saturated sample in air, too, with w_2 to designate its mass. The difference between w_1 and w_2 is rightly the mass of the kerosene that succeeded in entering the sample. Now, the pore volume present in the sample can be calculated through the equation written below, which utilizes ρ_o to represent the density of kerosene.

$$V_p = (w_2 - w_1)/\rho_o \tag{5.33}$$

This method has advantages such as the simple equipment and easy operation. Besides, as far as the convenience is concerned, in choosing available liquids for saturating, it is also feasible to employ formation water in stead of kerosene. When saturating a sample with kerosene, the operation should be completed speedily in order to prevent the kerosene from volatilizing, which may cause errors otherwise. Moreover, the water that can be used for saturating is not that distilled or fresh water, but exclusively the formation water, which does not cause expansions when encountered by some minerals. The porosity obtained from the liquid-saturating methods is usually specially called liquid-measuring porosity.

3. Measurement of Grain Volume V_S

Two commonly used methods will be introduced here.

(1) By Helium Porometer

Figure 5.19 is a schematic diagram for helium porometer. Assuming that the grain volume of the sample is V_s and the volume of the core chamber is V, the remaining volume left in the inhabited core chamber is $V-V_s$. To do the measurement, close the valve 1 and vacuumize the core sample. Open valve 2 and charge nitrogen, the standard gas cell, record pressure pk when pressure equilibrated. Open the valve 1, nitrogen will enter core chamber and pore of core by isothermal. Measure the final pressure in the system after expansion as p. according to the Boyel law, we have:

$$V_k P_k = P(V_k + V - V_s)$$

Thus, rock matrix volume can be calculated by formula:

$$V_s = V - \frac{V_k(P_k - P)}{P} \tag{5.34}$$

(2) By Solid Volumenometer

A solid volumenometer, as shown in Fig. 5.20, is composed of a base vessel and an upright graduated tube. When measuring: (1) Put the mashed cores (grains) into the base vessel; (2) Turn the upright tube upside down to render the opening upward, and inject kerosene into it until the liquid level reaches the scale point "5"; (3) Turn the opening of the upright tube to be downward and stick it into the base

Fig. 5.20 Solid
volumenometer

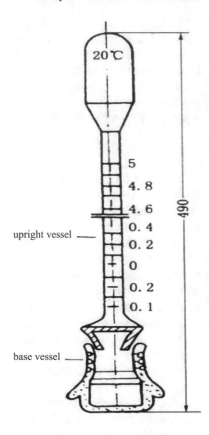

vessel so as to combine them together (Fig. 5.20). (4) Read the volume of the grains
V_s from the graduated upright tube.

Determination of rock porosity can be accomplished by many other methods
such as the NMR (nuclear magnetic resonance) and the resistivity method (see
Chap. 7). Readers can learn more about it in related books.

5.3.5 Factors Influencing Porosity of Rock

1. Size and Arrangement of Particles

Starting from ideal rock models composed of uniform spheres, we can learn
some regularities of porosity exhibited by various packing arrangements of uniformly sized spherical grains. As shown in Fig. 5.21, the porosity for the cubical
packing, the least compacted arrangement, can be computed easily: the unit cell is a
cube with sides equal to $2r$ where r is the radius of the constituent spheres;
therefore, the bulk volume of the unit cube is $V_b = (2r)^3 = 8r^3$; as there are eight
1/8 spheres in every unit cube, the total volume of the spheres, or the grain volume,

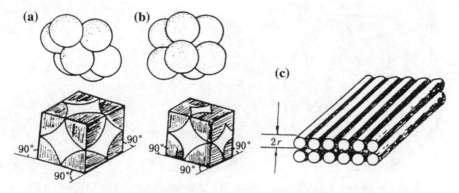

Fig. 5.21 Ideal rock models of various packing arrangements and their porosity (from graton and fraser). **a** Arrangement of cubic-packed uniform spheres ($\varphi = 47.64$ %), **b** arrangement of rhombohedral-packed uniform spheres ($\varphi = 25.96$ %), **c** Arrangement of cubic-packed uniform columns ($\varphi = 21.5$ %)

is $V_S = \frac{4}{3}\pi r^3$. Then, the porosity for the ideal rock model of cubical packing arrangement is:

$$\phi = \frac{V_b - V_S}{V_b} \times 100\% = \frac{8r^3 - \frac{4}{3}\pi r^3}{8r^3} \times 100\% = 47.6\%$$

Similarly, the computed porosity for the rhombohedral packing, the most compacted arrangement [Fig. 5.21(b)] turns out to be 25.9 %. Considering the geometric properties of basic rhombohedral bodies, scientists presented the calculation equation for the ideal arrangements of uniform spheres:

$$\phi = 1 - \frac{\pi}{6(1 - \cos\theta)\sqrt{1 + 2\cos\theta}} \tag{5.35}$$

Obviously, the porosity of ideal rock, which has nothing to do with the particle size, is a function of the packing patterns (the angle θ) only. When $\theta = 90°$, $\varphi = 47.6$ %; and when $\theta = 60°$, $\varphi = 25.9$ %.

The yields from the research on the various idealized models mentioned above are no more than the upper limits of porosity of the rocks treated as packings of grains. In other words, the naturally occurring rocks, composed of a variety of grains with different particle sizes and shapes, have porosity values surely smaller than these ideal ones.

Being broken, transported, cemented, and compacted before the parent materials were incorporated and solidified into rock, the real clastic rocks. The porosity of clastic rocks is significantly affected by the elements such as the mineral components of the detrital particles, the sorting level, the type and quantity of the cementing materials, and the post-diagenetic compaction.

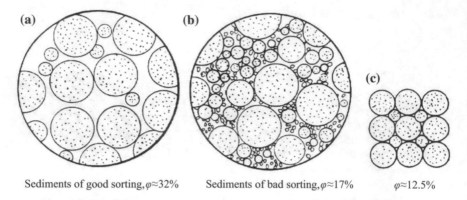

Sediments of good sorting,$\varphi \approx 32\%$ Sediments of bad sorting,$\varphi \approx 17\%$ $\varphi \approx 12.5\%$

Fig. 5.22 Impact of sorting degree on porosity. **a** Sediments of good sorting, $\varphi \approx 32$ %, **b** sediments of bad sorting, $\varphi \approx 17$ %, **c**. $\varphi \approx 12.5$ %

2. Sorting Property of Particles

Because the naturally occurring rocks are composed of a variety of particle sizes, the pores and throats among big particles may be filled up by small particles, leading to lower porosity and lower permeability of rock. Therefore, the sorting level of the particles has great effect on the porosity—better sorting property which means that the grains are more uniformly sized leads to greater porosity of rock, while worse sorting property causes lower porosity (Fig. 5.22).

Figure 5.23 shows the experimental relationship between the sorting coefficient and the porosity. When the sorting coefficient is close to 1, which means a quite even particle-size distribution, the porosity reaches its maximum. Furthermore, in the case that the value of sorting coefficient is above 1, greater sorting coefficient implies worse sorting property and lower porosity.

Besides, on the basis of the statistical regularity obtained from thousands of real rocks, it is recognized that the porosity is also related to the magnitude of the

Fig. 5.23 Relationship between sorting coefficient and porosity

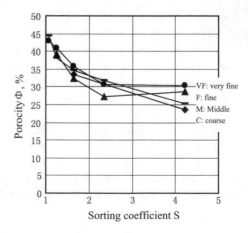

particle size—the larger the particle size, the lower the porosity. The reason why the fine-grained rocks have higher porosity than the rocks composed of coarse and well-rounded grains is that the detritus contained in a fine-grained rock usually exhibit worse roundness but greater angularity and that the detritus tend to be loosely incorporated in the case of grain-supported type of cementation.

As shown in Fig. 5.23, when the sorting coefficient is close to 1, or in other words the particle size is quite evenly distributed, the particle size has little effect on the porosity of rock.

3. Mineral Components and Cementing Materials of Rock

As we know, the quartz crystals are granular, the mica is plate-shaped, and some clay minerals are water-expandable. These facts indicate that the mineral components of rock influence the configurations of the particles. In addition, as feldspar has stronger lipophilicity and hydrophilicity than quartz, it can be coated a thicker liquid film when wetted by oil or water.

The composition and quantity of the cementing materials, as well as the type of cementation, are closely related to the oil-reserving properties of a rock. It is mud that dominantly comprises the cementing materials contained in clastic rocks, where a comparatively smaller amount of calcium and a very small amount of silica and iron also exist. Generally speaking, the loosely cemented sandstones with the loose type of mud cementation are endowed with good porosity, and, moreover, as the content of the cementing materials increases, the porosity decreases markedly. Table 5.9 shows the relationship.

Apart from the mud cementing materials, calcareous cementing materials are also found in several oil reservoirs of China, and the commonly encountered components of them include calcite and dolomite. The effect exerted by the calcareous cementing materials on the porosity of rock is much greater—when the content shared by them reaches a value higher than 3–5 %, the porosity will be reduced significantly.

4. Depth of Burial and Compaction

As both the reservoir temperature and subsurface hydrostatic pressure increase with depth of burial, the thick overburden rock layers, when a deeply buried interval is studied, inevitably cause the supporting rock grains to be aggregated closer together and to have plastic and irreversible motions relative to each other, which result in sharp porosity reduction. Figure 5.24 shows the relationship between porosity and depth. If the arrangement of grains has already been compacted to its utmost limit, local pressolution will occur to the solids at the contact points of the grains when the overburden pressure continues to increase. In such a case, the dissolved minerals, such as quartz, tend to recrystalline in the pore space, leading to further porosity reduction and even a complete elimination of the pores which means impermeability of rock.

Table 5.9 Relationship between sandstone porosity and mud content

Mud content, %	<2	2–5	5–10	10–15	15–20
Porosity, %	28–34	29–31	25–30	<25	<20

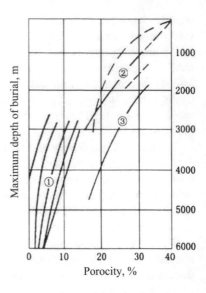

Fig. 5.24 Relationship between sandstone porosity and depth of burial (By Mayer-Curr 1978). ① muddy sandstone (containing mica), ② Jurassic–Cretaceous quartz sandstone, ③ Tertiary Period quartz sandstone

Figure 5.25 shows that, as for the carbonate rocks, trend toward lower porosity is also remarkable when their depth of burial increases. Born in specified environments (e.g., a specified stream environment) and correlating closely with biological processes, the carbonate rocks tend to contain multifarious forms of pores resulting from a variety of physical and chemical reactions that occur readily in company with the post-diagenesis environment changes. For example, parts of a carbonate rock are soluble under the action of formation water and minerals recrystallization at specified pressure and temperature, etc. Besides, the changes of

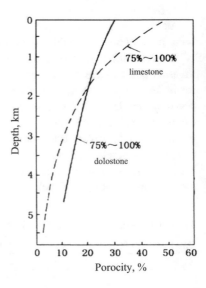

Fig. 5.25 Relationship between carbonate rock porosity and depth of burial (By J.W.Schmoker et al. 1982)

the environments to which the rocks are subjected also cause deformations of the initial pores that are developed during the deposition of materials. These "post-diagenesis environment changes," acting as influencing factors of porosity, include change of depositional environment, dolomization, corrosion, and tectonic stress actions. Carbonate rocks usually bear much more fractures produced by tectonic stress actions than that of clastic rock. Generally speaking, under the same stress conditions dolomite is where fractures are most developed, limestone comes second, and marlite is weakest developed with fractures.

Apart from the characteristics of lithology, the denseness and size (or scale) of the developed fractures are also associated with the thickness of rock. Generally, thick rock stratum bears wide-spacing but large-scale fractures, while thin rock stratum bears close-spacing but small-scale fractures.

5. Post-Diagenesis Actions

The effect of post-diagenesis actions on the porosity of rock lies in two aspects: firstly, more microfissures are developed as results of the action of tectonic stresses, increasing the porosity of rock; secondly, the porosity may be enhanced by the solution of rock grains and cementing materials by formation waters, but it is also possible to be reduced because the pores may be filled with the mineral precipitates from water.

5.4 Compressibility of Reservoir Rocks

5.4.1 Rock-Bulk Compressibility (Rock Compressibility)

The pore spaces enable a rock to be compressed, or in other words, to possess a certain elasto-plasticity. Under original conditions before development, rocks buried at depth are well in an equilibrium state, subjected to set of mutually balanced pressures including the overburden pressure (the external stress imposed by the overlying rocks), the reservoir pressure (the internal stress exerted by the fluids contained in the pores), and the pressure supported by the skeleton of rock matrix. After a reservoir has been developed and along with the production, its equilibrium relationship among the mentioned pressures will be broken, because the uplift of reservoir fluids and the consequent decrease of reservoir pressure will cause an increase of the differential pressure between the external and internal stresses, which implies that the load imposed by the overlying sediments entails deformations of the supporting rock grains and a tighter arrangement of them and, therefore, a decrease in pore volume. Figure 5.26 shows these details.

In order to illustrate the relationship between the reduction of pore volume and the reservoir pressure drop ΔP, the concept of "rock-bulk compressibility" is introduced here. The rock-bulk compressibility is defined as the fractional change in the bulk (apparent) volume of the solid rock material (grains) or equivalently the

Fig. 5.26 Deformation of sandstone skeleton of formation

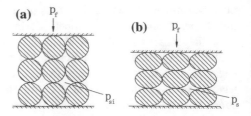

change in the pore volume contained in a unit bulk volume of rock, with a unit change in reservoir pressure. Expressed mathematically:

$$C_f = \frac{1}{V_b}\frac{\Delta V_p}{\Delta P} = -\frac{1}{V_b}\frac{\Delta V_p}{\Delta P_e} \qquad (5.36)$$

where

C_f Rock-matrix compressibility, MPa^{-1};
V_b Bulk (apparent) volume of rock, m^3;
ΔP Change or reservoir pressure, MPa;
ΔV_p Reduction of pore volume that happens to a reservoir pressure drop, m^3.

The shrinkage of pore volume resulting from reservoir pressure drops offers a driving force for the oil displacement which impels the fluids to flow toward the well bores. Therefore, the magnitude of the rock-bulk compressibility of a reservoir is also a measure of the elastic driving capacity of its rocks. This is rightly the reason why the rock-bulk compressibility is also called "elastic compressibility." Given that the value of C_f is already known, the change in pore volume that happens consequently due to a specified reservoir pressure drop can be calculated: $\Delta V_p = C_f V_b \Delta P$.

The rock-bulk compressibility generally ranges within 2–36×10^{-4} MPa^{-1}, with exceptions such as some abnormal high-pressure reservoirs that have greater values of this coefficient.

Some Euro-American countries also adopt the term "pore compressibility." It is defined as the fractional change in pore volume of the rock with a unit change in reservoir pressure and given by the following relationship:

$$C_p = \frac{1}{V_p}\frac{\Delta V_p}{\Delta P} = -\frac{1}{V_p}\frac{\Delta V_p}{\Delta P_e} \qquad (5.37)$$

where

V_p Pore volume in rock, m^3

The significances of other symbols are the same as stated above or before.

As noted that $V_p = V_b \varphi$, the bulk compressibility is related to the pore compressibility by the following expression:

$$C_f = C_p \varphi \tag{5.37'}$$

Based on the research on the rock-bulk compressibility C_p, a conversion formula, which correlates the laboratory-measured rock porosity under atmospheric conditions (ϕ_o) with the porosity under reservoir conditions (ϕ), is presented by predecessors:

$$\phi = \phi_0 e^{-C_P \Delta P_e} \tag{5.38}$$

where

ΔP_e Change of effective overburden pressure, MPa

$\Delta P_e = P_e - P_{e0}$, where P_e, the effective overburden pressure, is equal to the differential pressure between the overburden pressure and the reservoir pressure.

With the porosity obtained under experimental conditions where ΔP_e tends to be small-valued, this formula can be used to yield the porosity under reservoir conditions, where the value of ΔP_e is comparatively larger.

5.4.2 Total Compressibility

Thus far, we have learned the individual compressibility of oil, gas, water, and rock. In consideration of the elastic energy contained in the whole reservoir, the elasticity of all the formation fluids as well as the rock should be counted in. In such a case, a term called "total compressibility" or "comprehensive compressibility," symbolized as C_t, should be employed to represent the elasticity of the whole reservoir.

Provided that a reservoir pressure drop occurs in a formation, the reduction of the pore volume ΔV_p combined with the expansion of the fluid volume ΔV_L will impel a huge amount of reservoir fluids into the producing wells. That is to say, enormous elastic energy is stored in the reservoir. The following part, where the symbol V_b represents the volume of reservoir rock and the symbol ΔP represents the reservoir pressure drop, will give the derivation steps to obtain the volume of the produced oil ΔV_o that results from the combination mentioned above.

$$\Delta V_o = \Delta V_p + \Delta V_L$$

Because

$$V_L = V_b \cdot \phi$$

We can get

$$\Delta V_o = C_f V_b \Delta P + C_L V_b \phi \Delta P$$
$$= V_b \Delta P (C_f + C_L \varphi) \tag{5.39}$$

where

C_L Fluid compressibility, MPa^{-1};
ϕ Porosity of rock, fraction;
C_f Rock-bulk (rock) compressibility, MPa^{-1}.

If we set $C_f + C_L \varphi = C_t$ where C_t is the one named total compressibility, the physical significance of the total compressibility should be the fractional change in the total volume of the pore space and the fluids with a unit change in reservoir pressure. Expressed mathematically:

$$C_t = \frac{1}{V_b} \frac{\Delta V_o}{\Delta P} \tag{5.40}$$

When the three phases of oil, water, and gas coexist in a formation, the mathematic relationship among them is:

$$C_t = C_L \phi + C_f = (C_o S_o + C_g S_g + C_w S_w) \phi + C_f \tag{5.41}$$

where

S_o, S_g, S_w the saturation of oil, gas, and water fractions, respectively;
C_o, C_g, C_w, C_f the compressibility of oil, gas, water, and rock, MPa^{-1}, respectively.

The significances of other symbols are the same as stated above or before.

Note that when $C_L > C_f$, the contribution of the liquid compressibility to the elastic reserve is not necessarily larger than that of the rock compressibility. In fact, the judgment must be based upon the comparison between φC_L and C_f. This kind of sweeping generalization should be avoid and replaced by concrete analysis of concrete problems.

The total compressibility of formation comes from the combined affection of the elasticity both the rock and the fluids, and it is an important parameter reflecting the elastic reserve and the elastic energy stored in the studied formation. Given that the value of C_t, the formation volume, and the pressure drop existing in the formation are already known, the total volume of the oil and gas produced by virtue of the elasticity and expansion energy. Although the rock compressibility is usually quite small, the elastic energy and elasticity reserve is still considerable because the bulk volume of a formation is usually very huge.

5.5   Fluid Saturation in Reservoir Rocks

5.5.1   Definition of Fluid Saturation

When a reservoir rock is completely filled with a single kind of fluid, we can say that this rock is saturated with a fluid; and when there are more than one fluid coexist in a reservoir rock, the volumetric percentage of a component fluid is called the saturation of it.

1. Oil Saturation, Water Saturation, and Gas Saturation

Based on the definition given above, the oil saturation, water saturation, and gas saturation can be expressed as:

$$S_o = \frac{V_o}{V_p} = \frac{V_o}{V_b \phi} \tag{5.42}$$

$$S_w = \frac{V_w}{V_p} = \frac{V_w}{V_b \phi} \tag{5.43}$$

$$S_g = \frac{V_g}{V_p} = \frac{V_g}{\phi V_b} \tag{5.44}$$

where

S_o, S_w, S_g The oil saturation, water saturation, and gas saturation;
V_o, V_w, V_g The volume occupied by the oil, water, and gas in the pores of rock;
V_p, V_b The pore volume and the bulk volume of rock;
φ The porosity of rock, fraction.

Furthermore, the correlation among the three parameters, S_o, S_w, and S_g, can also be known:

$$S_o + S_w + S_g \equiv 1 \tag{5.45}$$

When there are only oil and water existing in rock, i.e., $S_g = 0$, the relationship between S_o and S_w is:

$$S_o + S_w = 1 \tag{5.46}$$

2. Initial Water Saturation—Connate Water Saturation

Before the development of oil reservoirs, the pores, partly occupied by water, is not 100 % saturated with oil. The so-called initial water saturation S_{wi} refers to the predevelopment ratio of the initial water volume V_{wi} to the pore volume V_p.

Proven by large amount of on-site coring analysis, all the reservoirs, involving those pure oil or pure gas reservoirs, contain a certain amount of stagnant water, which is usually called connate water. The positions where the connate water is distributed include the surface of the sand grains, the corners formed by contacting grains, as well as the microcapillary channels. The connate water contained in a formation rock was formed during the diagenetic process: originally, the sedimentary sandstone rocks born in a water environment were fully filled with water in their pores; as the crude oil is migrated to form reservoirs, part of the once trapped water was displaced by the oil while the other part of water was left in the pores, owing to the capillary effect and the absorption of the grain surface to the water; as a result, the water that survived the displacement of oil during the diagenetic process finally became the connate water.

Owing to differences in rock and fluid properties as well as conditions for migration, the reservoirs differ widely in connate-water saturation (usually within 20–50 %). Generally speaking, rocks such as coarse-grained sandstones, granular vuggy limestones, and all of the other rocks that contains macropores, have comparatively low connate-water saturation, while rocks like siltstones and low-permeability sandstones rich in mud have higher connate-water saturation. For example, the connate-water saturation in the M-Interval of the Laojun-Temple Oilfield of Yumen can be as high as 50 %. Figure 5.27 shows the relationship between the water saturation and the permeability of sandstones.

Considered from different perspectives, the initial connate-water saturation of a reservoir is also called "coexisting water saturation," "bound water saturation," "primary water saturation," "residual water saturation," "irreducible water saturation," "critical saturation," "equilibrium saturation," etc. Respectively explained, the name "residual water saturation," originated from the perspective of the genesis of the initial connate water, reflects the residual water trapped in the pores of a rock after the process of water-displacing oil which happened during formation of the rock; in terms of the state of being of the initial connate water, the name "bound water saturation," symbolized as S_{cw}, is the most reasonable one, while the initial connate water also deserves the name "coexisting water saturation" because it does coexist with the oil when a formation is newly opened. In a word, they are essentially the same matter with different names.

Fig. 5.27 Relationship between sandstone connate-water saturation and permeability

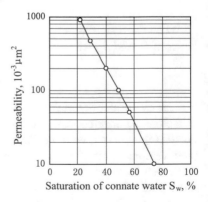

3. Initial Oil Saturation

The initial oil saturation of a formation is defined as the ratio of the oil volume contained in the formation (V_{oi}) to the pore volume present in rocks (V_p) under initial reservoir conditions. Expressed mathematically:

$$S_{oi} = \frac{V_{oi}}{V_p} \qquad (5.46)$$

The water saturation under this given condition is called the initial water saturation, which is usually symbolized as S_{wi}. With a known S_{wi}, we can get:

$$S_{oi} = 1 - S_{wi} \qquad (5.47)$$

The initial oil saturation is subjected to the predominant impact of the structures and surface properties of pores. Generally speaking, it is comparatively large-valued in cases like coarse-grained rocks, which implies that the rock is endowed with small specific surface, large pores and throats, good permeability, and low resistance for the oil and gas to displacing water.

The properties of crude oil also have definite influences on the fluid saturation. For example, regarding the high-viscosity-oil-bearing formations, they usually have high residual water saturation and low oil saturation because the impetus forces driving the crude oil are too weak to displace water and enter the pore spaces.

4. Current Fluid Saturations

After a formation has been developed for a period of time, the fluid saturation present currently in the formation is called the "current fluid saturation" also called "fluid saturation" for short.

5. Residual Oil Saturation and Remaining Oil Saturation

The residual oil refers to the oil hardly flushed out and left in the swept formation pores after flooding or some other oil production methods. The percentage accounted by its volume in the pore volume of rock is rightly the residual oil saturation, symbolized as S_{or}. Note that the residual oil, which still exist after flooding, is in a bound, non-flowing (stagnant) state.

The remaining oil refers to the oil that still remains in a reservoir after a certain development method, usually including two parts—the oil in the dead areas which are not swept by the oil-displacing agent (e.g., the injected water) and the oil that fails to be extracted out although lying within a swept area. How much remaining oil can be left in reservoir depends on many factors such as the geological conditions, the properties of crude oil, the types of the oil-displacing agents, the well pattern for production, and the technical skills. It should be mentioned that part of the remaining oil in a reservoir can still be depleted out through certain adjustment measures of the development as well as some stimulation treatments. The ratio of the remaining oil volume to the pore volume is called the remaining oil saturation.

For gas reservoirs, there are also concepts similar to those of the oil reservoirs. The repetitious details need not be given here.

[Example 5.1] In a reservoir, it is known that the oil-bearing area is $A = 14.4$ km^2, the effective thickness of oil formation is 10 m, the porosity is 0.2, the connate-water saturation is 0.3, the formation volume factor of crude oil is 1.2, and the relative density of crude oil is $d_4^{20} = 0.86$. Calculate the oil reserve in this reservoir.

Solution: According to the given conditions, the in situ volume of the crude oil in this reservoir is $V_o = (1 - S_{cw})\phi Ah$.

Then, the oil reserve (surface volume) is:

$$N = (1 - S_{cw})\phi Ah/B_o = [(1 - 0.3) \times 0.2 \times 14.4 \times 10^6 \times 10]/1.2$$
$$= 1.68 \times 10^7 \text{m}^3$$

Expressed in mass unit: $1.68 \times 10^7 \times 0.86 = 1445 \times 10^4$ (t).

5.5.2 Methods of Determining Fluid Saturation

There is many approaches to the problem of determining the fluid saturation of formation.

(1) Petrophysical methods: conventional method of core analysis (including retort method, distillation-extraction technique, and chromatography) and special method of core analysis (e.g., the determination of oil and water saturation based on the relative permeability curve or the capillary pressure curve).

(2) Logging methods: for example, the methods, such as the pulsed neutron capture logging and the nuclear magnetic logging, can be employed to determine the fluid saturation in the immediate vicinity of wellbore.

(3) Empirical statistical equations and charts: they can be used to give a rough estimation of the initial water and oil saturation.

The three methods introduced below are the most commonly followed ways for the determination of fluid saturation.

1. Solvent Extraction Method (Distillation-Extraction Technique)

Essentially, this method takes a sample and extracts the water from it to measure the water saturation, based on which the initial oil saturation can be obtained. Figure 5.28 shows the set of apparatus for a solvent extraction method. Firstly, take an oil-bearing sample and weigh it before putting it into the Millipore-baffle funnel (2). Then, heat the solvent contained in the bulb (1) to vaporize it and note that the utilized solvent should have a boiling point higher than that of water, for example, the methylbenzene (110 °C). Consequently, the water trapped in the sample is vaporized out and then condensed in the tube cooler (3) and collected in the graduated receiving tube (4), where the volume of the extracted water V_w can be read directly. As the pore volume present in the sample can be determined through

Fig. 5.28 Apparatus for the measurement of saturation in core (the extraction method)

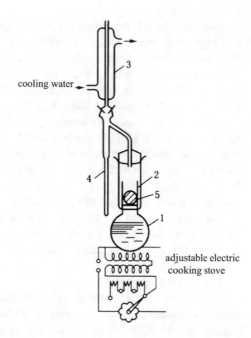

cooling water

adjustable electric cooking stove

those methods introduced previously, the water saturation can be obtained by definition.

The volume of water and oil in the sample can also be determined by virtue of their mass: measure the mass of the sample before and after a series of extracting, cleaning and drying processes, respectively, symbolized as w_1 and w_2, and, with the mass of the extracted water w_w obtained according to its volume, compute the volume of oil through the equation written below:

$$V_o = (w_1 - w_2 - w_w)/\rho_o$$

Then, the oil saturation is a direct determination:

$$S_o = \frac{w_1 - w_2 - w_w}{V_p \rho_o} \times 100\% \qquad (5.48)$$

where

ρ_o Oil density

The calculation equation for the gas saturation S_g is:

$$S_g = 1 - (S_w + S_o) \qquad (5.49)$$

The solvent extraction method has many merits such as complete cleaning of core, straightforward measuring principle, simple to operate, and accurate measurement of the water volume. As for the solvent, it should possess properties like high oil-washing ability, density smaller than water, and boiling point higher than water. For example, methylbenzene has a boiling point of 110 °C and a relative density of 0.867. Since the wettability of rock varies from sample to sample, different and specially adapted solvents should be chosen for different samples, which should be subject to the principle that no change in the rock wettability should occur. For example, a carbon tetrachloride solvent can be chosen for oil-wet rock, a proportionally mixed (e.g., 1:2, 1:3, 1:4) alcohol–benzene solvent can be chosen for water-wet rock; and a methylbenzene solvent can be chosen for neutral-wet rock and asphaltic crude oil. If there is crystallization water combined with the minerals, a solvent with a boiling point lower than water should be chosen to avoid extracting the crystallization water out of the sample.

The core is cleaned at the same time as the water contained in it is extracted. Thus, the extracting time should go on long enough for the purpose of complete cleaning. Taking the compacted rock sample as an example, the extracting time for it is at least 48 h.

2. Retorting Method

Figure 5.29 shows the set of apparatus for a retorting method. This method, also called evaporation method or pyrolysis method, takes a rock sample and heats the sample with a electric stove so as to vaporize the connate water out of it and then

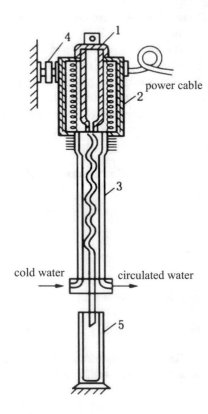

Fig. 5.29 Apparatus for the measurement of saturation in core (the resorting method)

power cable

cold water

circulated water

increase the temperature (50–650 °C) to further vaporize the oil. Both the vaporized water and oil are to be condensed in tube coolers and collected in metering tanks, where the volumes of them can be read. Then, as long as the pore volume contained in the pore is determined through related approaches, the oil saturation and water saturation can be obtained.

An error which may occur to the volume of the collected oil is that, in addition to the evaporation loss, the oil heated to high temperatures has a tendency itself to crack and coke, causing the retorted oil to be less than that really contained in the sample. As the error can be quite large, a fluid correction must be made on the sample data obtained from a retorting method. In laboratory, what is used most commonly to undertake the correction work is the relation curve of the real oil recovery against the oil in retort.

Note that the sample must not be overheated in the step of vaporizing connate water; otherwise, the temperature during this retorting process would be so high that the recovered water contains not only the interstitial water but also the crystallization water. In fact, the magnitude of the temperature in this step should be determined by virtue of the relation curve of the vaporized water volume against the temperature, and, usually, the first flattened interval of this curve represents the temperature needed to completely vaporize the connate water. Only after the connate water is vaporized out could the sample be heated up to 650 °C.

3. Chromatography

The method of chromatography is based upon the fact that water mixes in all proportions with alcohol. Procedure is as follows: weigh a sample and then immerge it in an alcohol solvent where the water contained in the rock will be fully dissolved into the alcohol; subsequently, use the chromatograph to analyze the alcohol that has already been mingled with water, and, according to concentration difference between the original alcohol and the later water-containing alcohol, calculate the water volume (V_w) from the sample; then, clean the sample by the solvent extraction method, weigh the dried sample, and compute the oil volume by the minusing method. Finally, with the pore volume that can also be determined, the water saturation and oil saturation of the sample can be obtained.

The key problem influencing the accuracy and reliability of the results from these methods presented above is whether the studied sample is adequately representative of the original fluid distribution and fluid contents present in the reservoir where the core comes from.

As a core is removed up to the surface, due to the pressure drop, the fluid originally contained in it may shrink, overflow, or be displaced and expelled. So, generally speaking, the oil saturation measured on the basis of core tends to be smaller than the reality in the underground formations where the core is removed from. Moreover, the error, which can be as high as 70–80 %, is dependent upon factors such as the oil viscosity and the solution GOR. Therefore, in practical application corrections against the errors caused by the shrinkage, overflow and displacement of the fluids must be made on the sample data obtained from these methods.

Chapter 6
Permeability of Reservoir Rocks

The porosity and the permeability are important parameters of particularly interesting to petroleum engineers. The porosity governs the storage capacity of rock or, in other words, the oil and gas contained in unit volume of rock. The permeability, which is a measure of the capacity of the medium to transmit fluids (oil, water, and gas) under some specified pressure differential, imposes a direct and great effect on the production of oil and gas. As most of the pores present in sandstones are interconnected, the fluids are able to flow through the porous medium, which is named "seepage flow." The laws governing seepage flow are quite different from that governing the flow in circular tubes which are usually discussed in hydromechanics. The flowing laws in porous medium and the permeability of reservoir rocks will be discussed in this chapter.

6.1 Darcy's Law and Absolute Permeability of Rock

6.1.1 Darcy's Law

Figure 6.1 shows the apparatus used in the famous Darcy's Experiment. In 1856, Henry Darcy investigated the flow of water downwards through a sand filter, a cylindrical sand column packed by uncemented uniformly sized grains. He found out some rules from his experiments:

As the water flows through the sand column, the volume flow rate of it is directly proportional to the cross-sectional area of the column (A), directly proportional to the pressure differential between the inflow and outflow faces of the column (ΔH or ΔP), while reciprocally proportional to the length of he column (L). Besides, it was also found that the flow rate did vary with the particle size packed in the sand filter

© Petroleum Industry Press and Springer-Verlag GmbH Germany 2017
S. Yang, *Fundamentals of Petrophysics*, Springer Geophysics,
DOI 10.1007/978-3-662-55029-8_6

Fig. 6.1 Schematic drawing
of Henry Darcy's experiment
on flow of water through sand

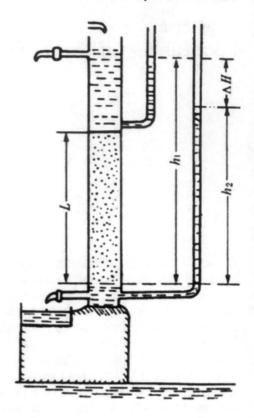

when the other conditions such as A, L, μ, and ΔP are given identically. Darcy interpreted his observations so as to yield a result in the form of equation, the famous Darcy's Law:

$$Q = K \frac{A \Delta P}{\mu L} \qquad (6.1)$$

where

Q Flow rate of the fluid through the sand column, cm^3/s;
A cross-sectional area of the sand column, cm^2;
L length of the sand column, cm;
μ viscosity of the fluid flowing through the sand column, mPa s;
ΔP pressure differential between the two ends of the sand column, 10^{-1} MPa;
K a constant of proportionality, named the absolute permeability of this porous medium, D.

In fact, the Darcy's Law is applicable to various porous mediums, including naturally occurring packs of cemented sands and sandstones. It is found that the value of K is characteristic of the pore structure of the studied porous medium and independent of the experimental fluids or the apparent geometric size of the medium. Therefore, K is called the absolute permeability of rock.

Darcy units are adopted in this equation. The petroleum industry adopts Darcy as the unit of permeability, which is symbolized as D and has a physical significance as follows:

A porous medium has a permeability of one darcy when a single-phase fluid with 1 mPa s viscosity that completely fills the voids of the medium will flow through it under conditions of viscous flow at a rate of 1 cm^3/s per square centimeter cross-sectional area under a pressure gradient of 1 atm/cm.

The calculation equation of K can be derived from Eq. (6.1):

$$K = \frac{Q\mu L}{A\Delta P} \quad (6.2)$$

Table 6.1 shows the units usually adopted in the Darcy's Law. Seen from the table, the permeability, which actually represents the area of the fluid-conducting channels contained in the rock, has the same dimensional units as area. Obviously, higher permeability indicates that the rock contains larger total area of channels through which the fluids can flow more easily.

Under most circulations, the unit millidarcy (mD) is more often used in petroleum industry, because that 1 darcy is too large a measure to use with the permeability of the oil and gas bearing reservoir rocks.

$1D = 1000\ mD = 1.02 \times 10^{-8}\ cm^2 \approx 10^{-8}\ cm^2 = 1\ \mu m^2$; $1\ mD = 10^{-3}\ \mu m^2$

Table 6.1 Units adopted in Darcy's Law

Parameter	Symbol	Dimension	Absolute unit		Mixed unit		
			CGS	SI	Darcy unit	Field unit	
						Metric system	English system
Length	L	L	cm	m	cm	m	ft
Mass	m	M	g	kg	g	kg	lb
Time	t, T	T	s	s	s	d	hr
Area	A, F	L^2	cm^2	m^2	cm^2	m^2	ft^2
Flow rare	q, Q	L^3/T	cm^3/s	m^3/s	cm^3/s	m^3/d	bbls/d
Velocity	v	L/T	cm/s	m/s	cm/s	m/d	ft/d
Density	ρ	M/L^3	g/cm^3	kg/m^3	g/cm^3	kg/m^3	lb/ft^3
Pressure	P	(ML/T^2)/ L^2	dyn/cm^2	N/m^2 (Pa)	atm	atm	lbf/in^2
Viscosity	μ	M/LT	g/cm s (P)	kg/ms (Pa s)	cp	cp	cp
Permeability	K	L^2	cm^2	m^2	D	D	mD

Table 6.2 Ranking of formations according to Permeability (from Russia)

Rank	Permeability		Evaluation
	μm^2	mD (or 10^{-3} μm^2)	
I	>1	>1000	Excellent permeability
II	1–0.1	1000–100	Good permeability
III	0.1–0.01	100–10	Mediate permeability
IV	0.01–0.001	10–1	Weak permeability
V	<0.001	<1	Impervious formation

Table 6.3 Ranking of oil-bearing formations according to permeability

Rank	Permeability
	mD (10^{-3} μm^2)
High permeability	>500
Mediate permeability	500–50
Low permeability	50–10
Ultra low permeability	<10

Table 6.4 Ranking of oil-bearing formations according to permeability

Rank	Permeability from well test interpretation	Gas permeability
	mD (or $10^{-3}\mu m^2$)	
High permeability	>50	>300
Mediate permeability	50–10	300–20
Low permeability	10–0.1	20–1
Ultra low permeability	< 0.1	<1

Permeability is one of the major indexes in reservoir evaluation, and it ranges between 5 and 1000 mD. A ranking of formations according to their permeability is made by a Russian and here shown in Table 6.2.

The Petroleum Industry Standard SY/T 6169-1995 of China ranks the permeability of oil-bearing formations into four levels which are shown in Table 6.3.

The Petroleum Industry Standard SY/T 6168-2009 of China ranks the permeability of gas-bearing formations into four levels which are shown in Table 6.4.

6.1.2   Measurement of Absolute Permeability of Rock

The measurement of the absolute permeability of rock is based on a form of the Darcy's Law, the Eq. (6.2). Assuming that the cross-sectional area A, the length L of the experimented core, and the viscosity μ of the employed fluid are already known. The flow rate Q and the pressure differential ΔP are needed to be measured for the determination of the absolute permeability K.

Since the absolute permeability is a natural property of rock, the following conditions must exist during the measurement:

1. The pore space is 100 % saturated with a single-phase, incompressible liquid, and only a steady state single-phase flow happens within the core;
2. Laminar flow;
3. No reaction between fluid and rock.

As counter examples for the condition (3), neither acid liquids nor distilled water can be used for permeability measurement where formation water (saline water) can afford, because that the last one mentioned fails to cause the expansion of the clay minerals.

The proportionality coefficient K in the Darcy's Law could not be a constant unless all of the three conditions are satisfied.

In comparison with liquids, gases (e.g., nitrogen, air) have many advantages such as vastitude source, low price, stable chemical property, availability, and convenience. Therefore, at times some gas, such as dried air and nitrogen, are also used for permeability measurement, but they do not obey Eq. (6.2) due to their compressibility. Related equations will be presented in Lesson 2.

6.1.3 Applying Conditions of Darcy's Law

The application of Darcy's Law must be limited to certain restrictions. As illustrated by the curve 1 in Fig. 6.2, the flow rate is not linear any longer as long as the seepage velocity exceeds an upper limit, called the critical seepage velocity, and under such a circumstance the flowing fluids are subjected to not only the viscous

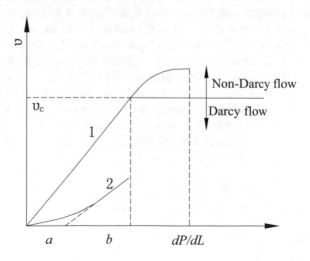

Fig. 6.2 Relationship between seepage velocity and pressure gradient

resistance, but also the inertial resistance. In a word, a change from linear flow to nonlinear flow happens to the fluid when its flow rate exceeds the critical seepage velocity, which means that the Darcy's Law is no longer applicable. Figure 6.2 shows that non-Darcy flow appears when the pressure exceeds the point b.

A parameter named "Reynolds number (Re)" is usually used to determine the critical seepage velocity. The definition of the Reynolds number is:

$$Re = \frac{v\rho\sqrt{K}}{17.5\mu\phi^{3/2}} \tag{6.3}$$

where

v seepage velocity, cm/s;
K permeability, D;
ρ density of fluid, g/cm^3;
μ viscosity of fluid, mPa.s;
φ porosity, fraction.

In the Darcy's Law, the seepage velocity is equal to the ratio of the volume flow rate to the cross-sectional area, i.e., $v = \frac{Q}{A}$. Since on every cross-section the grains contribute a part of area to A, the real velocity in the fluid-conducting channels is higher than the seepage velocity.

Empirically, the critical Reynolds Number for normal reservoirs is about 0.2–0.3. For example, when $Re = 0.2$, the critical seepage velocity can be obtained through Eq. (6.3):

$$v_c = \frac{3.5\mu\phi^{3/2}}{\rho\sqrt{K}} \tag{6.4}$$

Still another circumstance needs to be taken up here. When a fluid flows through a low-permeability compacted rock with a low velocity, a coat of hydration film may be formed on the particle surface due to the adsorption effect that occurs at the encounter of the two. This hydration film will not move until the pressure gradient is large enough, so, there is a "threshold pressure gradient" which is the lower limit for the liquid to flow. As illustrated by the curve 2 in Fig. 6.2, under the pressure gradients below this threshold the relationship between the flow rate and the pressure differential, which does not obey the Darcy's Law, is no loner linear.

As for the seepage flow of gas, the Darcy's Law in the form of Eq. (6.2) could not directly apply to its flowing regularities, either. For details, see the next lesson.

6.2 Gas Permeability and Slippage Effect

6.2.1 Calculation Equations for Perm-Plug Method

When liquid is being used as the measuring fluid, the volume flow rate (Q) is a constant at all the cross-sections of the core, because that liquid can be regarded as incompressible. However, the situation is different for gas—as the volume of gas varies obviously with temperature and pressure. The pressure gradient is gradually decreasing pressure along with the direction of flow and causes expansion of gas volume and consequently increasing volume flow rate. Figure 6.3 shows this case. Given that the seepage flow of gas within the core is a steady flow which does not change with time, the mass flow rate at every cross-section is kept at constant along with its flowing through the core. If we consider that the flowing happens under isothermal conditions, the following equations can be obtained according to the Boyle–Mariotte's Law:

$$QP = Q_o P_o = \text{constant} \quad \text{or} \quad Q = \frac{Q_o P_o}{P} \tag{6.5}$$

where

P_o atmospheric pressure, 10^{-1} MPa;
Q_o volume flow rate of gas at atmospheric pressure (volume flow rate at the exit-end), cm^3/s;
Q, P volume flow rate and pressure at an arbitrary cross-section of the core.

Taking a small length dL of the core to study, the differential form of Darcy's Law can be obtained:

$$K = -\frac{Q\mu}{A} \cdot \frac{dL}{dP} \tag{6.6}$$

Fig. 6.3 Pressure distribution in core for linear seepage flow

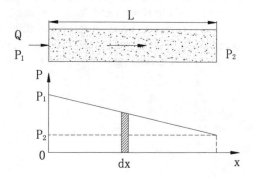

Note that dP and dL have opposite direction; so, a minus sign should be added to the right side of the equation to assure the positive value of K.

Substitute Eq. (6.5) into the equation given above:

$$K = -\frac{Q_o P_o \mu}{A} \frac{dL}{P dP} \tag{6.7}$$

By performing separation of variables and integration operation, we can get the following equations.

$$\int_{P_1}^{P_2} K P dP = -\int_o^L \frac{Q_o P_o \mu}{A} \cdot dL \tag{6.8}$$

$$K_g \frac{P_2^2 - P_1^2}{2} = -\frac{Q_o P_o \mu}{A} \cdot L \tag{6.9}$$

$$K_g = \frac{2 Q_o P_o \mu L}{A(P_1^2 - P_2^2)} \tag{6.10}$$

where

K_g gas permeability, μm^2;
Q_o volume flow rate, cm^3/s;
P_o atmosphere pressure, $10^{-1} MPa$;
A cross-sectional area of core, cm^2;
μ viscosity of gas, mPa.s;
L   length of core, cm;
P_1, P_2 the absolute pressures, respectively, at the inflow face and the outflow face of the core, 10^{-1} MPa.

This is the calculation equation for gas permeability, which is reciprocally proportional to the pressure square difference $(P_1^2 - P_2^2)$ between the two ends of the core.

Figure 6.4 shows a typical experimental flowchart for the determination of permeability by the perm-plug method. The gas supplied by the high-pressure nitrogen cylinder firstly goes through the pressure-regulator by which the upstream pressure is kept stabilized. Then, it flows through the core, where a pressure differential is formed between the two inflow and outflow faces. By measuring the pressure differential and the outflow rate of the gas, the permeability of the core can be calculated according to Eq. (6.10).

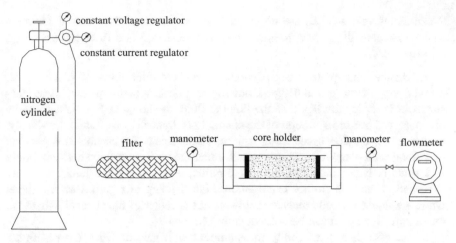

Fig. 6.4 Typical experimental flowchart of perm-plug method

6.2.2 Slippage Effect

Some new phenomena of interesting to engineers occur as amounts of measurements by the perm-plug method are performed on different cores and under different pressure differentials. Experimental results show the following:

1. Different values of K_g are observed when different mean pressures, which is defined as the average of upstream flowing and downstream flowing pressure $[\overline{P} = (P_1 + P_2)/2]$, are exerted on the same core permeated by the same gas.

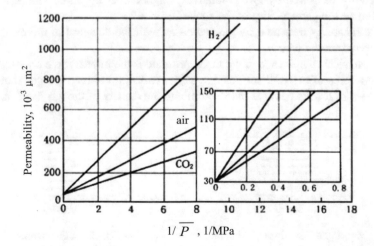

Fig. 6.5 Permeability to different gases under different average pressures (by Klinkenberg)

2. Different values of K_g are observed when different gases are used for measurements on the same core under the same mean pressure $\overline{P}$. Figure 6.5 shows these curves.

Klinkenberg interpreted these phenomena in view of the flow-rate distributions, respectively, of the gas and liquid flowing in pores. When a permeability measurement is made with liquid as the flowing fluid, as shown in Fig. 6.6a, the flow rate observed at a cross-section of the channel is elliptically distributed. Obviously, in such a case the flow speed along with the center line of the channel is larger than that along with pore walls. This is because that the molecular force between liquid and solid is larger than that between liquids, and thereby the liquid viscosity exhibited at the wall surface is extremely high, leading to a zero flow rate there, while the liquid viscosity reaches its lowest at the center of the channel, where the maximum flow speed can be consequently observed.

In the case of a permeability measurement with gas, or when the perm-plug method is performed, different flow-speed distribution occurs. On one hand, the molecular force between gas and solid is much weaker than that between liquid and solid, so that there are still a part of gas molecules at the wall surface keeping in motion; on the other hand, as the gas molecules keep colliding and making momentum exchange with each other, the gas molecules at the wall surface would be carried with the flux to flow directionally. This is the so-called slippage effect of gas, also named Klinkenberg effect, after discovered by Klinkenberg.

The slippage effect can explain all the phenomena mentioned above:

1. Permeability measured by gas is relatively higher than by liquid

When measuring with liquid, the channel space for flowing is narrowed by the stagnant liquid film attached on the wall surface. However, when measuring with gas, the gas nearby the wall surface also joins in the directional flowing due to the slippage effect. Therefore, thanks to more pore space available for flowing, the permeability measured by gas is relatively higher than by liquid and is a better reflection of the permeability of the rocks for study.

2. Permeability measured by gas increases with the decrease in the mean pressure exerted on the core

The physical significance of the mean pressure is the impact force imposed upon every unit area of the wall surface due to the collision between the gas molecules and the solid. It rests with the momentum and the density of the gas itself. A lower

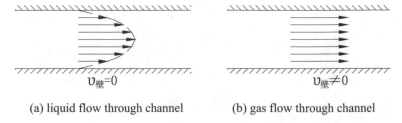

(a) liquid flow through channel   (b) gas flow through channel

Fig. 6.6 Gas slippage effect. **a** Liquid flow through channel, **b** gas flow through channel

mean pressure implies that the gas has smaller density and that the molecules have lower frequency to collide with each other, which enhances the flowing of the gas and leads to more serious slippage effect and thereby higher permeability of the rock.

Conversely, if the mean pressure over the core is raised, the slippage effect will gradually go down and thereby the permeability will decrease. To say further, if the mean pressure is raised up to infinite, the flow properties of the gas will approach to that of liquid, and the force between the gas and solid will become so strong that the gas film attached to the wall surface will not move any more. In such a case the permeability will approach a constant K_∞ which is close to the permeability measured by liquid and therefore named the "equivalent liquid permeability" or "Klinkenberg permeability."

In 1941, Klinkenberg presented the mathematical expression for the perm-plug gas permeability in view of the slippage effect:

$$K_g = K_\infty \left(1 + \frac{b}{\overline{P}}\right) \qquad (6.11)$$

where

K_g permeability measured by perm-plug gas;
K_∞ equivalent liquid permeability;
$\overline{P}$ mean pressure;
b constant for a given gas in a given medium, named "slippage coefficient" or "slippage factor." It depends on the mean free path of the gas and the size of the openings in the porous medium.

For the flowing of the given gas in a single capillary tube:

$$b = \frac{4C\lambda\overline{P}}{r} \qquad (6.12)$$

where

r radius of the capillary tube (equal to the radius of the pores in rock);
C constant of proportionality approximated to 1;
λ mean free path of the gas molecules at the given mean pressure.

According to the kinetic molecular theory presented in the General Physics, we can know that:

$$\lambda = \frac{1}{\sqrt{2}\pi d^2 n} \qquad (6.13)$$

where

d molecular diameter, decided by gas type;

n molecular density, dependent upon the mean pressure $\overline{P}$.

3. Permeability varies in value when using different gases for measurement

The slippage effect also has something to do with the properties of the used gas. As different gases (e.g., H_2, air, and CO_2) are different in molecular weight, molecular diameter, and free path, they have different slippage coefficient b, too. The smaller the molecular weight of the used gas, the greater the value of b, indicative of more serious the slippage effect. Figure 6.5 shows the permeability variance of different gases with the changing of pressure.

4. The difference between the permeability measured by gas (K_g) and by liquid (K) varies in value when the measurement is performed on different cores

Only when the pore radius approaches the mean free path of gas molecules, which increases the frequency of collision between gas molecules and solid walls, can the microscopic mechanism in the pore space acts remarkably enough to cause obvious macroscopic performance, the slippage effect of gas. Because of their small pore radius, the compacted cores have quite great slippage coefficient and bear serious slippage effect when permeated by gases. This phenomenon is illustrated by the steep straight lines shown in Fig. 6.7. Oppositely, since in high-permeability cores the diameter of the pore openings is much larger than the free path of the gas molecules, the gas can flow easily through these cores, and, therefore, the slippage

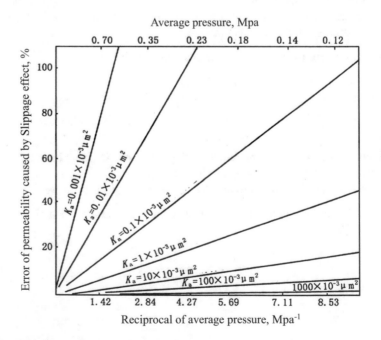

Fig. 6.7 Chart for the correction of Klinkenberg effect

effect, as illustrated by the straight lines with gentle slope in Fig. 6.7, is found to be insignificant.

In conclusion, the slippage effect of gas is more significant in low-permeability rocks, especially the compacted rocks measured at low pressure. So, it is generally specify that those rocks with permeability lower than 100 mD (0.1 μm^2) should undergo a correction for their measured Klinkenberg permeability.

In lab, the Klinkenberg permeability is measured and corrected in this way:

Firstly, on the basis of a group of permeability measurements under different mean pressures ($\overline{P}$) to get a series of gas permeability, K_g. Secondly, draw the $K_g - 1/\overline{P}$ relationship curve which turns out to be a straight one as described by Eq. (6.11). Then, get the value of K_∞, the intercept on the K_g coordinate axes, which is exactly the absolute permeability of the rock shown Fig. 6.5

Apart from real measurements, empirical equations and charts can also be employed for the work of correction. Figure 6.7 is one of the most commonly used correction charts.

With respect to the effect of the gas pressure on the gas permeability measurements, in routine core analysis, the usual measurement of permeability is made with air at a mean pressure of approximately $P = 1$ atm, and the results can be used as criteria for the permeability evaluation of rocks. Section 6.4 gives details of the measurements.

6.3 Factors Affecting the Magnititude of Rock Permeability

Compared with the porosity, the permeability of rock is influenced in a more complicated way by the factors such as the environment for the formation of rocks, the diagenesis effect, and the rock structures.

6.3.1 Deposition

1. Effect of rock skeleton construction and rock structure

Many properties such as particle size, sorting, cementation, and bedding all affect the rock permeability. Figure 6.8 got from laboratory researches on sandstones shows that conditions such as loose cementation, fine particle size, and bad sorting lead to lower permeability.

Different rock structures, for example, cross-bedding, wave-bedding, and graded-bedding, have significant effects on the permeability of rock. As the depositional rock strata with positive rhythm usually have finer and finer particle size from bottom down to top, the permeability upwards becomes lower and lower correspondingly, which is the reason why the lower parts of a formation are usually

Fig. 6.8 Relationship
between permeability and
particle size, sorting
coefficient

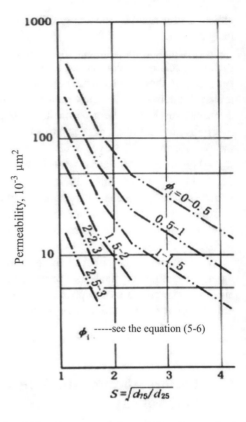

Fig. 6.8 Relationship between permeability and particle size, sorting coefficient

prematurely waterlogged during waterflood development. Meanwhile, the vertical permeability and the horizontal permeability of the same sandstone stratum are greatly different, either.

For carbonate rocks, these bearing only initial pores usually have higher horizontal permeability in comparison with their vertical permeability; while these reduced pores are also developed tend to have higher vertical permeability in comparison with their horizontal permeability. Therefore, routine core analysis must be concerned with the direction in which the core is drilled to realize which permeability the core is representative of the horizontal or the vertical.

2. Effect of pore structure of rock

Equations that pertain to the effect of pore structures on rock permeability are presented by scientists (Eqs. 6.62–6.67) give derivation details):

$$K = \frac{\phi^3}{2\tau^2 S_S^2 (1-\phi)^2} \tag{6.14}$$

$$K = \frac{\phi \times r^2}{8\tau^2} \tag{6.15}$$

Fig. 6.9 Relationship among permeability, porosity, and particle size (Chilingarian 1964). *1* Coarse and extremely coarse grain, *2* coarse and moderately coarse grain, *3* fine grain, *4* muddy grain, and *5* clay

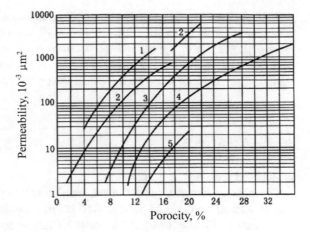

where

S_S the specific surface that take the rock matrix volume as denominator;
τ the tortuosity of the rock pore.

Seen from Eq. (6.14), the permeability K has a first-power relation with the porosity φ and a second-power relation with the pore radius r as well as the specific surface S. This equation clearly indicates that the permeability of rock depends on not only the porosity, but also the parameters (e.g., the pore radius and the specific surface) that reflect the rock structures.

The specific surface of a rock, one of the influencing factors for the permeability, is also decided by the particle size and the pore radius. Generally speaking, finer

Fig. 6.10 Relationship among permeability, porosity, and particle size (Timmerman 1982). ① clean sand, ② well sorted fine sand, ③ extremely well sorted fine sand, ④ well sorted and very fine sand, ⑤ moderate sorted and very fine sand, and ⑥ bad sorted sand

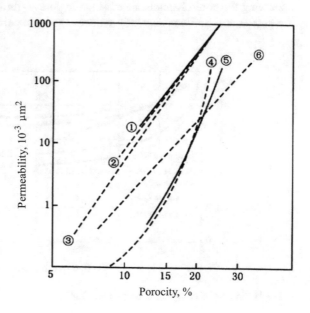

particle size and smaller pore radius imply greater specific surface and lower permeability of rock. Besides, Figs. 6.9 and 6.10 shows that the interconnectivity, tortuosity of the pore space, and the coarseness of the internal pore surface also have effect on the rock permeability. Furthermore, a low permeability usually also signifies that the rock has a high connate water saturation.

6.3.2   Diagenesis

1. Effect of reservoir static pressure
    Compaction experiments on clean, dried samples prove that K_p/K_i (K_p is the permeability at the current effective overburden pressure, and K_i is the permeability at the initial effective overburden pressure) intimately associated with the overburden pressure P—when a higher pressure is disposed on the sample, the measured permeability will become lower; and, as shown in Fig. 6.11, once the overburden pressure surpasses a specified value (e.g., 20 MPa), the permeability will decrease sharply. What's more, the permeability of the clay sandstones tends to be more impressionable to the increase or decrease in the overburden pressure.
    According to relevant reports, the permeability of the recent sedimentary sandstones can be as high as 10 or even 100 D, while that of those compacted and thereby solidified sandstones seldom exceeds 1 or 2 D. In fact, most of the compacted sandstones possess a quite small permeability ranging from a few to a few hundreds millidarcy—even some in great depths just own a permeability lower than 1 mD—and, as proved by experiments, their low permeability should be ascribed to the compaction effect, which exerts a profound influence on the formation by narrowing the pore-channels, increasing the pore-to-throat ratio and the tortuosity of the pore space, and consequently reduces the rock permeability sharply.

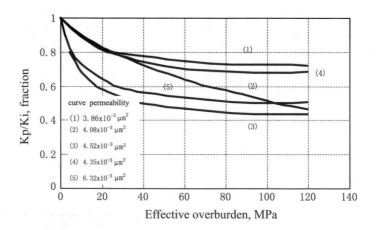

Fig. 6.11 Impact of compaction on permeability

2. Effect of cementation

During either the early or the late diagenesis, the precipitated cementing materials occupy a part of the conducting channels, increase the pore-to-throat ratio, and raise the degree of roughness of the pore walls, all of which lead to a lower permeability. Figures 6.12 and 6.13 show these relationships.

3. Effect of erosion

The corrosion effect does increase the rock porosity to some extent, but with regard to the permeability, its variation tendency resulting form the corrosion effect depends on chance—it may be increased and may not because the secondary pores developed from the corrosion effect are usually irregular in shape and great in the pore-to-throat ratio and the tortuosity.

Fig. 6.12 Relationship between permeability and porosity, cemention type

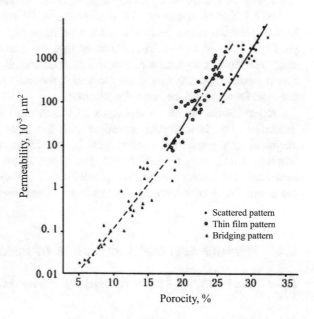

Fig. 6.13 Relationship between permeability and mud content

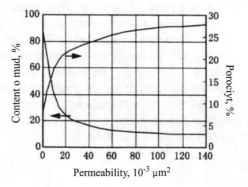

6.3.3   Tectogenesis and Others

The fractures and fissures developed from tectogenesis, or tectonic actions, enhance both of the porosity and the permeability of formations, especially the carbonate formations. In some cases, they may alter an impermeable carbonate to a highly or moderately permeable one. With the tectonic actions, more or less, it may also follow that the reservoir fluids reacts with the rocks chemically or physically, and that the rock permeability is changed as a result. For example, clay minerals expand when confronting incompatible formation water and become big enough to plug up the channels, leading to lower permeability; and, as another example, harmful migrations of mud or absorptions of the asphalts, waxes, and colloids onto the pore walls may be caused by a flowing high-viscosity crude oil.

Byplakof et al. found when the experimented temperature is raised, the over-burden pressure exerts less effect on the permeability of rock, especially at comparatively low pressures. The reason is that the expansion of the rock and the contained fluids counteracts part of the compaction effect when the temperature is raised, which means that the degree to which the permeability is decreased with the pressure increase will be naturally lowered.

Besides, there are also some cases in which the rock permeability is reduced because of the development operations and formation damage. For instance, the chemical or physical reactions caused by the incompatibility of the extraneous working fluids, the permeability changes triggered by inappropriate development measures or executions, etc. Consequently on all of these factors mentioned above, the permeability of reservoir rocks is not fixed and rigid.

6.4   Measurement and Calculation of Permeability

6.4.1   Laboratory Measurement of Permeability

1. Permeability measurement of routine small core

Small cores refer to the cores approximately 2.5 or 3.8 cm in diameter and 4–6 cm in length. A variety of apparatuses have been invented for the laboratory measurements of permeability, all of which are based on the Darcy's Law. The basic procedure for measurement is as follows: firstly, let a fluid (liquid or gas) flowing through the core and begin to measure necessary parameters, including the pressures, P_1 and P_2, respectively, at the inflow and outflow surfaces (or the pressure differential) and the volume flow rate Q; secondly, obtain the viscosity of the used fluid by experiments or resorting to empirical methods; then, compute the permeability according to Eqs. (6.2) or (6.10). A typical flowchart for the experiment is shown in Fig. 6.4.

2. Perm-plug method for the permeability measurement of routine small core

The most commonly used method currently in China is a perm-plug method using a flow metering-tube. It also called the gas-driving-water method. As shown in Fig. 6.14, the apparatus for such a method consists of two parts: (1) the core holder used to clamp the core and (2) the flow metering tube stuck into the water in the sink.

The top-end of the core is open to atmosphere so P_1 is held as a constant equivalent to the atmospheric pressure; when evacuating the tube with (3) the air-suction ball, the water-surface is sucked swiftly from (4) the sink up to a certain level in the tube; subsequent to the evacuation operation, the outside air begins to enter the core and then flows through it, and finally arrives at the tube due to the pressure differential over the core, the bottom-end of which is subjected to a negative or subatmospheric pressure caused by, as a result of the action of the hydrostatic head, the drop of the former water-surface in the tube; obviously, the gas pressure at the outflow surface of the core is a variable, depending on the hydrostatic head h formed by the water column in the tube. The descending speed of the water-surface is associated with the permeability of the core—the higher the permeability, the greater the descending speed (vice versa). If the distance $(h_0 - h_x)$ that the water-surface dropped and the time (T) of this process are recorded, the permeability K can be calculated by equation:

Fig. 6.14 Flow metering tube for the measurement of gas permeability. *1* core holder, *2* flow metering tube, *3* gas sucker, *4* water container

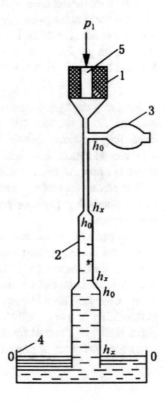

$$K = \frac{B\mu L}{TA} \times 10^6 \qquad (6.16)$$

where

K the permeability of reservoir rocks, mD;
B a constant for a given flow metering tube, cm³/atm;
L length of the core, cm;
μ air viscosity, mPa s;
A cross-sectional area of the core, cm²;
T time consumed by the water surface to drop form h_o to h_x, s.

The constant B is dependent upon not only the apparatus structure, but also the local condition of pressure. In addition to the real measurement, B can also be obtained from the equation written below:

$$B = C\{-2F[\ln(CP_o - h_x) - \ln(CP_o - h_o)]$$
$$+ \frac{V * + 1033.6P_oF + h_oF}{CP_o}\left[\ln\frac{CP_o - h_x}{h_x} - \ln\frac{CP_o - h_o}{h_o}\right]\} \qquad (6.17)$$

where

F cross-sectional area of the flow metering tube, cm²;
P_o the local atmospheric pressure, atm;
$V*$ the residual volume left above h_o, cm³;
h_o the height from the 0–0 water-surface (in the sink) to the h_o scale mark, cm;
h_x the height from the 0–0 water-surface (in the sink) to the h_x scale mark, cm;
C constant, $C = 2067.2$.

See from the figure, the flow metering tube is composed of three differently sized sections. Each section will contribute a distinctive length of the $(h_o - h_x)$ when measuring the same sample. So, the differently valued B of them should be taken into consideration and which section should be chosen for the measurement depends upon the permeability of the rock.

The advantages of this method lie in the simplicity of its apparatus, the rapidity of the measuring process, and its suitability for the in situ measurements during field exploration.

3. Permeability measurement of full-diameter core

For the heterogeneous formations interspersed with fractures, vugs or solution cavities, conglomerates, etc., the small cores are not representative any more. Therefore, it is the full-diameter cores that should be used to serve the permeability measurement. With a Hassler-type core holder, both the horizontal and the vertical permeability of a core, which, respectively, represent the horizontal and the vertical permeability of the real formations, can be measured and obtained.

1) Measurement of horizontal permeability

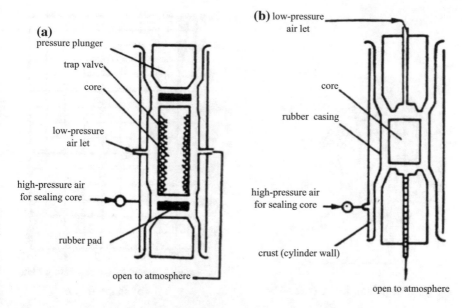

Fig. 6.15 Hassler-type full-diameter core holder. **a** Horizontal flow, **b** vertical flow

A diagram of the Hassler-type core holder is given in Fig. 6.15. The housing of this core holder is a long steel cylinder in which there is a rubber sleeve to form an enclosed annular space with the outer cylinder. As seen from the diagram, since there are ports allowing high-pressure gases to enter or exist the annular space, confining pressure can be applied to the central core sample by injecting gas or liquid into the annular space, which also help enclose and seal off the core. With the upper and lower end plugs, pressures can also be applied to the core from the two ends. Besides, there is also a pair of inlet and outlet located on the cylinder side as the entrance and exit of the measuring low-pressure air.

Subsequent to putting a core into the holder, it is time to exert pressure onto the plugs, with the aid of hydraulic press, until the rubber pads seal off the two ends of the core, and then to exert confining pressure onto the circumference of the core so that to seal the side surface of the core. When measuring, the gas consecutively goes through the inlet, the filter screen, the core sample, the filter screen on the other side, and finally the outlet. Figure 6.16 shows the flow direction.

The calculation of the horizontal permeability still obeys the Darcy's Law on the premise that an extra shape-factor be introduced to the equation. This shape-factor depends on the core diameter and the angle sheltered by the arcuate filter screens, which are put partly locally over the core circumference for gas distribution. The value of E can be obtained from electric models, displacement experiments, etc. The Darcy's Law after correction turns out to be the following:

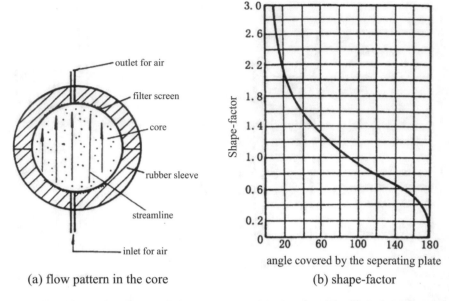

(a) flow pattern in the core   (b) shape-factor

Fig. 6.16 Flow pattern and shape-factor when the horizontal permeability of core. **a** Flow pattern in the core, **b** shape-factor is measured

$$K = E \cdot \frac{2Q_oP_o\mu}{L(P_1^2 - P_2^2)} \times 10^{-1} \qquad (6.18)$$

where

K	horizontal permeability, μm^2;
Q_o	gas volume flow rate through the core at the atmospheric pressure P_o, cm^3/s;
μ	gas viscosity, mPa s;
L	length of the core, cm;
P_1, P_2	pressures at the inflow and the outflow ends of the core, MPa;
E	shape factor, depending on the core diameter and the angle sheltered by the arcuate filter screens. It can be obtained by resorting to the curve shown in Fig. 6.16. For example, when the angle $\theta = 90°$, $E = 1$.

2) Measurement of vertical permeability

For the measurement of the vertical permeability of a core, the measuring gas needs to flow through the core from one end to another and the couple of filter screens should be put at the two ends instead of the side surface of the core. And, correspondingly, the inlet and outlet openings for the gas are located on the two end plugs. The measurement process and calculation equation are same as that for the routine small cores.

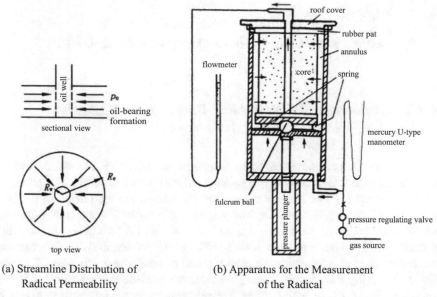

(a) Streamline Distribution of
Radical Permeability

(b) Apparatus for the Measurement
of the Radical

Permeability of Full-Diameter Cores

Fig. 6.17 Measurement of radical permeability

3) Measurement of radical permeability

The apparatus for the measurement of rock radical permeability is shown in Fig. 6.17. Drill a hole at the center of a full-diameter core and then put the hollow-hearted core into the core chamber where the center hole of it needs be connected with the inlet opening of the upper plug. Note that the lower end of the core must be closely clung to the thick rubber pad under it, so that gas leakage can be avoided.

During measurement, clump the core tightly by pressurizing the plugs and then connect it with the gas supply; measure and determine the pressures at the inflow and the outflow ends of the core, P_1 and P_2, and the gas volume flow rate flowing through the core. According to the Darcy's Law for radical flow, we can know that:

$$K = \frac{Q_o P_o \mu \ln\left(\frac{d_e}{d_w}\right)}{\pi h (P_1^2 - P_2^2)} \times 10^{-1} \tag{6.19}$$

where

K	gas permeability, μm^2;
d_e	outside diameter of the core, cm;
μ	liquid viscosity, mPa s;
d_w	diameter of the center hole of the core, cm;

h height of the fore, cm;
P_0 atmospheric pressure;
P_e, P_w pressures at the inflow and the outflow ends of the core, MPa.

6.4.2 Evaluation of Permeability Based on Well-Logging Data

In field practices, sometimes the formation permeability can also be determined by well-logging curves, an indirect means to avoid costly core drillings. The principle of this method: (1) The connate water saturation S_{wi} is closely related to the permeability of the core sample. Generally speaking, a rock of small rock pore size tends to be endowed with high S_{wi} and low K. (2) So to the sandstones, the formation with a higher porosity usually has a better permeability. (3) We can use the information of a formation's porosity φ and connate water saturation S_{wi} from the well-logging data for evaluating and determining the permeability of it. As shown in Fig. 6.18, in well-logging interpretation the equation written below is usually employed for permeability evaluation.

$$K = \frac{c\phi^a}{S_{wi}^b} \qquad (6.20)$$

In the equation, a, b, and c are constants dependent on the rock pore structure and the fluid properties. For example, Timur once offered that $a = 4.4$, $b = 2.0$, and $c = 0.136$, upon which the equation becomes the following:

$$K = \frac{c\phi^a}{S_{wi}^b} \qquad (6.20)$$

The unit of K is darcy.

The error of the permeability obtained from the evaluation based on the well-logging data is rather great and sometimes can be as high as 50 %. So, comparison with lab core tests is needful when this method is used for determining the permeability profile formation of a formation

Fig. 6.18 Evaluation of permeability based on Φ and S_{wi}

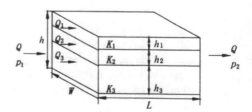

6.4.3   Calculation of Permeability Based on Mean Pore Radius (r) and Porosity (φ)

Founded on the model of bundles of capillary, the following equation can be got (Lesson 6.6 gives details of the derivation):

$$K = \frac{\phi \times r^2}{8\tau^2} \tag{6.23}$$

where

K permeability, μm^2;
r pore-channel radius of rock, μm;
φ porosity, fraction;
τ tortuosity, fraction.

Generally speaking, the pore-channel radius of real oil-bearing reservoir rocks falls between 1 and 10 μm, and the tortuosity of their pore space is about $\tau = 1$–1.4. Hence, making use of the above mathematical relationship, we can obtain the mean permeability of formation.

6.4.4   Methods for the Calculation of Average Permeability

Complicating matter is the fact that the data obtained from measurements usually come in enormous quantities—many countries specify in their core analysis standards that at least three cores should be extracted from every single meter of the reservoir intervals, so there are thousands of poroperm data from even a single well, and, what puts the analysis into a sea of troubles, these data vary within a considerable wide range. Therefore, a mean value of the permeability, referred as the mean permeability, should be determined to give a better description of the studied formation. The mean permeability should be representative for the formation characteristics concerned with the fluid-conducting property and meaningful enough to be used for supporting the reservoir evaluation as well as the reserve calculations. Currently, the most commonly used methods include the ones listed below.

1. Arithmetic averaging method

Assuming that the permeability and porosity at the ith of the total n sampling points are, respectively, K_i and φ_i, we can easily get the mean permeability by the arithmetic averaging method:

$$\overline{K} = \frac{\sum\limits_{i=1}^{n} K_i}{n} \tag{6.23}$$

$$\overline{\phi} = \frac{\sum_{i=1}^{n} \phi_i}{n} \tag{6.24}$$

2. Weighted Averaging Method

The permeability of the ith measured core sample is K_i, and the effective thickness and oil-bearing area of the layer where the core came from are h_i and A_i. According to the thickness of each oil layer, the mean permeability yielded from weighted averaging is as follows:

$$\overline{K}_h = \frac{\sum K_i h_i}{\sum h_i} \tag{6.25}$$

And according to the area of each oil layer, the mean permeability is:

$$\overline{K}_A = \frac{\sum K_i A_i}{\sum A_i} \tag{6.26}$$

And according to the volume of each oil layer, the mean permeability is:

$$\overline{K}_V = \frac{\sum K_i A_i h_i}{\sum A_i h_i} \tag{6.27}$$

Since the weighted averaging method has taken into consideration the permeability K_i of each core and the effective thickness, the oil-bearing area of the represented oil layer, the results yielded come nearer the truth.

Moreover, the mean values of other physical-property parameters such as the porosity and the original oil saturation. can also be calculated by these methods.

Apart from the two mentioned, there are many other ways to calculate the mean permeability $\overline{K}$. For example:

$$log\overline{K} = (logK_1 + logK_2 + \cdots + logK_n)/n \tag{6.28}$$

$$\overline{K} = \sqrt[n]{K_1 \times K_2 \times \cdots K_n} \tag{6.29}$$

$$\frac{n}{\overline{\overline{K}}} = \frac{1}{K_1} + \frac{1}{K_2} + \cdots \frac{1}{K_n} \tag{6.30}$$

The subsequent discussion will be focused on the averaging methods according to the physical process of the underground seepage flow.

3. Permeability of combination layers in parallel connection

(1) Linear seepage flow

Consider the case where the flow system is comprised of layers of porous rocks with different permeability (K_1, K_2, K_3) as shown in Fig. 6.19. The average permeability K can be computed as follows:

Fig. 6.19 Liner flow, parallel combination of beds

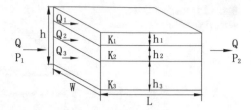

Since the three layers are connected in parallel, the pressure differentials over the length of them are identical; and thence, the total volume flow rate is equal to the sum of the three individual volume rates:

$$Q = Q_1 + Q_2 + Q_3 \tag{6.31}$$

According to Darcy's Law:

$$Q_1 = \frac{K_1 W h_1 (P_1 - P_2)}{\mu L}; \ Q_2 = \frac{K_2 W h_2 (P_1 - P_2)}{\mu L}; \ Q_3 = \frac{K_3 W h_3 (P_1 - P_2)}{\mu L} \tag{6.32}$$

Taking K as the equivalent permeability, the total volume flow rate can be written as:

$$Q = \frac{K W h (P_1 - P_2)}{\mu L} \tag{6.33}$$

Considering that the total thickness of the formation is equal to the sum of the individual thickness of each layer, that is $h = h_1 + h_2 + h_3$, we can substitute Eqs. (6.32) and (6.33) into Eq. (6.31) and get ; $Kh = K_1 h_1 + K_2 h_2 + K_3 h_3$
Then

$$K = \frac{K_1 h_1 + K_2 h_2 + K_3 h_3}{h_1 + h_2 + h_3} = \frac{\sum\limits_{i=1}^{n} K_i h_i}{\sum\limits_{i=1}^{n} h_i} \tag{6.34}$$

(2) Radial seepage flow

Consider the case where the formation is composed of three porous mediums with different permeability (K_1, K_2, K_3), and the flow system is a radial seepage flow as shown in Fig. 6.20. Here, the average permeability K can be computed as follow:

The same as the case of linear flow, the total thickness of the formation is: $h = h_1 + h_2 + h_3$

And the total volume flow rate is:

Fig. 6.20 Radical flow,
parallel combination of beds

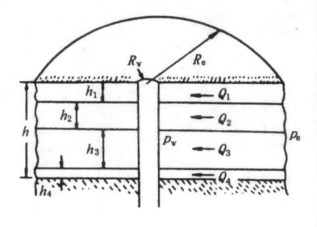

$$Q = Q_1 + Q_2 + Q_3$$

According to the Darcy's Law for radical-flow system, we know that:

For the whole formation: $Q = \frac{K2\pi h(P_e - P_w)}{\mu \ln R_e / R_w}$

For each individual layer:

$$Q_1 = \frac{K_1 2\pi h_1 (P_e - P_w)}{\mu \ln R_e / R_w} \qquad Q_2 = \frac{K_2 2\pi h_2 (P_e - P_w)}{\mu \ln R_e / R_w} \qquad Q_3 = \frac{K_3 2\pi h_3 (P_e - P_w)}{\mu \ln R_e / R_w}$$

Therefore

$$\frac{K2\pi h(P_e - P_w)}{\mu \ln R_e / R_w} = \frac{K_1 2\pi h_1 (P_e - P_w)}{\mu \ln R_e / R_w} + \frac{K_2 2\pi h_2 (P_e - P_w)}{\mu \ln R_e / R_w} + \frac{K_3 2\pi h_3 (P_e - P_w)}{\mu \ln R_e / R_w}$$

The equation above can be simplified to: $Kh = K_1 h_1 + K_2 h_2 + K_3 h_3$

Then, by generalizing,

$$K = \frac{K_1 h_1 + K_2 h_2 + K_3 h_3}{h_1 + h_2 + h_3} = \frac{\sum_{i=1}^{n} K_i h_i}{\sum_{i=1}^{n} h_i} \qquad (6.35)$$

It is noted that the same calculation appear in the radial-flow network as in the linear system.

4. Permeability of combination layers in series connection

(1) Linear seepage flow

Another possible combination for flow system is to have the values of different permeability (K_1, K_2, K_3) arranged in series. In the case of linear flow, as shown in Fig. 6.21, the fluid under a pressure differential $P_1 - P_2$ flows vertically through the three layers of which the stretching length are, respectively, L_1, L_2, and L_3. The average series permeability for the total volume can be evaluated as follows:

Fig. 6.21 Liner flow, series combination of beds

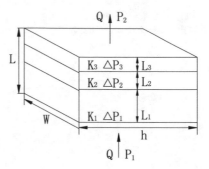

Firstly, in such a series combination, it can be known that the volume flow rates through every layer are identical; and the total pressure differential ΔP is equal to the sum of the three individual pressure differential: $\Delta P = \Delta P_1 + \Delta P_2 + \Delta P_3$, and the total stretching length of the formation is: $L = L_1 + L_2 + L_3$.

According to Darcy's Law, the total pressure differential and the three individual pressure differentials can be obtained:

$$\Delta P = \frac{Q\mu L}{KWh}; \quad \Delta P_1 = \frac{Q_1\mu L_1}{K_1 Wh_1};$$

$$\Delta P_2 = \frac{Q_2\mu L_2}{K_2 Wh_2}; \quad \Delta P_3 = \frac{Q_3\mu L_3}{K_3 Wh_3}$$

Because that $\Delta P = \Delta P_1 + \Delta P_2 + \Delta P_3$,

$$\frac{Q\mu L}{KWh} = \frac{Q_1\mu L_1}{K_1 Wh_1} + \frac{Q_2\mu L_2}{K_2 Wh_2} + \frac{Q_3\mu L_3}{K_3 Wh_3}$$

Then, with the consideration that $h = h_1 = h_2 = h_3$, substitute $Q = Q_1 = Q_2 = Q_3$ and $L = L_1 + L_2 + L_3$ into the given equation:

$$K = \frac{L_1 + L_2 + L_3}{\frac{L_1}{K_1} + \frac{L_2}{K_2} + \frac{L_3}{K_3}} \tag{6.36}$$

(2) Radial seepage flow

Radical seepage flow in a system of series connection occurs when a contaminated zone cause by mud invasion is formed around the wellbore. As shown in Fig. 6.22, the seriously contaminated zone has a radius of R_1 and a permeability of K_1; the lightly contaminated zone has a radius of R_2 and a permeability of K_2; while the uncontaminated zone has a permeability of K_3. According to the Darcy's Law for radical-flow system, we obtain pressure differential over each formation interval:

For the seriously contaminated zone ($R_w \rightarrow R_1$): $\Delta P_1 = \frac{Q\mu\ln(R_1/R_w)}{K_1 2\pi h}$

For the lightly contaminated zone ($R_1 \rightarrow R_2$): $\Delta P_2 = \frac{Q\mu\ln(R_2/R_1)}{K_2 2\pi h}$

For the uncontaminated zone ($R_2 \rightarrow R_e$): $\Delta P_3 = \frac{Q\mu\ln(R_e/R_2)}{K_3 2\pi h}$

Fig. 6.22 Radical flow,
series combination of beds

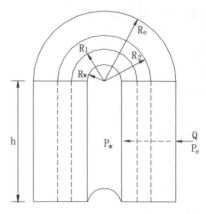

Taking Q and $\overline{K}$, respectively, as the volume flow rate through the formation and the average permeability, we can write the total pressure differential in such a form: $\Delta P = \frac{Q\mu\ln(R_e/R_w)}{\overline{K}2\pi h}$.

As the three zones are in series connection, we can know that $\Delta P = \Delta P_1 + \Delta P_2 + \Delta P_3$.

Substitute differential pressures into the equation given above, we can get:

$$\Delta P = \frac{Q\mu\ln(R_e/R_W)}{\overline{K}2\pi h} = \frac{Q\mu\ln(R_1/R_W)}{K_1 2\pi h} + \frac{Q\mu\ln(R_2/R_1)}{K_2 2\pi h} + \frac{Q\mu\ln(R_e/R_2)}{K_3 2\pi h}$$

Due to the continuous flow, which implies $Q = Q_1 = Q_2 = Q_3$, and considering that the viscosity of the fluids flowing through the zones are identical, we can get the average permeability in a simplified form:

$$\overline{K} = \frac{\ln\frac{R_e}{R_w}}{\frac{1}{K_1}\ln\frac{R_1}{R_w} + \frac{1}{K_2}\ln\frac{R_2}{R_1} + \frac{1}{K_3}\ln\frac{R_e}{R_2}} = \frac{\ln\frac{R_e}{R_w}}{\sum_{i=1}^{n}\frac{1}{K_i}\ln\frac{R_i}{R_{i-1}}} \qquad (6.37)$$

6.5   Permeability of Naturally Fractured and Vuggy Rocks

Naturally occurring fractures and vugs, or solution cavities, are particularly developed and quite prevalent in carbonate formations where the fracture system acts not only the storage space for oil and gas, but also the conduits for fluid flow. And, in some cases, many clastic rocks, e.g., the compacted sandstones may also comprise fractures.

Due to the complexity and randomness of the fractures developed in real formations, it is rather difficult to ensure the drilled cores, to be representative of the general view of the formation. Sometimes downhole televisions can help us catch a glimpse of the distribution and scale of the fractures within a local area of the studied formation. However, so far there still has not been a perfect or excellent method to directly observe and measure the permeability of underground fractures, and, therefore, most of the researches on the permeability of fractured rocks are based on plenty of simplified conditions and simplified models, for example, the pure-fractured rock model and the double-porosity medium model. Here, the permeability of the pure-fractured rock and the permeability of the fracture-pore double-medium rock will be introduced.

6.5.1   Permeability of pure-fractured rocks

1. Porosity of pure-fractured rocks

The research on the permeability of a fractured rock necessitates a preceding description of the fracture characteristics contained in the rock. As shown in Fig. 6.23, it is a model where the fracture of uniform width penetrates throughout a core. We can use the fracture density n and the fracture width b to describe its characteristics. The fracture density is defined as the ratio of the total length of the fractures at core end surface to the cross-sectional area of the core. Expressed mathematically:

$$n = \frac{l}{A} \tag{6.38}$$

where

n fracture density, cm^{-1};
l total length of the fractures at core end surface, equal to the sum of the fracture lengths, cm;

Fig. 6.23 Description of fracture

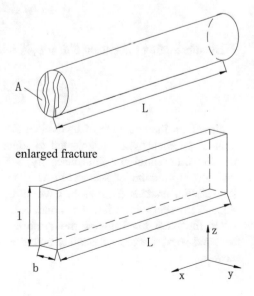

A cross-sectional area of the fractured core, cm^2.

Then, the fracture porosity φ_f of the core is:

$$\phi_f = \frac{lbL}{AL} = nb = \frac{lb}{A} \tag{6.39}$$

where

φ_f fracture porosity, fraction;
b fracture width, cm.

2. Permeability calculation of pure-fractured rocks

The flow within a fracture can be regarded as the flow happening between a couple of parallel plates. According to the Boussinesq Equation, the flow rate running through every unit length of the fracture exclusively in the Z direction is:

$$q = \frac{b^3}{12\mu}\frac{dP}{dx} \tag{6.40}$$

where

q flow rate running through every unit length of the fracture, cm^3/s;
μ kinetic viscosity of liquid;
dP/dx pressure gradient.

As the total length of the fractures at the core end surface is l, the total flow rate running through the end surface of the core is:

$$Q = l \cdot q = \frac{lb^3}{12\mu}\frac{dP}{dx} \tag{6.41}$$

By substituting the equation $lb = \varphi_f A$ into Eq. (6.41), we can get:

$$Q = \frac{A \cdot \phi_f \cdot b^2}{12\mu}\frac{dP}{dx} \tag{6.42}$$

Based on the principle of equivalent flow, if another rock possesses identical external geometry and is subjected to identical seepage-flowing conditions (e.g., pressure differential and fluid viscosity) with the fractured rock, according to the equivalent principle of seepage-flowing resistance, there must be an equivalent permeability enables the rock to have an equivalent porous resistance against the seepage flowing within it and to conduct an equivalent flow rate.

On the other hand, taking the symbol K_f as the effective permeability of the fractured rock, we can get another equation of the flow rate based on the Darcy's Law:

$$Q = \frac{K_f A}{\mu} \frac{dP}{dx} \tag{6.43}$$

As Eqs. (6.42) and (6.43) are equal to each other, we can get:

$$K_f = \frac{\phi_f b^2}{12} \tag{6.44}$$

where

b fracture width, cm;
φ_f fracture porosity, fraction;
K_f fracture permeability, cm^2.

If we use the Darcy units, the above equation will be altered into:

$$K_f = \phi_f b^2 \cdot \frac{10^8}{12} = 8.33 \times 10^6 b^2 \phi_f \tag{6.45}$$

where

K_f fracture permeability, μm^2.

Seen from the equation, the permeability of a fracture is governed to a large degree by the width and the extending degree of it. For example, even a fracture opening of a certain number of microns has a considerable permeability—on the order of 1 darcy. Besides, the width of fractures usually varies widely from 10 μm to 2 cm.

6.5.2 Permeability of fracture-pore double-medium rocks

For this kind of rocks, although both the matrix pores and the fractures have capacity for storing and conducting fluids, the matrix pores play a dominant role as the storage space, and the fractures play a dominant role as the conduits for the seepage-flow.

The permeability of a fracture-pore rock is equal to the sum of the matrix permeability and the fracture permeability. Expressed mathematically:

$$K_t = K_m + K_f \tag{6.46}$$

where

K_t rock permeability, also named total permeability, μm^2;
K_m matrix permeability, μm^2;
K_f fracture permeability, μm^2.

Fig. 6.24 Description of vugs

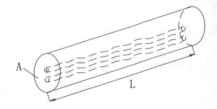

6.5.3   Permeability of vuggy rocks

Similar to the process of getting the fracture permeability, we can calculate the vug porosity (Fig. 6.24):

$$\phi_h = \frac{N\pi r^2 L}{AL} = \frac{N\pi r^2}{A}$$

(6.47)

where

φ_h porosity of vuggy rock;
r vug radius, cm;
L length of the vuggy core sample, cm;
N number of the vugs at the core cross-section;
A cross-sectional area on the vuggy core sample, cm^2.

According to the Poiseuille's Law, the flow rate running through a single vug is:

$$q = \frac{\pi r^4}{8\mu}\frac{\Delta P}{L}$$

(6.48)

where

q flow rate through a single vug, cm^3/s;
μ kinetic viscosity of the fluid, dyn.s/cm^2;
$\Delta P/L$ pressure gradient, dyn.

So, if there are totally N strands of vugs, the total flow rate through the whole cross-section of the rock is:

$$Q = Nq = \frac{N\pi r^4}{8\mu}\frac{\Delta P}{L} = \frac{\phi_h A r^2 \Delta P}{8\mu L}$$

(6.49)

Taking the symbol K_h as the effective permeability of this vuggy rock, the equation written below can be got on the basis of the Darcy's Law:

$$Q = \frac{K_h A}{\mu} \frac{\Delta P}{L} \qquad (6.50)$$

According to the equivalent principle of flow resistance, Eq. (6.50) is equal to Eq. (6.49). So, we can get:

$$K_h = \frac{\phi_h r^2}{8} \qquad (6.51)$$

where

r vug radius, cm;
φ_h porosity of the vuggy rock, fraction;
K_h permeability of the vuggy rock, cm^2.

If we use the Darcy units, the above equation will be altered into:

$$K_h = 12.5 \times 10^6 \phi_h r^2 \qquad (6.52)$$

where

r vug radius, cm;
K_h permeability of the vuggy rock, μm^2.

6.6   Ideal Models of Rock Structure

In consideration of the complexity of the pore structure in sandstone rocks, people have been trying to build ideal models to simply the real rocks. These models are based on which the porosity, permeability expressions become easier to be derived and the quantitative relations among those rock physical-property parameters turn to be easier to built up. They can be summarized as follows: (1) the soil model, (2) the capillary buddle model, also named the model of bundles of capillary tubes, and (3) network model.

The soil model, which had already been introduced in the Chap. 5, is mainly used in the calculations for porosity, but not convenient for that for permeability. This lesson will focus on the model of bundles of capillary tubes.

6.6.1   Model of Bundles of Capillary Tubes

1. Model of bundles of capillary tubes

As shown in Fig. 6.25, the pore space contained in the real rock is composed of irregular pores and channels; but the model of bundles of capillary tubes simplifies it into an ideal rock where the pore space is composed of uniformly sized, parallel capillary tube bundles

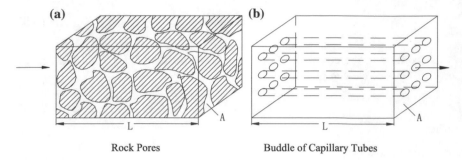

(a) Rock Pores

(b) Buddle of Capillary Tubes

Fig. 6.25 Model of bundles of parallel capillary tubes

Before further discussion, the following symbols need to be introduced firstly:

L length of the core sample, cm;
A cross-sectional area of the core sample, cm^2;
n capillary number per unit area of the cross-sectional area, number/cm^2;
V_s volume of the rock skeleton, cm^3;
a the total internal surface area of all the capillary tubes, cm^2.

Thence, the porosity of this hypothetical rock is:

$$\phi = \frac{V_e}{V_b} = \frac{nA(\pi r^2)L}{AL} = n\pi r^2 \tag{6.53}$$

And the specific surface of this hypothetical rock is:

$$S = \frac{a}{V_b} = \frac{nA(2\pi r)L}{AL} = n(2\pi r) = n\pi r^2 \cdot \frac{2}{r} \tag{6.54}$$

Based on Eqs. (6.53) and (6.54), we can get:

$$r = \frac{2\phi}{S} \tag{6.55}$$

The specific surface taking the skeleton volume as the denominator:

$$S_S = \frac{a}{V_S} = \frac{a}{(1-\phi)V_b} = \frac{S}{1-\phi} = \frac{2n\pi r}{1-\phi} \tag{6.56}$$

The specific surface taking the pore volume as the denominator:

$$S_P = \frac{a}{V_P} = \frac{a}{\phi V_b} = \frac{S}{\phi} = \frac{2n\pi r}{\phi} \tag{6.57}$$

So

$$S = S_S(1 - \phi) = S_P \phi \tag{6.58}$$

2. Law of capillary seepage flowing

Consider the case where a fluid with a viscosity of μ is flowing (in the manner of laminar flow or viscous flow) through a capillary tube with a length of L and an inside radius of r_o, under the pressure differential is $(P_1 - P_2)$. Provided that the fluid is able to wet the interior wall of the capillary tube, the flow velocity will be kept as zero at the tube interior wall and reaches its maximum value at the center line of the tube; what's more, the flow viscosity is constant at the circles of points which are equidistant from the center line—as shown in Fig. 6.26, the fluid particles with the same flow velocity form into a cylinder layer. The viscous force between adjacent cylinder layers is:

$$F = \mu A \frac{dV}{dx}$$

where

A area of the cylinder layer with a distance r away from the center line, cm^2;
dV/dx velocity gradient, cm/(s cm);
F viscous force, dyn.

So, taking the cylinder layer that is at a distance of r from the center line as an example, the viscous force of it is:

$$F_r = \mu A \frac{dV}{dx} = \mu(2\pi r L) \frac{dV}{dr}$$

As the driving force acting on this liquid cylinder is $(P_1 - P_2)\pi r^2$ (dyn), the viscous force must be equal to the driving force supposing that the fluid have no acceleration:

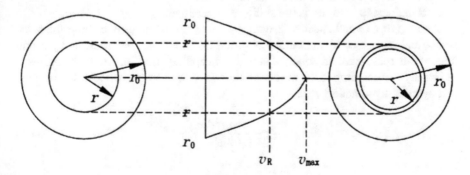

Fig. 6.26 Velocity distribution of viscous fluid flowing in capillary tube

$$\mu(2\pi rL)\frac{dV}{dr} + \pi r^2(P_1 - P_2) = 0$$

$$dV = -\frac{(P_1 - P_2)rdr}{2\mu L}$$

Then, we can easily get the integral form of the above equation:

$$V = -\frac{(P_1 - P_2)r^2}{4\mu L} + C_1$$

When $r = r_0$, $V = 0$. By substituting this known relationship, we can obtain the integral constant C_1:

$$C_1 = r_o^2(P_1 - P_2)/4\mu L$$

$$V = \frac{(r_0^2 - r^2)(P_1 - P_2)}{4\mu L} \tag{6.59}$$

This equation gives the moving velocity of the cylinder layer at a fixed distance r from the center line. Taking a whole cross-section of the capillary tube into consideration, we can realize that the velocity exhibits a parabolic variation with respect to r. Therefore, the seepage flow rate through the capillary tube can be obtained by superimposing the flow rate of all the moving liquid cylinders:

$$q = \int_0^q dq = \int_0^{r_0} VdA = \int_0^{r_0} \frac{(P_1 - P_2)(r_0^2 - r^2)}{4\mu L} \times 2\pi rdr$$

After integration:

$$q = \frac{\pi r_0^4(P_1 - P_2)}{8\mu L} \tag{6.60}$$

This equation, which describes the flow rate of the viscous seepage flow through a capillary tube, is known as the Posenille's Law.

3. Relationship between permeability and pore radius

Assume of a hypothetical rock with the same geometric size, bearing the same fluids, and impressed by the same pressure differential with real rock, and per unit area of this rock, there are totally n strands of capillary tubes of which the radius is uniformly r. According to the the Posenille's Law, we can get the flow rate running through this hypothetical rock:

$$Q = \frac{nA\pi r^4 \Delta P}{8\mu L} \tag{6.61}$$

With the permeability K of rock, we can get another form of flow rate based on the Darcy's Law:

$$Q = \frac{KA\Delta P}{\mu L}$$

According to the equivalent principle of seepage flowing, the two flow rates are equal:

$$\frac{KA\Delta P}{\mu L} = \frac{nA\pi r^4 \Delta P}{8\mu L}$$

Since the porosity of the hypothetical rock is $\phi = nA\pi r^2 L/(AL)$, the above equation can be reduced to the below one:

$$K = \frac{\phi \cdot r^2}{8} \tag{6.62}$$

where

r mean pore radius of real rock, cm;
K permeability of real rock, cm^2.

As the porosity of rock is subject to a small variation while the mean pore radius varies significantly (1–10 μm), it is the pore radius that plays a dominant role in deciding the permeability of rock. According to Eq. (6.62), the mean pore radius can be computed:

$$r = \sqrt{\frac{8K}{\phi}} \tag{6.63}$$

where

K permeability, cm^2;
r mean pore radius of real rock, cm.

4. Relationship between permeability and specific surface
Substitute Eq. (6.55) into Eq. (6.63):

$$K = \frac{\phi \cdot r^2}{8} = \frac{\phi}{8} \times \frac{4\phi^2}{S^2} = \frac{\phi^3}{2S^2} \tag{6.64}$$

Substitute Eq. (6.57) into Eq. (6.64):

$$K = \frac{\phi^3}{2S^2} = \frac{\phi^3}{2S_S^2(1-\phi)^2} = \frac{\phi}{2S_\phi^2} \qquad (6.65)$$

5. Corrections

In consideration of the disparities between ideal models and real rocks, people introduce a correction coefficient to resolve contradictions to some extent. Replacing the constant 2 in Eq. (6.65) with another constant ζ, we can get the get the generic expression of the Gaocaini-Equation, which describes the relationship between permeability and specific surface:

$$K = \frac{\phi^3}{\zeta S^2} = \frac{\phi^3}{\zeta S_S^2(1-\phi)^2} = \frac{\phi}{\zeta S_\phi^2} \qquad (6.66)$$

The constant ζ is called the "Gaocaini coefficient" and assigned a value 5 in the initial derivation (1927).

As shown in Eq. (6.66), the permeability is in inverse proportion to the specific surface. So, the mudstones, bearing small pores and thereby great specific surface, have low permeability.

Furthermore, another imparity between ideal model and real rock is that the pore space within the latter is not alike direct tubes as shown in Fig. 6.25 but bound to run into twists and turns. In order to give a corrective change, in the year 1932, Kalman introduced a parameter named "tortuosity coefficient" that altered Eq. (6.64) into:

$$K = \frac{\phi \cdot r^2}{8\tau^2} = \frac{\phi}{8} \times \frac{4\phi^2}{\tau^2 S^2} = \frac{\phi^3}{2\tau^2 S^2} = \frac{\phi^3}{2\tau^2 s_s(1-\phi)} \qquad (6.67)$$

If we use the Darcy units and assigning $\tau = 1.0$, equation above becomes:

$$K = \frac{\phi^3}{2S^2} \times 10^8 \qquad (6.68)$$

$$S = 7000\phi\sqrt{\frac{\phi}{K}} \qquad (6.69)$$

where

S specific surface taking the bulk volume of rock as denominator, cm^2/cm^3;
φ porosity of rock, fraction;
K permeability of rock, μm^2.

If assigning $\tau = 1.4$, Eq. (6.67) becomes:

$$K = \frac{\phi^3}{2 \times 1.4^2 S^2} \times 10^8 \qquad (6.70)$$

$$S \approx 3600\phi\sqrt{\frac{\phi}{K}} \qquad (6.71)$$

Generally speaking, there are certain statistical regularities about the relationship between rock permeability and porosity: (1) permeability increases to some extent with the increase in porosity; (2) permeability decreases sharply with the decrease in the particle size of rock grains, that is to say, when the grains become finer, the mean pore radium becomes smaller, and thence, the permeability becomes lower (Fig. 6.10).

The model of bundles of capillary tubes, based on which we discussed in this lesson the correlations between permeability and pore radius, and between permeability and specific surface, is a simple but useful model explaining seepage flowing in rocks.

6.6.2   Network Model

Seeing that there are still some gaps between the structures of the capillary bundle model and real rocks, people proposed network models, using a network composed of large numbers of capillary tube groups to simulate real rock pore space. Therefore, the network approaches closer to the reality.

In brief, network model is a spatial network structure consisting of lines, including both straight and curve lines, points, and it can be one-dimensional, two-dimensional or three-dimensional.

In the case of one-dimensional network model, as shown in Fig. 6.27, the pore diameters varies randomly or regularly, being equivalent to a capillary tube with variable diameter along with the direction of flow.

Two-dimensional network, as shown in Fig. 6.28, taking on a shape in some kind of lattice pattern, is formed by crisscrossed column and row lines that are spaced irregularly or symmetrically and is endued with a number of pores, among which the interconnectivity are randomly determined. In such a network wherein the pore space is modeled as lines, the points of intersection of the lines, called notes, represent the junctions of the real rock pores; the line segments between every two notes represents the throats in real rock, and the coordination number is

Fig. 6.27 Diagram of one-dimensional model

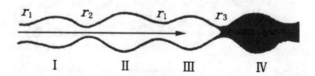

I   II   III   IV

the number of the conduits attached to a note. What should be noticed is that there are no dead pores where all the openings to the conduits are plugged up. The pore size, involving the pore diameter and the pore length, can be designed by people. However, in such a two-dimensional network model only one phase can be continuous while the other phase must be discontinuous. A system of two coexisting continuous phases can be established in a three-dimensional network model.

The proposal of network model is based on full reorganization of the complexity of pore structure. People discovered long ago that a rock pore was usually interconnected with more than one throats, and the pores and the throats is not evenly developed in size no matter investigated in the two-dimensional plane or in the three-dimensional space. With respect to the combination of the pores and the throats, they pair up with each other in infinite variety of forms.

Feith (1956) firstly built up a two-dimensional model by combing the model of bundles of capillary tubes and the ideal soil model. He used regularly shaped cylinder tubes to represent pore space, and the tube-diameter was randomly distributed as pores do. The tubes, designed differently in length and diameter according to specified proportion, were arranged and connected with each other to form a two-dimensional network, where each tube communicated with a certain number of neighboring tubes. Then, Darling developed Feith's model to a three-dimensional one, so as to give a more realistic description of the pore structure of rock.

In order to further clarify the use of mercury injection curve to explain the geometrical characteristic of the pores and the throats involved in rock, Wardlaw (1976) designed a transport two-dimensional model wherein the detailed process of the intrusion of mercury into the pore system, as show in Fig. 6.28c, can be directly observed by naked eyes. In this model, all the intergranular pores are equal in size, but the channels commutating with each pore are of assorted sizes ranging within the six designed values: "6" is the largest one, and "1" is the smallest one. Seen from the model where mercury is being injected only from the left side, the mercury flows along the major throat channels. This example indicates that network models are able to more vividly simulate the seepage flow in porous medium as well as phenomena such as microfingering.

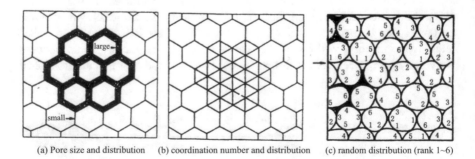

(a) Pore size and distribution   (b) coordination number and distribution   (c) random distribution (rank 1~6)

Fig. 6.28 Diagram of two-dimensional model (by Wardlaw et al. 1978)

Fig. 6.29 Model of
three-dimensional network

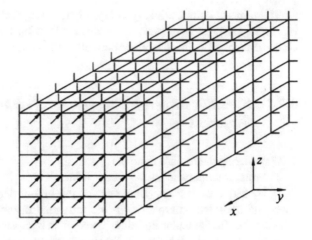

Figure 6.29 shows an ideal three-dimensional network model. Each segment of line in the model represents a single pore. At each note, it is hard to specify which of the connected pores perform inflow or outflow. That is to say, the flow occurring in the model is rather complicated. In fact, as the pores involved in such a model are of various irregular shapes and distributed in various irregular forms, it is almost impossible to establish an analytical relationship like that for the model of buddle of capillary tubes. One of the methods to solve this problem is to give a statistical description, and another is to realize the quantitative forecast for the seepage-flow behavior by the aid of computer programming.

In recent years, with the development of the computer technology and the mathematical statistics, the network model has been further developed. However, limited to the space of this book, the models cannot be discussed in depth. The readers may refer to the relevant literature.

6.7   Sensibility of Sandstone Reservoir Rocks

Due to the presence of certain minerals of sensibility, reservoir rocks may suffer sharp decrease in permeability when subjected to changes of reservoir conditions, which are inevitable during development. Therefore, in order to prevent this kind of reservoir damage, the regular manner in which these vulnerable minerals exhibit their sensibility should be studied and grasped. Before the development of a reservoir, a variety of evaluations and experiments of the mineral sensibility should be conducted to predict the degree to which the reservoir rocks may be damaged when encountered by the extraneous fluids. What's more, special care and necessary measurements should be given to deal with the problem of mineral sensibility during some operations that may cause reservoir damage.

The mineral composition, quantity, distribution, and occurrence of the cementing materials present in rocks have direct effect on the reservoir sensibility. The characteristics of some sensitive minerals will be discussed here.

6.7.1   Sensitive Minerals Present in Sandstone Cementing Materials

1. Water-expandable clay minerals

   1) Structural characteristics of clay minerals

"Clay mineral" is a generic name for the highly dispersed hydrated phyllosilicates and hydrated amorphous silicate minerals. In terms of the structure of clay mineral, the fundamental building block of it is crystal—the lateral stretching of crystals shapes into the crystal sheet; the vertical overlapping of crystal sheets shapes into the crystal layer; and the lateral stretching plus the vertical overlapping of crystal layers shapes into the clay mineral. The four levels involved in the structure of clay mineral can be illustrated as a run-through as follows:

$$\text{crystal} \xrightarrow{\text{laterallystretch}} \text{crystal-piece} \xrightarrow{\text{vertivallyorerlap}} \text{crystal-layer}$$
$$\xrightarrow{\text{laterallystretch} + \text{vertivallyorerlap}} \text{clay mineral}$$

Most of clay minerals are composed of silicon-oxy tetrahedron sheets and alumina-oxy octahedral sheets, and, depending on clay types, how strongly the adjacent sheets are combined together is different from one to another. Among the clay minerals, the sturdiest combination between crystal sheets lies in kaoline, where the sheets are combined by the bonding of OH^- group and O atom; while the feeblest combination lies in montmorillonite, where the sheets are combined by the bounding between O atoms—this is rightly the reason why the water molecules and the cations carried by the water can easily enter the space between the montmorillonite crystal sheets. Besides, the high-valent metal cations within the crystal lattices tend to be replaced by some low-valent metal cations—for example, Al^{3+} may be replaced by Mg^{2+}, and Si^{4+} may be replaced by Al^{3+}—and the charge change caused by this behavioral attribute of clay minerals, coupled with the broken bonds at the edges and corners of the crystal layers, consequently lead up to the charge imbalance of the crystal layers, which enables a certain amount of cations (exchangeable cations) to be absorbed and attach onto the edges and corners and into the space between layers. Apparently, the property of being water-expandable has something to do with the entrance of water into the space between weakly combined crystal layers and, simultaneously, the hydration layer formed due to the absorbed cations.

2) Main crystal, crystal sheet, and crystal layer of clay minerals
(1) Silicon-oxy tetrahedron

The silicon-oxy tetrahedron unit consists of a central silicon ion Si^{4+} surrounded by four equidistantly and closely spaced oxygen ions O^{2-} in the shape of a tetrahedron. The silicon ion shares its charge equally between the four oxygen ions, leaving each oxygen ion with an excess charge of negative one. As any group of three points of them can define a plane, there are totally four planes, living up to its name of "Tetrahedron" (Fig. 6.30).

(2) Crystal sheet of silicon-oxy tetrahedron

Because that the radius of silicon ion ($0.39 \overset{o}{A}$, $1 \overset{o}{A} = 10^{-10}$ m) is much smaller than that of the oxygen ions ($1.40 \overset{o}{A}$), the silicon atom occupies the space opening (interstice) between the oxygen atoms and shares four valence electrons with the four oxygen atom; and, to describe quantitatively, the distance between the oxygen ions (O–O) is $2.61 \overset{o}{A}$, the distance between the oxygen and silicon ions is $1.80 \overset{o}{A}$, and the space available for a tetrahedron unit to display its ions is $0.55 \overset{o}{A}$. As shown in Fig. 6.30, each tetrahedron shares the 3 underside vertex oxygen ions with its adjacent tetrahedrons forming a two-dimensional plane, the crystal sheet of silicon-oxy tetrahedron.

However, the apical oxygen ions, being at another plane, do not link with each other. So, the whole crystal sheet of silicon-oxy tetrahedron can be expressed chemically as $n(Si_4O_{10})$. Now let us consider the array of the tetrahedrons: the bottom surfaces of these polyhedrons are joined together to form a big flat plane by bonding the three corner oxygen ions of each of them, leaving the fourth apical

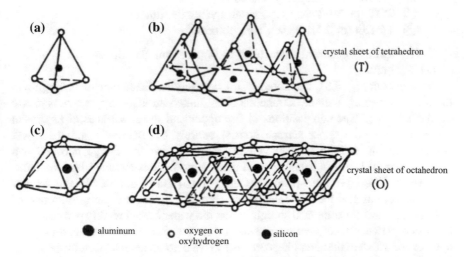

Fig. 6.30 Crystal structure of clay mineral. **a** Single silicon-oxy tetrahedron, **b** silicon-oxy tetrahedron sheet, **c** aluminum (oxyhydrogen)-oxy octahedrons, and **d** aluminum(oxyhydrogen)-oxy octahedrons sheet

vertexes staying on the same side of the plane and thereby the tetrahedrons "pointing" in the same direction—in this manner, the general effect of this array appears as a collection of hexagons linked up into a single stretch, which is called tetrahedral crystal sheet and symbolized as T.

(3) Aluminum-oxy octahedron

As shown in Fig. 6.30, the aluminum-oxy octahedron unit consists of a central aluminum ion Al^{3+} (or magnesium ion Mg^{2+}), surrounded by O^{2-} (or OH^{-}) in the shape of a three-dimensional geometrical polyhedron that has totally eight surfaces.

(4) Crystal sheet of aluminum-oxy octahedron

The sheet is composed of two staggered slices of aluminum-oxy octahedrons with aluminum ions or magnesium ions emplaced between them. As shown in Fig. 6.30, the O^{2-} or OH^{-} of an octahedron are always shared with its adjacent octahedrons, making the upper and lower ranks coordinated together. Obviously, in this manner the octahedrons can be assembled on the same plane, and the sheet can spread infinitely following the same pattern. In consideration of the different metal ions emplaced at the center, the octahedron with aluminum ion is called dioctahedron, while the octahedron with magnesium ion trioctahedron. So, the crystal sheet can be chemically expressed as $n[Al_4(OH)_{12}]$ or $n[Mg_6(OH)_{12}]$, and usually symbolized as O.

(5) Crystal layer

A crystal layer, as basic structural unit of crystal, is a superposition of crystal sheets—a sheet of tetrahedron coupled with a sheet of octahedron, two sheets of tetrahedron with a sheet of octahedron between them, etc.

Clays can be categorized depending on the way that tetrahedral (T) and octahedral sheets (O) are packaged into layers.

① TO (or 1:1 Type): e.g., Kaolinite;
② TOT (or 2:1 Type): e.g., montmorillonite, illite;
③ TOT,O (or 2:1:1 Type): e.g., chlorite.

3) Physical and chemical characteristics of common clay minerals

(1) Kaolinite

As shown in Fig. 6.31, kaolinite is a 1:1 type phyllosilicate mineral belonging to the triclinic system, with one tetrahedral sheet linked through oxygen atoms to one octahedral sheet. The compositions of the upper and lower surfaces of its crystal layer different—the upper surface consist entirely of hydroxyls and the lower surface consists entirely of oxygen ions. Because that the charge between two adjacent crystal layers is balanced, and there is no cations attracted into the interlayer space. The layers, one piled atop another and linked not only by the van der Waals force, but also the hydrogen bonds formed between a certain proportion of OH^{-} groups and are combined so tightly that the water cannot readily permeate into the space between them, and even if the layers are surface hydrated and propped open to some extent that propping force is too weak to contend against the power of cohesion layers of kaolinite. Therefore, kaolinite is a comparatively stable and non-expansive clay mineral which is difficult to be hydrated.

Fig. 6.31 Structure of kaolinite

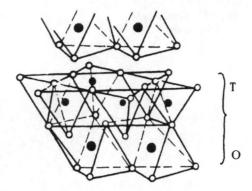

However, the kaolinite is weak to make a stand against mechanical force. In spite of a certain amount of hydrogen bond force, the combination of the kaolinite crystal layers is very weak because that the powerless van der Waals force still plays a dominant role between the layers. As a result, the kaolinite is of low hardness and cleavage character. When facing the swashing of flowing water, the kaolinite, owing to the crack of the cleavages, tends to shatter into scaly particles that may migrate in the formation and decrease the permeability.

Halloysite, another kind of kaolinite that grows from natural hydration, has sheets of water molecules among the laminated crystal layers, making the distance between the layers become greater. What should be mentioned is that these molecules do not associate with interlayer cations, which is different from the case of montmorillonite. Although this kind of clay mineral also readily break and migrate before hydration, the water molecules attached to it during hydration can be removed away in an environment of 60 °C plus or longtime drying. So, in dealing with a core sample of halloysite, the drying temperature should be kept at 60 °C, and the relative moisture should be kept at 45–50 %.

(2) Montmorillonite and illite

Figure 6.32 shows the structure of a montmorillonite crystal layer. The structural unit of the montmorillonite, crystal layer, consists of two tetrahedron sheets with a A1–O–OH octahedron sheet between them, which can be expressed briefly as TOT. The characteristic of this structure is that the high-valent metal ions (Al^{3+}, Si^{4+}) within the polyhedrons can be partly replaced by some low-valent metal ions (Mg^{2+}, Ca^{2+}, Na^+, etc.). For example, the trivalent Al^{3+} may be replaced by the bivalent Mg^{2+}, and the quadrivalent Si^{4+} may be replaced by the monovalent Na^+. Result of the exchange of cations: (1) the lost positive charges cause the clay to be negatively charged and, for charge balance, need some exchangeable cations that can be attached onto the appearance surface of the crystal or into the interlayer space between the crystal layers; (2) the uneven charge distribution resulting from the exchange of cations avail further exchange of cations to happen more easily.

Besides, as each crystal layer of montmorillonite faces its adjacent layer with only O ions, there is only Van de Waals force between them to combine the two together, and, thanks to the weak combination, those exchangeable cations, in

Fig. 6.32 Structure of
montmorillonite

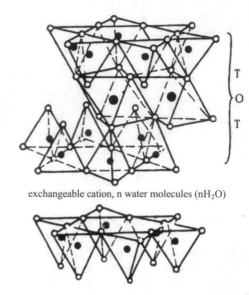

exchangeable cation, n water molecules (nH_2O)

companion with which there are also abundant water molecules and some other polar ions, are allowed an entrance into the interlayer space of the clay. Once a montmorillonite comes into contact with water solution, the physical and chemical properties of it will change more or less, depending on the ion composition and concentration of the water solution. The most important consequent change is that, as along as there are small-radius monovalent exchangeable cations present in the water solution, the water will continue entering the interlayer space and giving occasion for further hydration. In this case, the interlayer distance will be increased by 13–20 times. The hydrate and thereby swelled, cracked montmorillonite crystal layers usually split along with the cleavages and go apart from each other to take on a bended, thin, and scaly shape of appearance. As this shape of appearance is incapable of impeding coagulation, the montmorillonite usually migrate with the flowing fluids as agglomerates, no matter what kind of an environment (acid or alkaline) it is in.

Although illite belongs to the same category of clay minerals with montmorillonite, it has much smaller cation exchange capacity than the latter one by virtue of a kind of strong bond in it structure, which is formed in the presence of K^+ and plays a quite important role in preventing water from entering the interlayer space. And, as a result, illite is an inexpansible clay mineral. As shown in Fig. 6.33, the reason why illite does not contain interlayer water is that the size of potassium ion is equivalent with the size of the interstice present in the ditrigonal network, and, consequently, it can not enter the interstice space carrying with accessory water. In addition, because the combing force of the potassium-to-oxygen (K–O) bond is stronger than that of the hydrogen bond, the crystal layers of an illite are combined much more tightly, and it is very difficult for it to exchange its interlayer potassium ions with the cations in the outside solutions.

Fig. 6.33 Structure of illite

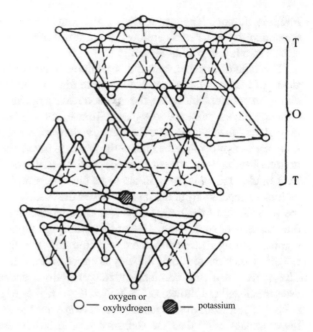

○— oxygen or oxyhydrogen   ◉— potassium

(3) Chlorite

Chlorite is also inexpansible because of its strong combination force between crystal layers. It is noteworthy that there are usually considerable amounts of Fe^{2+} and Fe^{3+} in chlorite, and these ions will react with acids and produce precipitates which may hurt the formations. So, engineers have tangible reasons to take it seriously.

4) Mechanism of the instability of clay minerals

When encountering water, clay minerals may suffer some changes because of their instability, for example, expansion, dispersion, and flocculation.

(1) Expansion of clay minerals

The change that happens to the clay mineral contacted with water can be divided into two stages:

① Surfacial hydration (also called interlayer expansion)

During this process, the main impetus is the absorption energy of the water layered at the clay surface; and, the absorbance of water, the orientation, and thickness of the oriented water-film are all decided by the hydratability of the interlayer cations and the surface electric charge density at the clay surface. As proven by researches, in the case that the exchange cations of a montmorillonite are Ca^{2+}, Mg^{2+}, or H^+, the attraction among clay crystal layers becomes stronger, the water-film becomes thinner, and the orientation of the water molecules become more regular; while in the case that the exchange cation of a montmorillonite is Na^+, the attraction among clay crystal layers becomes weaker, the water-film becomes thicker, and the orientation of the water molecules become more irregular. That is to say, sodium montmorillonite has greater expandability than calcium montmorillonite. The sequence of different clay minerals according to their expandability is listed as follows: montmorillonite > mixed layer clay containing expansible layers > illite > Kaolinite.

The relationship between the superficial hydratability and the mineral composition of clay can be interpreted with the aid of "exchangeable cation." When the clay is in contact with water, the interlayer exchangeable cations trapped in the clay structure have a tendency to separate from it, leaving behind the negatively charged clay sheets that would be consequently dispersed because of the repulsive force between each other and then allow the water to invade in. Note that the kaolinite has no interlayer cations, so it does not expand when dispersed in water. When the same kind of clay minerals is put in the same environment, the tendency of the interlayer cations to separate from the layer is under the control of the law of mass action and the magnitude of the ionic valence, and dependent upon the adsorption energy of the exchangeable cations. Generally speaking, higher ionic valence of cations signifies stronger attraction force with the clay structure and weaker tendency to escape, which decides that the clay is more difficult to be hydrated and dispersed. The sequence of the common interlayer cations according to their tendency to escape is listed as follows: $Li^+ > Na^+ > K^+ > Rb^+ > Mg^{2+} = Ca^{2+} = Sr^{2+} = Ba^{2+} = Cs^{2+}$. Seen from the sequence, the reason why the calcium montmorillonite has lower expansibility than the sodium montmorillonite is that Ca^{2+} has weaker tendency than Na^+ to escape from the interlayer space of montmorillonite, owing to its greater absorption energy than the latter one.

② Osmotic hydration (outside surface hydration)

If the salinity of the extraneous fluid is lower than that of the formation water, the higher ionic concentration at the clay outside surface will attract the water molecules contained in the extraneous fluid to approach the appearance surface of the clay, which will subsequently form an oriented water-film there and increase the electrostatic force of the electric double layer. As the electric double layers around the particle surfaces repel each other because of their alike electric charge, the clay particles tend to push each other away, making the clay trending to expand. Therefore, the dominant factor that controls the formation of the hydrated shell covering the clay surfaces is the osmotic equilibrium state of the semipermeable membrane formed around the clay particles. In other words, it is the osmotic hydration that leads up to clay expansion.

In conclusion, the type of the clay mineral and the composition of the exchangeable cations, as well as the ionic composition and concentration of the aqueous medium are the main influencing factors for the surface hydration and osmotic hydration of clay minerals. Taking the montmorillonite as example, when it comes across a low-salinity extraneous water, it will firstly undergo the surface hydration, and the expansion of the crystal layers will make the clay volume doubled; subsequently, what's going on to occur is the osmotic hydration, which will make the clay volume increase manyfold. In fact, montmorillonite is the most expansive clay mineral.

(2) Flocculation and dispersion

Flocculation and dispersion are another two important characteristics of the clay-water system. When the clay particles in the aqueous medium have a tendency to assemble and to form into conglomerates, we can say that the clay is in a state of flocculation, and when these conglomerates split and disassemble, we can say that the clay is in a state of dispersion. The flocculation and dispersion can be explained by the theory of electric double layer which is shown in Fig. 6.34.

Figure 6.34 shows that in an aqueous solution containing electrolytes there are coexist positively charged cations and negatively charged anions (OH⁻), by virtue of the electro ionization. When this kind of solution is encountered by clay, the cations in it will be exchanged with the cations trapped in the clay crystal structure, which is known as "cation exchange"; at the same time, the OH⁻ will be absorbed onto the appearance surface of clay where the excess positive charges lie, coat the clay surface, and form a negative layer of OH⁻ there, and, then, an equivalent amount of positive charges will naturally approach the negative layer to balance the

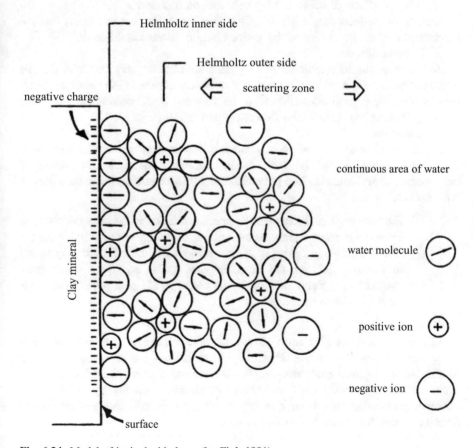

Fig. 6.34 Model of ionic double layer (by Fink 1981)

charge, and finally form a positive layer of cations outside it. If the balancing cations are far away from the clay, the clay tends to be dispersed by the repulsion force of the negative charges coating the clay particles; while if the balancing cations are near to the clay, the powerful attraction between the positive and negative charges will cancel the repulsion force between the clay surface-negative layers and consequently lead to the flocculation of the particles.

As proved by researches, when the exchangeable cations of the clay or the electrolytes are Ca^{2+}, Mg^{2+}, or H^+, the clay prefers to undergo flocculation, and when the exchangeable cation is Na^+, the clay prefers to undergo dispersion, and, what's more, an increase in the electrolyte concentration in the aqueous solution will enhance the tendency of the clay to be stable or flocculate.

In normal cases, the clay is in an equilibrium state with the equation water, which is referred to as the state of flocculation. However, in the case that the extraneous fluid has rather low salinity or contains dispersive cations like Na^+, the clay minerals will be forced to undergo dispersion and migration that do great harm to the formation.

5) Potential effects of different clay minerals on formations

Seen from the preceding analysis of the clay instability, we can know that the mechanisms of the instability of the various clay minerals are different.

(1) Montmorillonite

The damage to formation caused by montmorillonite clay cements can be attributed to its extremely great sensibility to water. Especially, the sodium montmorillonite may expand 600–1000 % in its volume when encountered by water, resulting in sharp decrease in the formation permeability.

(2) Kaolinite

As the most common and most abundant clay cement present in sandstone formations, kaolinite usually pack the intergranular space in the shape of book-pages, vermiform, etc. The potential effect of kaolinite on formations lies in two aspects:

① Fill the original intergranular space to make it become much finer intercrystal pores which makes little contribution to the permeability.

② The kaolinite aggregates with poor adhesion to the solid are easy to fall into pieces and separate from the rock grains under the shear stress caused by the fluids flowing in the pores, resulting in pore/throat clogs by the kaolinite minerals.

(3) Illite

Illite is the most complex form of clay minerals. Illite can be divided into two categories in terms of its cause of formation: in a scale-like conformation and in a fibrous hair or streaked conformation. The former one is usually distributed across grain surface, influencing the permeability of formation mainly by reducing the effective pore radius; the latter one has much more complicated impact on the formation, and the impact lies in three aspects:

① The fibrous, hair-like illites in the pore space staggered so that the original intergranular pores are transformed into a large number of fine pores, leading to much more complicated pore structure and thus a significant reduction in the reservoir permeability.

② These illites have considerably large specific surface area and a great amount of adsorbed water, so the rock has high connate water saturation.

③ Subjected shear stress due to the fluids, the illite breaks easily and is ready to be transported to the throat constrictions where clogs may be caused.

(4) Chlorite

Chlorite is often adsorbed on the grain surfaces in a willow-like form or by filling the pores in the form of button-like aggregates, also growing in the pore spaces in a bridge-like pattern. Chlorite present in petroleum reservoirs is an iron-rich clay mineral. Chlorite dissolves in presence of acid and releases iron until the acid is exhausted, with colloidal ferric hydroxide precipitation left in the formation. As these three-valent iron colloid particles are comparatively large in size, it is easy for them to plug the pore throat and cause formation damage.

6) Swelling capacity of clay

The "swelling capacity" refers to the degree of expansion of clay. It is the ratio of the expansion volume of clay to its original volume. The swelling capacity of clay depends not only on the properties of clay, but also the characteristics of water. Usually, as shown in Fig. 6.35, fresh water causes the most formidable expansion of clay.

Determination of the degree of clay swelling usually resorts to the clay dilatometer shown in Fig. 6.36. It consists of micrometering tube and a semipermeable diaphragm funnel below which is the liquid (e.g., water) awaiting to be

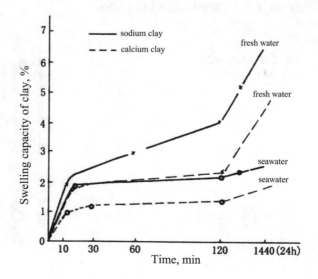

Fig. 6.35 Relationship between expansion of clay and water quality, time

Fig. 6.36 Clay dilatometer

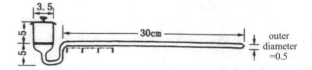

tested. A filter paper is laid upon diaphragm, and in test, a certain amount of clay (about 0.1 g) is placed on the filter paper. The volume of the absorbed liquid is read from the metering tube, while the expansion volume of the clay is read from the graduated funnel.

2. Sulfate cements and their dehydration property

The sulfate minerals involved in reservoir rocks are mainly gypsum ($CaSO_4 \cdot nH_2O$) and anhydrite ($CaSO_4$). One of the most striking features of sulfate minerals is their dehydration at high-temperature, which affects the physical-property analysis of reservoir rocks in the measurement of saturations. Sulfate content more than 5 % in reservoir rock is enough to significantly affect the value of the irreducible water saturation.

When gypsum is heated, there will be crystal water precipitating from the gypsum as long as the temperature reaches a certain level (e.g., 64 °C), although very slowly. When the temperature exceeds 80–100 °C, the crystal water will precipitate quickly. As the temperature continues to rise, the precipitation of the cystal water will fold in speed. Figure 6.37 gives the relation curve between the dehydration of gypsum and temperature.

Therefore, the water saturation measured from in-lab tests such as an extraction by the toluene with a boiling point of 105 °C is usually higher than it really is. What's more embarrassing is that sometimes the total fluid saturation obtained from these tests exceeds 100 %. For example, it is been confirmed that the dehydration of gypsum and clay minerals is one of the reasons for the overestimated water saturation in the Yumen oilfield of China.

Fig. 6.37 Dehydration property of gypsum

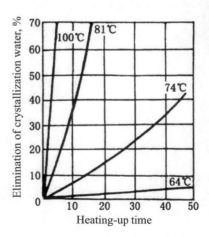

In the core analysis process, two measures can be taken to prevent the error caused by the dehydration of gypsum at high temperatures: ① release the oil and water out of the core by the aid of the centrifugal force generated from the rapidly rotating centrifuge; ② Prevent the gypsum from dehydrating by using a mixed solution of chloroform and methanol prepared according to a quality proportion of 13:87, which has a boiling point (53.5 °C) lower than the dehydration temperature of gypsum.

3. Limy cements and their characteristics

Limy cement is mainly composed of carbonate minerals, such as limestone (CaCO3), dolomite (CaMg $(CO_3)_2$), sodium (Na_2CO_3), potassium salt, (K_2CO_3), and siderite ($FeCO_3$). The common carbonate minerals present in sandstone are calcite $CaCO_3$ and dolomite CaMg (CO_3) $_2$. These carbonate minerals in sedimentary rocks can be divided into two categories: primary and epigenetic. The content of carbonate rocks vary greatly among different oil fields but ranges between 9.5 and 21 % in the cement materials. A carbonate reservoir is wholly formed by the carbonates. But there are also cases of rather small content of carbonates, for example, quartz sandstone. The content of the carbonates not only affects the physical properties of the hydrocarbon bearing rocks, but also impacts the effects of the stimulation treatments.

1) Significance of the study of limy cement

(1) Analysis of the content of the carbonate minerals in rock, especially the content of the epigenetic carbonate minerals, is an available approach to obtain information about the activity patterns of a formation's hydrodynamic field and about the characteristics of the age when the formation was born.

Studies have shown that there is a good corresponding correlation between the carbonate content and the porosity of a formation. Because of the intense activities of underground water, formations suffer intrusion of acidic and thus erosion of the carbonate minerals involved in them. The parts where water flows with comparatively greater rate usually have more carbonate minerals to be dissolves and less to be left, finally with large porosity developed; on the contrary, the parts where water flows with smaller rate have more carbonate minerals to be left and thus small porosity developed.

(2) One of the characteristics of carbonate minerals is to react with the acid, and that is why they are called acid-reacting minerals. Most of the carbonate minerals react with acid with no precipitation of sediments, so they will not cause harm to formation but lead to increase in permeability. In production engineering, sometimes acid is deliberately crowed into the formation to realize effective innovation of formation, so as to improve the production capacity of oil wells and the injection capacity of water wells.

The nature of such processes is to make use of the reaction between acids and salts, as a result of which the carbonate minerals are dissolved and thus the porosity and permeability of formation is increased. Whether an acidification should be adopted for a formation depends primarily upon the carbonate content of it. For the conventional sandstone formations, these with carbonate content of greater than 3 % are allowed to be modified by acidification.

(3) Protection of formation

The so-called acid-sensitive minerals refer to the minerals that generate sediments as a result of their reaction with acids, which leads to plugs of pore-channels and lowered permeability of formation. Acid-sensitive minerals in cement include the iron-rich chlorite, the pyrite (FeS_2), and the siderite ($FeCO_3$).

Siderite and acid react to produce sediments. Their reaction is:

$$FeCO_3 + 2HCl = FeCl_2 + H_2O + CO_2 \uparrow$$
$$2FeCl_2 + 3H_2O + 3O = 2Fe(OH)_3 + 2Cl_2 \uparrow$$

Iron-rich chlorite and acid react and resultantly cause expansion of clay. Besides, as the iron-rich chlorite is dissolved in acid, the Fe^{+2} will exist as colloidal precipitation when the pH value ranges from 5 to 6, which causes formation damage. Then, Fe $(OH)_3$, a flaky crystal which is relatively large in size, tends to plug the pore-throats of rock. The related reactions are listed below:

$$3FeO \cdot Al_2O_3 \cdot 2SiO_2 \cdot 3H_2O + 6HCl = 3FeCl_2 + 2Al(OH)_3 + 2SiO_2 \cdot 3H_2O$$
$$2FeCl_2 + 3H_2O + 3O = 2Fe(OH)_3 + 2Cl_2 \uparrow$$

In order to prevent the precipitation of Fe^{2+} and Fe^{3+} in formation, additives such as iron chelating agent and deoxidizers can be added to the working liquid of acidification, so that the formation is protected.

2) Measurement of the carbonate content

Measurement of the carbonate content in formation should be in accordance with the Oil Industry Standard SY/T 5336-2006. The reaction is:

$$CaCO_3 + 2HCl = CaCl_2 + H_2O + CO_2 \uparrow$$

Determination the carbonate content in rock sample is based on the volume of CO_2 released from the reaction and the corresponding pressure. According to the volume of CO_2, the content of all the carbonates is converted $CaCO_3$ content. The formula of computation is:

$$C = \frac{V\rho}{0.44W} \times 100\% \tag{6.72}$$

where

C content of $CaCO_3$ in rock sample, %;
V volume of CO_2 released from the reaction between the sample and HCl, cm^3;
ρ density of CO_2 at room temperature and atmospheric pressure, g/cm^3;
W weight of rock sample, g.

If there is $MgCO_3$ or $CaMg(CO_3)_2$ contained in rock, the chemical reactions are:

$$MgCO_3 + 2HCl = MgCl_2 + H_2O + CO_2 \uparrow$$
$$CaMg(CO_3)_2 + 4HCl = CaCl_2 + MgCl_2 + H_2O + CO_2 \uparrow$$

The content of carbonate minerals differs greatly among different sandstone formations. Some formations do not contain carbonates at all, while some others may have carbonate content higher than 20 %.

6.7.2   Evaluation for Formation Sensibility

During all of the links involved in reservoir exploration and development—well drilling, cementation, completion, perforation, stimulation treatments, work over operations, water injection operations, etc.—the formations inevitably come into contact with extraneous fluids and the solid particles contained in them. The term "formation damage" refers to the alteration of virgin characteristics of a formation, usually caused by the exposure to extraneous incompatible fluids and leading to decreased permeability as well as reduced production capacity. So, taking into account all of the sensibilities of the payers, system evaluations must be taken to protect the oil or gas-bearing formations so as to bring their potential of production into full play. Here, we will focus on the evaluation techniques for clastic formations.

Formation sensibility consists of velocity sensibility, water sensibility, acid sensibility, salt sensibility, etc. The sensitivity characteristics of the minerals contained in formation and the potential harms of them are shown in Table 6.5. The evaluation of formation sensibility includes petrologic analysis, conventional core analysis, special core analysis, and some core flow tests.

Petrological analysis of rock and conventional core analysis are conducted to yield information about the original state of formation, involving the lithology, mineral composition, pore distribution, rock permeability, cement composition, clay content, rock type, etc., so that the potential factors of formation damage can be speculated and the possible degree of damage can be qualitatively judged. Based on them, core flow tests are performed to measure the degree of the various sensitivity of formation, including the speed sensitivity, the water sensitivity, the salt sensitivity, and the acid sensitivity. Then, compatible fluids are found for the formation on the basis of system evaluation and finally, after a comprehensive study, design proposals are given for the drilling, the well completion, and the stimulation treatments.

The following section will present the core flow test for sensitivity evaluation. The test procedure is shown in Fig. 6.38.

1 Evaluation test for velocity sensibility

In formation, uncemented or weakly cemented clay and debris particles of diameter <37 μm are collectively called formation fines. They flow in the pore

Table 6.5 Sensibility of reservoir minerals (by Shen Pingping 1995)

Sensitive mineral	Degree of sensibility	Potential sensibility	Conditions for the occurrence of sensibility	Method to inhibit sensibility
Montmorillonite	Maximum	Water sensibility	Fresh water system	High salinity fluid
	Moderate	Velocity sensibility	Fresh water system, high flow rate	Acid treatment
	Moderate	Acid sensibility	Acidification	Inhibitor for acid sensibility
	Moderate	Alkaline sensitivity	Chemical flooding	Addition of inhibitor
Illite	Maximum	Velocity sensibility	High flow rate, fresh water system	Low flow rate
	Moderate	Block of small pore	Hydrofluoric acidification	High salinity fluid, anti-swell agent
	Minimum	K_2SiF_6 precipitation		Inhibitor for acid sensibility
Kaolinite	Moderate	Velocity sensibility	High flow rate, high PH value High transient pressure	Stabilizing agent, low flow rate Low transient pressure
	Moderate	$Al(OH)_3$ precipitation	Acidification	Inhibitor for acid sensibility
	Moderate	Alkaline sensitivity	Chemical flooding	Addition of inhibitor
Chlorite	Maximum	$Fe(OH)_3$ precipitation	Oxygen-rich system	Deoxidizer
	Moderate	MgF_2 precipitation	High PH hydrofluoric acidification	Inhibitor for acid sensibility
	Minimum	Velocity sensibility	high flow rate, high pH value	Low flow rate
Carbonates	Moderate	CaF_2 or MgF_2 precipitation	Hydrofluoric acidification	Acid preflush before hydrofluoric treatment Inhibitor for acid sensibility
Ferrodolomite	Moderate	$Fe(OH)_3$ precipitation	High PH value, oxygen-rich system	Inhibitor for acid sensibility, deoxidizer
Pyrite	Moderate	Sulfide precipitation	Fluid containing Ca^{2+}, Sr^{2+}, Ba^{2+}	Anti-scale agent

(continued)

Table 6.5 (continued)

Sensitive mineral	Degree of sensibility	Potential sensibility	Conditions for the occurrence of sensibility	Method to inhibit sensibility
Siderite	Minimum	Fe(OH)$_3$ precipitation	High PH value, oxygen-rich system	Inhibitor for acid sensibility, deoxidizer
Mixed layer clay	Moderate	Velocity sensibility	High flow rate	Low flow rate
	Moderate	Water sensibility	Fresh water system	High salinity fluid, anti-swell agent
	Minimum	Acid sensibility	Acidification	Inhibitor for acid sensibility
Feldspar	Minimum	Fluorosilicate precipitation	Hydrofluoric acidification	Inhibitor for acid sensibility
Uncemented quartz grains	Moderate	Velocity sensibility	High flow rate	Low flow rate
			High transient pressure	Low transient pressure

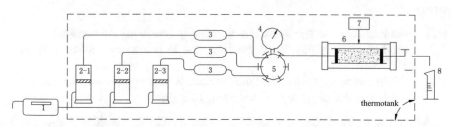

1-pump   2-1formation water   2-2 sub formation water   2-3 deionized water   3-filter   4-pressure gauge

5-six-way valve   6-core holder   7-confining-pressure pump   8-thermotank

Fig. 6.38 Evaluation test for formation sensibility

space with fluids and tend to accumulate at throat constrictions, resulting in "bridge block" there and thus reduced formation permeability.

Evaluation test for velocity sensitivity is aimed at discovering the relationship between the variation of the formation permeability and the seepage-flowing velocity. If the studied formation is sensitive to velocity, the critical flow rate (V_c), the rate at which the formation begins to exhibit abnormal phenomena due to over-rapid flow, must determined in the test. The critical flow rate provides a basis for the determination of the injection rate and the reasonable production capacity of oil wells.

The steps of an evaluation test for velocity sensibility are as follows: formation water is injected into the experimental core with different flow rates in turn (0.1, 0.25, 0.5, 0.75, 1.0, 1.5, 2.0, 3.0, 4.0, 5.0, 6.0 ml/min), and the permeability of the core at each injection rate is measured and determined in steady state. The permeability and the injection rate data are plotted on the vertical axis to form a relation curve. As shown in Fig. 6.39, the abrupt change of the curve corresponds to the critical velocity V_c after which the permeability decreases sharply with the increase in flow rate, indicating that the rock is being sensitive to the flow rate. Since the critical flow rate has been found out, the designed gap between the subsequently experimented flow rates is allowed to be larger. If there is no critical flow rate measured, the test is ended at the flow rate of 6.0 cm^3/min.

The build up of "Bridge block" depends upon not only the size and quantity of the formation fines, but also the size of the throats involved in the formation. At large flow rates, some particles may be exported from the rock sample with the fluid, so the core shows gradual increase in permeability later in the test (Fig. 6.39).

Velocity sensitivity is quantitatively evaluated according to a variety of indexes. Here two of them will be introduced.

1) Permeability damage index

The permeability damage index due to velocity sensibility is defined as:

$$D_{KV} = \frac{K_L - K_{Lmin}}{K_L} \tag{6.73}$$

where

D_{KV} permeability damage index due to velocity sensibility, dimensionless;

K_L original permeability when the flow rate is below the critical flow rate, 10^{-3} μm^2;

K_{Lmin} minimum permeability when the flow rate is larger than the critical flow rate while smaller or equal to 6 cm^3/min, 10^{-3} μm^2.

As shown in Table 6.6, the level to which a velocity-sensitive formation is damaged in terms of permeability can be evaluated according to the permeability damage index.

2) Velocity sensibility index

The velocity sensibility index is the ratio of the permeability damage index due to velocity sensibility to the critical flow rate of the core:

Fig. 6.39 Curve of evaluation test for water sensibility (by Shen Pingping 1995)

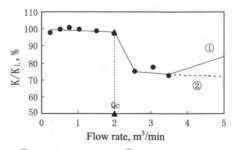

①-with fines washed out ②-without fines washed out

Table 6.6 Degree of permeability damage due to velocity sensibility

Degree of permeability damage	Permeability damage index
Severe	$D_{KV} \geq 0.70$
Moderate to severe	$0.70 > D_{KV} \geq 0.50$
Moderate to poor	$0.50 > D_{KV} > 0.30$
Poor	$0.30 \geq D_{KV} > 0.05$
Negligible	$D_{KV} \leq 0.05$

$$I_V = \frac{D_{KV}}{V_c} \tag{6.74}$$

where

I_V velocity sensibility index, d/m;
V_c critical rate, m/d.

Taking into consideration both the critical flow rate of the core and the permeability damage index, the velocity sensibility index offers evaluation for velocity sensibility of formation more reasonably. Table 6.7 gives the details.

2. Evaluation test for water sensibility

The phenomenon of water sensibility refers to the decrease in permeability caused by the clay expansion, dispersion, and migration after the entrance of some extraneous fluids that are incompatible with the formation. Before coming into contact with the extraneous fluids (e.g., the injected water), the clay minerals under reservoir conditions have been in a relative equilibrium with the formation water. However, once the clay minerals encounter the extraneous fluids that possess low salinity, they will expand and lead the formation permeability to be lowered.

American scholar Moore found that, generally speaking, the layers bearing a clay content of 1–5 % have the best physical properties, while the layers bearing clay content more than 5–20 % have poor physical properties. Especially, if water-sensitive clay minerals are involved in a formation, blocks of the rock pores and throats may be suffered given that the clay content is too high.

Evaluation of the use of water-sensitive index of core samples of the water-sensitive. Water sensitivity index is defined as follows:

$$I_W = \frac{K_L - K_W}{K_L} \tag{6.75}$$

Table 6.7 Evaluation of velocity sensibility with the velocity sensibility index

Degree of velocity sensibility	Velocity sensibility index
Severe	$I_V \geq 0.70$
Moderate to severe	$0.70 > I_V \geq 0.40$
Moderate to poor	$0.40 > I_V > 0.10$
Poor	$I_V \leq 0.10$
Negligible	$I_V \leq 0.05$

where

I_W Water sensitivity index, dimensionless;

K_W permeability to de-ionized water, $10^{-3}\mu m^2$;

K_L liquid permeability of the core void of chemical and physical reactions such as hydrous expansion, $10^{-3}\mu m^2$.

According to the grading standards of the Marathon Oil Company, the degree of water sensibility is ranked as shown in the Table 6.8.

3. Evaluation test for salt sensibility

It was found that sensitivity for water-sensitive formations a drop in salinity will result in heightened crystal layer expanding and aggravated expansion of clay minerals, which directly lowers the permeability of formation. Therefore, evaluation tests must be carried out to shed light on the variation regularity of permeability following the gradual decrease in the salinity of formation water or the invasion of the extraneous fluids of low salinity. Based on such tests, the critical salinity for a formation can be determined.

The principle of the evaluation test for salt sensitivity is as follows: salt waters of different salinities are injected into the core following an order from the high salinity to the low salinity, and for each experimented salt water, the permeability is measured. The K/K_{ws} and the permeability data are plotted on the vertical axis to form a relation curve as shown in Fig. 6.40. K_{ws} is the permeability to the water with the highest salinity. The salinity at the sudden change of the curve is rightly the critical salinity, denoted by C. The determination of the critical salinity provides a reference for the preparation of the fluids used for field operations.

The salt sensitivity is a measure for the formation's ability to tolerate low-salinity fluids, and the critical salinity C (unit: mg/L) represents the formation's degree of salt sensibility. Table 6.9 shows the evaluation for the degree of salt sensibility.

For the same formation, seen from the table, the critical salinity for the compound salt water (e.g., standard salt) is usually lower than that for the mono salt water (e.g., NaCl). The so-called compound salt water is prepared by adding some salts into the distilled water to establish specified mass concentrations (7.0 % NaCl, 0.6 % $CaCl_2$, 0.4 $MgCl_2 \cdot 6H_2O$ and 92 % distilled water).

Table 6.8 Ranking of degree of water sensibility

Degree of water sensibility	Water sensibility index
Negligible	$I_W \leq 0.05$
Poor	$0.05 < I_W \leq 0.30$
Moderate to poor	$0.30 < I_W \leq 0.50$
Moderate to severe	$0.50 < I_W < 0.70$
Severe	$0.70 \leq I_W < 0.90$
Extraordinarily severe	$I_W \geq 0.90$

Fig. 6.40 Curve of evaluation test for salt sensibility (by Shen Pingping 1995)

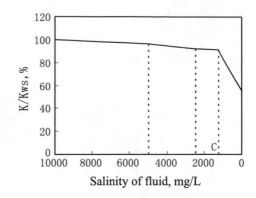

Table 6.9 Evaluation for degree of salt sensibility

Salinity sensibility evaluated by standard salt water (compound salt)		Salinity sensibility evaluated by NaCl salt water	
Degree of salinity sensibility	Critical salinity (mg/L)	Degree of salinity sensibility	Critical salinity (mg/L)
Poor	$C \geq 1000$	Poor	$C \leq 1000$
Moderate to poor	$1000 < C < 2500$	Moderate to poor	$1000 < C < 3000$
Moderate	$2500 \leq C \geq 5500$	Moderate	$3000 \leq C \geq 7000$
Moderate to severe	$5500 < C < 8000$	Moderate to severe	$7000 < C < 10,000$
Severe	$8000 \leq C < 25,000$	Severe	$10,000 \leq C < 30,000$
Extraordinarily severe	$C \geq 25,000$	Extraordinarily severe	$C \geq 30,000$

4. Evaluation test for acid sensibility

The phenomenon of acid sensitivity refers to the decrease in permeability as a result of the reactions between the acid-sensitive minerals in formation and the extraneous acid liquid, which produce gels, precipitates, or disbanded fines.

Evaluation tests for acid sensitivity must be performed before the acidification of a formation is carried out, so as to learn whether the acid liquid will damage the formation and the degree of the potential damages.

Acid liquid of a volume of 0.5–1.0 PV is injected into the core, and the reactions between this and the acid-sensitive minerals in the rock are permitted adequate time before the residual acid is discharged. In the test, the permeability of the core is measured, respectively, before and after the injection of acid.

Evaluation test for acid sensitivity, aimed at simulating the damage-prone parts of a formation and determining its degree of acid sensibility, is different from the evaluation test for acidification effect. Therefore, the former one requires a small amount of acid liquid (0.5–1.0 PV) to be injected while the latter one requires a much larger amount (more than 5 PV). This is because that, when such an amount of acid liquid is injected into the core in a evaluation test for acid sensitivity, the pH value of the system increase rapidly as the acids are being consumed in reactions, allowing the precipitates to form themselves rather readily.

Chapter 7
Other Physical Properties of Reservoir Rocks

With the deepening of our discuss about the physical properties of the underground formations, some other characters in addition to the already introduced basic parameters (e.g., the particle-size composition, the surface ratio, the porosity, the compressibility, the saturation, and the permeability) should be involved in this book, so as to present a more overall introduction of the reservoir rocks. Here, the knowledge to be introduced includes the rock electrical conductivity, thermal properties, and radioactivity.

7.1 Electrical Conductivity of Fluids-Bearing Rocks

The electrical conductivity of rocks is closely related to the rock lithologic characters, the oil-storing properties, and the saturation. Results from the studies about the electrical conductivity of rock help in charactering rock, segmenting the different layers of oil, water, and gas, providing more information about the formation physical parameters such as permeability and porosity, and many more other aspects. The property of electrical conductivity of rock is already used in laboratory measurement of fluid saturations and resistivity logging.

7.1.1 Electrical Conductivity and Resistivity of Material

The resistivity of a material is the reciprocal of conductivity and is commonly used to define the ability of a material to conduct current. The resistivity of a material is defined by the equation as follows:

© Petroleum Industry Press and Springer-Verlag GmbH Germany 2017
S. Yang, *Fundamentals of Petrophysics*, Springer Geophysics,
DOI 10.1007/978-3-662-55029-8_7

$$\rho = \frac{rA}{L} \tag{7.1}$$

where

ρ resistivity, Ω m;
r resistance, Ω;
A cross-sectional area of the conductor, m^2;
L length of the conductor, m.

The electrical resistivity ρ is a material property being independent of the geometrical shape of the conductor. As a more characteristic and accurate property of each material than the resistance, resistivity is useful in comparing various materials on the basis of their ability to conduct electric current.

Porous rock is comprised of rock fragments and void space. According to their different principles of electrical conductivity, the porous rocks can be divided into two categories: rocks of ionic conduction and rocks of electronic conduction.

The rocks of ionic conduction, for example, the sedimentary rocks, are capable of conducting electric current because of the positive and negative ions contained in the solutions present in the interconnected pores and channels. Their electrical resistivity depends on the property and concentration of the conducting solutions.

The rocks of electronic conduction, for example, most of the igneous rocks, are capable of conducting electric currents by virtue of their freely moving electrons owned by the rock minerals. Their electrical resistivity depends on the property and amount of the conducting minerals.

Most of the rock formations involved in the petroleum exploration and development belong to the category of rocks of ionic conduction. Sedimentary rocks are comprised of mineral solid grains which are referred to as the rock skeleton. The solids, with the exception of certain clay minerals, are nonconductors of electricity. The electrical properties of a rock depend on the geometry of the voids and the fluids with which these voids are filled. Among the fluids of interest in petroleum reservoir, the oil and gas are nonconductors, while the formation water containing dissolved salts is a conductor. Current is conducted in water by movement of ions and can thus be termed electrolytic conduction. Therefore, a sedimentary rock's capacity to conduct electricity depends on the electrical conductivity of the formation water involved in the rock pore system.

Table 7.1 shows a list of information about the resistivity of the commonly encountered minerals and rocks. It is seen that different minerals and rocks are significantly different in resistivity; with exception of the metallic minerals, the dominant rock forming minerals, such as quartz, feldspar, mica, and calcite, have very high resistivity. From the perspective of the origin of rocks, most of the igneous rock (e.g., basalt and granite) have high electrical resistivity, while the sedimentary rocks have comparatively low electrical resistivity.

Table 7.1 Electrical resistivity of some rocks and minerals

Name	Resistivity (Ω M)	Name	Resistivity (Ω M)
Clay	$1-2 \times 10^2$	Anhydrite	$10-10^6$
Mudstone	5–60	Quartz	$10^{12}-10^{14}$
Shale	10–100	White mica	4×10^{11}
Unconsolidated sandstone	2–50	Feldspar	4×10^{11}
Compacted sandstone	20–1000	Oil	$10-10^{16}$
Oil-/gas-bearing sandstone	2–1000	Calcite	$5 \times 10^3-5 \times 10^{12}$
Coquinoid limestone	20–2000	Graphite	$10^{-6}-3 \times 10^{-4}$
Limestone	50–5000	Magnetite	$10^{-4}-6 \times 10^{-2}$
Dolomite	50–5000	Iron pyrites	10^{-4}
Basalt	$500-10^5$	Copper pyrites	10^{-3}
Granite	$500-10^5$		

Most of the discovered oil and gas reservoirs in the world are buried in sedimentary rocks; so, the variation regularity of the resistivity of the sedimentary rocks will be our focus in the following discussions.

7.1.2 Relationship Between Electric Resistivity of Rock and Formation Water Properties

The resistivity of formation water depends upon the chemical composition of the dissolved salts, the concentration, and the temperature.

1. Relationship between the Formation Water Resistivity and the Chemical Composition of Salts

The water present in oil and gas reservoirs mainly includes $NaCl$, KCl, Na_2SO_4, $MgSO_4$, and $CaSO_4$. As these salts have different degrees of ionization and their ions have different valances as well as different mobility rates, the formation water has different resistivity when containing different salts. With higher degree of ionization, larger percentage of the salt compounds can be dissociated into ions, rendering the formation water to have lower electrical resistivity. With higher ionic valence which indicates the ions carry more charges, the formation water has lower electrical resistivity. With higher mobility rate, the ions move faster, and therefore, the formation water has lower electrical resistivity. Table 7.2 shows the resistivity of several kinds of salt solutions with a concentration of 10 g/L at 18 °C.

The formation water can be considered as $NaCl$ solution, since $NaCl$ is usually the major salt containing in formation water. If formation water contains other kind

Table 7.2 Electrical resistivity of several kind of salt solutions ($T = 18$ °C)

Salt Solution	NaCl	KCl	CaCO$_3$	Na$_2$SO$_4$	MgSO$_4$
Electrical resistivity (Ω m)	0.55	0.7	0.47	0.95	1.4

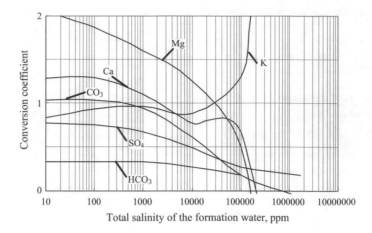

Fig. 7.1 Chart of conversion coefficient for equivalent sodium chloride salinity

of salts besides NaCl, an equivalent salt content can be calculated and be used to calculate the electrical resistivity of the formation water. Figure 7.1 shows the conversion factor curves for each ion. X-axis is the total salinity of formation water which represents salt content, in Mg/L, while Y-axis is the conversion factor for each ion. According to the total salinity of formation water, conversion factor for each ion can be found on the corresponding curves on the chart. Then the equivalent salt content can be calculated by summing the product of the salinity of each ion and its conversion factor.

[Example 7.1] According to the analysis of the reservoir water sample, the salinity of calcium ion is 460 mg/L, the salinity of sulfate ion 1400 mg/L, and the salinity of sodium ion and chloride ion is 19,000 mg/L. Calculate the total salinity and the equivalent sodium chloride salinity.

Solution:

(1) The total salinity is $460 + 1400 + 19,000 = 20,860$ (mg/L)
(2) Read the related coefficients in Fig. 7.1: the coefficient for Ca is 0.81, and the coefficient for SO_4^{2-} is 0.45. The equivalent sodium chloride salinity is yielded through the following equation, where each salinity is multiplied by the corresponding coefficient.

$$460 \times 0.81 + 1400 \times 0.45 + 19000 = 20000 \, (\mathrm{mg/L})$$

2. Relationship between Electric Resistivity of Formation Water and Solution Concentration and Temperature

When the salt concentration of the formation water is raised, the number of ions in the solution is increased, and consequently, the solution progresses in electric conductivity, leading to lower electric resistivity. When the temperature of the

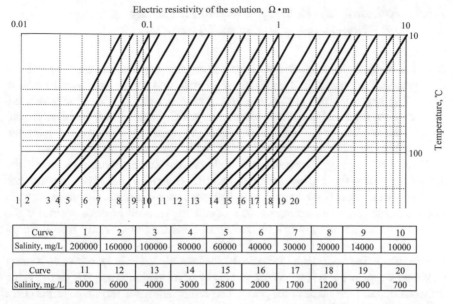

Curve	1	2	3	4	5	6	7	8	9	10
Salinity, mg/L	200000	160000	100000	80000	60000	40000	30000	20000	14000	10000

Curve	11	12	13	14	15	16	17	18	19	20
Salinity, mg./L	8000	6000	4000	3000	2800	2000	1700	1200	900	700

Fig. 7.2 Relationship between the electric resistivity of the NaCl solution and the concentration and temperature

solution increases, the mobility rate of the ions becomes larger and the solution becomes more conductive for the electric current, whereupon the electric resistivity is lowered. Besides, increase in temperature results in concomitant increment of the solubility of some salts contained in the formation water, causing rise of the ion concentration and reduction of the electric resistivity. According to the results yielded from in-lab measurement, the relationship between the electric resistivity of sodium chloride solution and its concentration and/or temperature has shown in Fig. 7.2.

[Example 7.2] It is known that the salinity of a NaCl solution is 2×10^4 mg/L; calculate its electric resistivity at 100 °C.

Solution: The 8th curve given in the chart is the one for the salinity of 2×10^4 mg/L. Read the corresponding coordinates at the point D, where the curve meet the horizontal line of 100 °C. The horizontal coordinate, 0.11 Ω m, is rightly the solution's electric resistivity at 100 °C.

Seen from discuss given above, the electric resistivity of formation water is inversely proportional to the salt concentration and has negative correlation with temperature—the electric resistivity of formation water decreases approximately by 2 % when the temperature is increased by 1 °C. Besides, note that the electric resistivity of formation water varies with the kind of salts contained in it.

As the rock's electric resistivity is directly proportional to the electric resistivity of formation water contained in it, it can be deduced that the rock's electric resistivity is reversely proportional to the electric resistivity of formation water, as well as the temperature.

7.1.3 Relationship Between Electric Resistivity of Water-Bearing Rocks and Porosity

A rock's capacity to conduct electricity dominantly depends on the pore volume (i.e., porosity) contained in a unit volume of rock and the resistivity of formation water. The larger the porosity, or the lower the resistivity of formation water, the more capable the rock to conduct electricity, which means lower resistivity; conversely, the smaller the porosity, or the higher the resistivity of formation water, the less capable the rock to conduct electricity, which means higher resistivity.

Given that there are cementing materials present among sandstone grains, the cementing materials, having relatively higher electricity resistivity against the electric current, makes the current path become more tortuous. Therefore, consolidated sandstones, being analogous to a conductor reduced in cross-section while increased in length, usually is less capable of conducting electricity than the consolidated sandstone and thus has higher electric resistivity.

Seen from the analysis given above, the electric resistivity of water-bearing sandstones depends upon the electric resistivity of formation water, the porosity, the cementation, and the shape of the pores. The cementation and the shape of the pores are decided by rock lithology.

In order to find the relationship between the electric resistivity of rock and the porosity, as well as the electric resistivity of formation water, we can intentionally take the lithology as constant. Proven by experiments, the electric resistivity of a rock completely saturated with formation water (R_o) is directly proportional to the electric resistivity of formation water (R_w), and that is to say, the ratio R_o/R_w is a constant. This constant, being independent of the resistivity of the formation water contained in the rock, is decided only by the porosity, cementation, and pore shape of the rock.

R_o/R_w gives one of the most fundamental concepts in considering the electrical properties of reservoir rocks, and it is called the formation factor F:

$$F = \frac{R_o}{R_w} \tag{7.2}$$

where

F formation factor, or relative resistance, dimensionless;
R_o the resistivity of the rock when 100 % saturated with water having a resistivity of R_w, $\Omega\,m$;
R_w the resistivity of the formation water, $\Omega\,m$.

The relationships between the electrical properties and other physical prosperities of the rock are illustrated by the following derivations. Consider a cube shown in Fig. 7.3a of salt water which has a cross-sectional area A, a length L, and a resistivity R_w. If an electrical current is caused to flow across the cube through an area A and a length L, the resistance of the tube can be determined by the equation below:

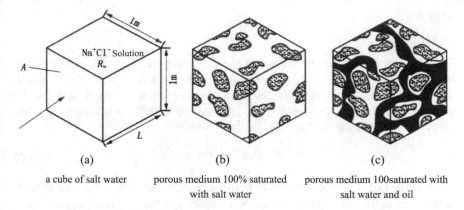

(a)	(b)	(c)
a cube of salt water	porous medium 100% saturated with salt water	porous medium 100saturated with salt water and oil

Fig. 7.3 Three kinds of unit body (the *white regions* represent saline water; the *black regions* represent oil; and the scattered patterns represent rock)

$$r_1 = \frac{R_w L}{A} \tag{7.3}$$

Figure 7.3b shows a cube of porous rock of the same dimensions of the cube shown in Fig. 7.3a and 100 % saturated with salt water of resistivity R_w. Considering the solids to be nonconducting, the electrical flow must thence be conducted through the water-filled pores. The cross-sectional area available for conduction, which equals the actual or effective cross-section of the water-filled pores, is now A_a, and the path length of current flow, the average length that an ion must traverse in passing through the channels, is, obviously, longer than L and represented by the symbol L_a. Then, the resistance r_2 of such a cube can be determined:

$$r_2 = \frac{R_w L_a}{A_a} \tag{7.4}$$

With R_o as the resistivity of the rock when 100 % saturated with water having a resistivity of R_w, by definition we know that: $R_0 = \frac{r_2 A}{L}$

Substitute Eq. (7.4) into the equation above:

$$R_o = \frac{R_w L_a A}{L A_a}$$

Therefore

$$F = \frac{R_0}{R_w} = \frac{L_a/L}{A_a/A} = \frac{\sqrt{\tau}}{A_a/A} \tag{7.5}$$

where

τ tortuosity.

It can clearly be seen that there is a relationship of some kind between the electrical conductivity and the physical-property parameters of rock. The formation factor F is a parameter describing the rock porosity and geometrical shape of the pore space. Different pore-structure models have been built up by researchers, but all of them can be generalized into the form below:

$$F = C\phi^{-m} \tag{7.6}$$

The constant C is under the effect of the tortuosity. Since $C = L_a/L$, its value is necessarily equal to or larger than 1. Theoretically speaking, C falls between 1 and 2. The value of m depends upon the shrink of the pore space or the decrement in the number of channels available for the electric current.

Winsaner presented a way to determine the tortuosity by introducing the time consumed by an experimental ion transported through the porous medium:

$$\left(\frac{L_a}{L}\right)^{1.67} = F\phi \tag{7.7}$$

In 1942, Archie proposed the formula as follows:

$$F = \phi^{-m} \tag{7.8}$$

where

φ porosity, fraction;
m cementation factor.

For cemented rocks, m value is 1.8–2.0, and for uncemented rocks, m value is 1.3.

Figure 7.4 shows the comparison of the various relation formulas for formation factor.

In the system of log–log coordinates as shown in the figure, the vertical axis represents the formation factor (F) while the horizontal axis represents the porosity (φ). Since there is an approximate linear relation between F and φ, the equation correlating them can be yielded according to the straight-line segment of the curve:

$$F = \frac{R_0}{R_w} = \frac{a}{\phi^m} \tag{7.9}$$

where

R_0 the resistivity of the rock when 100 % saturated with water having a resistivity of R_w, Ωm;
m cementation factor, varying with the degree of cementation.

Fig. 7.4 Comparison of the
various relation formulas for
formation factor

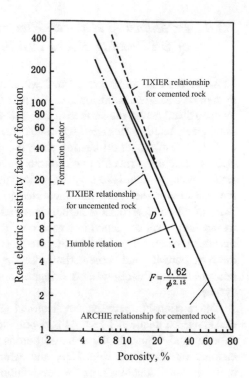

R_w electric resistivity of formation water, Ω m;
a proportionality factor, varying in value with the lithology of rock;
φ intercommunicating porosity of rock, %.

Figure 7.5 shows the relation curve between F and φ, which is based on relevant data obtained from some oil fields of China. Seen from the figure, F and φ obey a linear relationship between each other when put in the system log–log coordinates.

Fig. 7.5 Example relation
curves between for F and ϕ

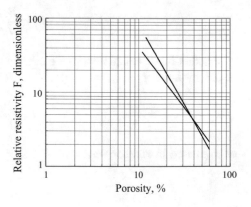

7.1.4 Relationship Between Electrical Resistivity of Oil-Bearing Rocks and Saturations

Consider the case in which oil and water are simultaneously present in the pore system of water-wet sandstone. Given that the pores are interconnected, the water tends to dwell across the surface of the grains while the oil is centrally located in the pore space, being surrounded by water. Due to the high electrical resistivity of oil (10^9–10^{16} Ω m), in an oil-bearing rock, the paths capable of being followed by the electric current are much more tortuous than that in a water-bearing rock with the same lithology—that is to say, the rock with oil is analogous to a conductor being reduced in cross-section while increased in length. Therefore, the electrical resistivity of oil-bearing rocks is higher than that of water-bearing rocks, and the larger the amount of oil contained in a rock (i.e., the higher the oil saturation), the higher the rock's electrical resistivity. So, the electrical resistivity of rock, in addition to the rock porosity and cementation, also depends on the geometrical shape of the pores as well as the oil–water distribution and saturation conditions within the pore system.

The relationship between the electrical resistivity of rock and the oil saturation can either be deduced by simplified pore-structure models or be computed from experimental datum. If the cube of porous rock contains both water and hydrocarbons as shown in Fig. 7.3c, the water is still the only conductor. The cross-sectional area available for conduction is reduced further to A_a', and the path length is changed to L_a'. In a similar manner to the foregoing examples, the resistance of the cube is:

$$r_3 = \frac{R_w L_a'}{A_a'} \tag{7.10}$$

The resistivity of a partially water-saturated rock is defined as $R_t = \frac{r_3 A}{L}$. So, by substituting r_3 into it, we can get:

$$R_t = \frac{R_w L_a'}{L A_a'} \tag{7.11}$$

Another fundamental notion of electrical properties of porous rocks is that of the resistivity index I:

$$I = \frac{R_t}{R_w} \tag{7.12}$$

or expressed in another way:

$$I = \frac{A_a/A_a'}{L_a/L_a'} \tag{7.13}$$

Equation (7.13) indicates that the resistivity index is a function of the effective path length of the current and the effective cross-sectional area, depending on not only the pore-structure parameters, but also the saturations.

Taking the form of ratio, the resistivity index has eliminated the influence from the electric resistivity of formation water, the porosity of rock, and the shape of rock pores. For a rock with specified lithology, the resistivity index depends only on the oil saturation of the rock.

The resistivity index can be determined experimentally. Firstly, choose a representative rock core and measure its electric resistivity R_o when 100 % saturated with water; then, introduce oil into the core gradually and measure the variation of the core's resistivity R_t at different oil saturations, based on which the resistivity index can be obtained.

Experiments on various rocks that are different in lithology have yielded results proving the same rule. The empirical formula is listed below:

$$I = \frac{R_t}{R_o} = \frac{b}{S_w^n} = \frac{b}{(1 - S_o)^n} \tag{7.14}$$

where

S_o oil saturation;
n saturation exponent;
S_w water saturation;
b coefficient.

The exponent n and the coefficient b depend on the rock lithology and can be determined experimentally. When the values of n and b have already been known, the oil and gas saturation of the formation can be thence obtained from the relation equation or relation curve established between I and S_o.

Hydrocarbons cannot conduct electricity. So, it is obvious that the higher the hydrocarbon saturation of a formation, the higher the electric resistivity of it.

The resistivity index formula, an empirical equation correlating the electric resistivity of rock and the water saturation, is of significant importance in electric logging. It is proven by it that the resistivity index is a function of the water saturation and the electric current path.

$$I = C'S_w^{-n} \tag{7.15}$$

$$I = R_t/R_o$$

where

C' tortuosity function;
n saturation exponent.

Based on analysis and summary on a large amount of experimental data, Archie presented a formula in the form given below:

$$I = S_w^{-2} \tag{7.16}$$

William also proposed his formula for cemented rocks:

$$I = S_w^{-2.7} \tag{7.17}$$

The relationship between electric resistivity and water saturation is shown in Fig. 7.6, where a comparison between the Archie's Equation and some other relevant data has been presented.

According to the experimental results, it is found that I and S_o approximately have a linear relation between each other in a system of log–log coordinates wherein the vertical coordinate represents the resistivity index I while the horizontal coordinate represents the oil saturation S_o. Figure 7.7 shows some examples from three different oil fields.

Fig. 7.6 Relationship between resistivity index and water saturation

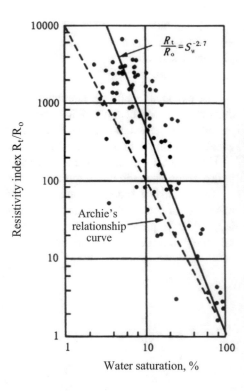

Fig. 7.7 Relationship between oil saturation S_o and resistivity index I

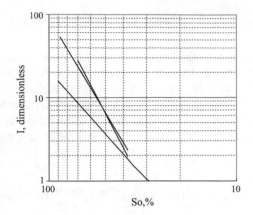

7.1.5   Measurement of Electrical Resistivity of Rocks

Figure 7.8 shows the technological process and schematic circuit diagram for the four-electrode test for electrical resistivity. Changing the saturations of oil and water in the core through a steady state displacement process, we can measure the corresponding the rock resistivity at different values of S_w.

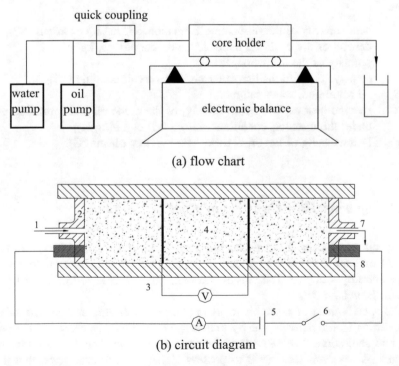

(a) flow chart

(b) circuit diagram

Fig. 7.8 Technological process and schematic circuit diagram for the four-electrode test for electrical resistivity. *1* inlet for fluid, *2* gland at core end, *3* rubber cylinder, *4* core, *5* electric power source, *6* switch, *7* outlet for fluid, and *8* electrode

7.2   Thermal Properties of Reservoir Rocks

The thermal properties of reservoir rocks refer to such rock characteristics as specific heat, heat capacity, heat conductivity coefficient, thermal diffusion coefficient, etc. Thermodynamic calculations are involved in many aspects of the development of oil & gas reservoirs, for example, the heat-transfer process taking place in the wellbore as well as the formation, the oil extraction measures especially in the engineering of oil recovery by heating, the EOR methods, etc.

7.2.1   Heat Capacity of Oil-Bearing Formations

The heat capacity is defined as the quantity of heat necessary to raise the temperature of a unit mass of a substance by 1 °C. The unit of that is kJ/(m^3 °C). The heat capacity of an oil-bearing formation depends on the heat capacity of the dried rocks and the saturation conditions of the fluids (oil, water). It is calculated through the equation below:

$$M = \varphi(\rho_o S_o C_o + \rho_w S_w C_w) + (1 - \varphi)(\rho Cp)_R \qquad (7.18)$$

where

M heat capacity of the oil-bearing formation, kJ/(m^3 K) or kJ/(m^3 °C);
ρ_o density of the crude oil under reservoir conditions, kg/m^3;
φ porosity of the formation, %;
ρ_w density of the formation water under reservoir conditions, kg/m^3;
S_o, S_w oil saturation, water saturation, %;
C_o, C_w specific heat capacity, respectively, of the crude oil and formation water under the reservoir conditions, kJ/(kg °C) or kJ/(kg K);
$(\rho Cp)_R$ heat capacity of the dried rock, kJ/(m^3 K) or kJ/(m^3 °C).

7.2.2   Specific Heat Capacity of Rock

The specific heat capacity of a material, also called specific heat for short, is the amount of heat that must be added to a unit mass of the material to raise its temperature by 1 degree. The unit of that is kJ/(kg °C) or kcal/(kg °C) or (British system) Btu/(1 bm °F).

The specific heat of reservoir rocks is affected by factors such as the mineral composition is, the porosity, the heat capacity of the fluids contained in the pores, the fluid saturations, the salinity of the formation water, and the reservoir temperature. A reservoir rock tends to possess higher specific heat given that it has

Table 7.3 Specific heat capacity, heat conductivity coefficient, and thermal diffusion coefficient of dried rocks and rocks saturated with water

Type of rock	Density (g/cm^3)	Specific heat capacity [kJ/(kg K)]	Heat conductivity coefficient [W/(m K)]	Thermal diffusion coefficient (10^4 m^2/s)
Dried rock sample				
Sandstone	2.08	0.766	0.877	1.97
Siltstone	1.92	0.854	0.685	1.50
Shale	2.32	0.804	1.043	2.00
Limestone	2.19	0.846	1.701	3.29
Fine grained sandstone	1.63	0.766	0.626	1.80
Coarse grained sandstone	1.75	0.766	0.557	5.66
Rock sample saturated with water				
Sandstone	2.27	1.055	2.754	4.13
Siltstone	2.11	1.156	2.612	3.84
Shale	2.39	0.888	1.687	2.85
Limestone	2.39	1.114	3.547	4.79
Fine grained sandstone	2.02	1.419	2.751	3.45
Coarse grained sandstone	2.08	1.319	3.071	4.01

larger porosity, higher water saturation, and higher temperature. This regularity is more distinct under the condition of low salinity. Generally, the specific heat capacity of sandstones is 0.837–1.315 kJ/(kg °C) or kJ/(kg K); the specific heat capacity of dried core is 0.766–0.846 kJ/(kg °C); and the specific heat capacity of the core saturated with water is 0.888–1.419 kJ/(kg °C). Table 7.3 gives for more information about the specific heat capacity of various rocks.

Generally speaking, the specific heat capacity of mudstones is greater than that of calcareous sandstones. Besides, the specific heat of rocks also depend the condition of temperature—when holding the pressure as constant, the specific heat increases slightly with the increase in temperature.

7.2.3 Heat Conductivity Coefficient

The heat conductivity coefficient is defined as the quantity of heat that passes through a unit area of a substance in a given unit of time due to a unit of temperature gradient. The unit of that is kJ/(h m °C) or W/(m °C). The mathematic expression of that is:

$$\lambda = \frac{Q_\text{h}L}{\Delta t \cdot A \cdot \Delta T} \tag{7.19}$$

where

λ the heat conductivity coefficient, W/(m K);
Δt time consumed by the thermal conduction, s;
A cross-sectional area of the core, m^2;
ΔT temperature difference between the two core ends, K;
Q_h quantity of heat transmitted during the time Δt, J;
L the length of the core, m.

The heat conductivity coefficient of rock is mainly decided by the mineral compositions of the rock, the porosity, and the water saturation. Proven by experiments, the heat conductivity of mineral grains is many times superior to that of water, which is approximately 30 times that of the heat conductivity of air. Related data is listed in Fig. 7.4. In addition, the heat conductivity of oil-bearing formations ranges between 1.3–1.6 kcal/(m h °C) = 130.4–160.5 kJ/(d m °C).

7.2.4 Thermal Diffusion Coefficient

Reflecting the relationship between the rock's capacity of heat transmission and the rock's heat storage capacity during the unsteady state heat conduction, the thermal diffusion coefficient is defined as:

$$\alpha = \frac{\lambda}{C\rho} \tag{7.20}$$

where

α the thermal diffusion coefficient, m^2/s;
λ the heat conductivity coefficient, W/(m K);
C specific heat capacity, J/(kg K);
ρ density of rock, kg/m^3.

The larger the thermal diffusion coefficient α, the quicker the heat obtained by a part of the rock can diffuse in the whole rock formation, which indicates that the rock tends to reach uniformity in temperature more readily.

7.3 Acoustic Characteristics of Reservoir Rocks

Acoustic waves are caused by mechanical vibration. By the interaction between the particles, the sound waves propagate from the focus to the other part of the medium in all directions. Sound can be transmitted as both longitudinal and transverse

Table 7.4 Acoustic velocity of longitudinal in commonly allodium

Medium	Acoustic velocity (m/s)	Medium	Acoustic velocity (m/s)
Air	330	Mud stone	1830–3962
Methane	442	Permeable sandstone	2500–4500
Oil	1070–1320	Salt rock	4600–5200
Water, cement	1530–1620	Compact limestone	7000
Loose clay	1830–3962		

waves in medium, and the acoustic velocity of longitudinal wave is about 1.73 times of that of the transmitted wave when they are propagating in most of natural rocks. The advantage of the longitudinal wave in velocity of propagation makes it more convenient and more widely used for the study of the laws of acoustic propagation in rocks. Table 7.4 shows the acoustic velocity of longitudinal wave in some commonly media and sedimentary rocks.

The acoustic velocity of a rock varies according to its elasticity and density. The density of rock is a dominant factor in deciding the acoustic velocity. It is found that the larger the rock's density, the higher the acoustic velocity of the wave transmitted by it. Taking the corresponding relation between the density and the porosity of rock into consideration, a conclusion can be drawn: the larger the porosity, the smaller the rock's density and the lower the acoustic velocity. This characteristic of wave the technology of acoustic logging is based on this acoustic phenomenon. In acoustic logging, a transmitter located in the borehole emits a pulse of mechanical energy which is recorded by one or more receivers located in the borehole some distance away from the transmitter. Based on some empirical equations, the porosity of the tested formation can be calculated according to the acoustic signals detected by the receivers.

In the recent years, such simulation technologies, as artificial earthquake, in-well ultrasonic wave, etc., have come into service; the propagation principle of acoustic wave has given birth to many new physical stimulation approaches applicable to the development of reservoir.

Part III
Mechanics of Multi-Phase Flow in Reservoir Rocks

The preceding two parts of this book have respectively delineated the physical properties of reservoir fluids and reservoir rocks, and this part, as further discussion of the multiphase flow mechanism of reservoir fluids, will deal with the distribution and behavior of multiphase flow in reservoir rocks.

In general, such a simplified case as single-phase system is seldom found in actual reservoirs; on the contrary, two-phase systems of immiscible oil and gas and even three-phase systems of oil, water, and gas are more commonly involved in real reservoirs. A combination of oil and water is a multicomponent, anisotropic, and unstable system.

Reservoir rocks, actually being porous media with extremely great specific surface and highly dispersed pores of all kinds of shapes, have tremendously large area of contact with the fluid contained in it, even if there is only a single phase present. Now, let us take our idea a step further. In the case that there are two or three phases coexisting in the pore space, many more interfaces would be formed between water and rock, oil and rock, oil and water, oil and gas, gas and water, etc. The total area of these interfaces would be tremendously large and maybe a variety of physical and chemical reactions that take place between the fluids as well as between the fluids and the rock surfaces. The neighboring molecules of the two phases interact with each other and give rise to many problems with regard to interfacial properties; for example, the disappearance of the oil-water interface during immiscible displacement and the additional capillary pressure existing at the oil-water interface. In a word, the distribution and seepage flowing of the multi-phase fluids in the pore space of reservoir rocks is a problem of great complexity.

As the interfacial phenomena are closely associated with the forces taking place between the molecules of materials, we should begin our study from the micro point of view and then go on with some deeper discussions, including the physical properties of the multiphase fluid in formations (e.g., the wettability of rock and the capillary force), the microscopic seepage-flowing mechanism (e.g., the relative permeability and various capillary resistances), the characteristics of the oil and water distribution in the rock pore space, the regularities of the residual oil distribution and the influencing factor for it, the approaches to eliminate or minimize the additional resistances, the regularities that govern the macroscopic production of oil wells, the optimum ways to develop reservoirs, and so on.

Chapter 8
Interfacial Phenomena and Wettability of Reservoir Rocks

Established by usage, a property that is caused by the intermolecular forces within a phase is usually called "bulk property," while a property that is caused by the intermolecular forces at the interface between two phases is called "interfacial property."

The substance of interfacial phenomenon is a part of the course of physical chemistry. To start with, we should introduce some physical and chemical properties regarding the interfaces between reservoir fluids, for example, the interfacial tension, surface adsorption, surface wettability, and interfacial viscosity, and subsequently, we will discuss the interfacial phenomena occurring between reservoir fluids and rocks.

8.1 Interfacial Tension Between Reservoir Fluids

8.1.1 Free Surface Energy of the Interface Between Two Phases

The bounding surface between any two phases can be called "interface," and when one of the two phases is gas, the interface is also directly called "surface." For example, the bounding surface separating the solid or liquid from the gas could be called the solid surface or liquid surface, while bounding surface between the solid and the liquid, or that between two liquids, must be called interface.

Here, the water surface is the starting point of our discussion. It is found that the molecules at the water surface and those staying remote from the water surface are

© Petroleum Industry Press and Springer-Verlag GmbH Germany 2017
S. Yang, *Fundamentals of Petrophysics*, Springer Geophysics,
DOI 10.1007/978-3-662-55029-8_8

Fig. 8.1 Diagram of forces
on surface and inner
molecules of liquid

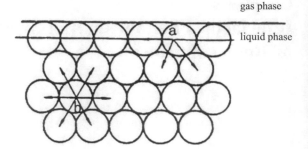

tolerating different forces. Each water molecule that is remote from the interface is pulled equally in all directions by the neighboring water molecules, resulting in a net force of zero and thus a state of equilibrium. However, a molecule at the water surface, e.g., the one marked as "a" in Fig. 8.1, is subjected to a set of unbalanced forces involving not only an attraction force from the gas molecules lying immediately above the surface, but also an attraction force form the water molecules lying below the surface. As the attraction from the gas is much smaller than that from the water, the resultant force points to the inside of the water phase and is perpendicular to the water surface. As a result, the representative molecule "a" sustains a tendency to move inwards deeper in the water phase, and the whole water surface has a tendency to squeeze itself together until it has the lowest surface area possible. Obviously, a surplus of energy is preserved by the water-surface molecules under the unbalanced forces, and this kind of energy is referred to as the "free surface energy."

A certain amount of work is required to move water molecules form within the body of the liquid to the water surface, and this work will be finally transformed into the free surface energy of the water molecule it has moved.

In conclusion, the free surface energy of the interface between two phases is caused by the unbalanced molecular force field at the interface. The characteristics of the free surface energy include the following:

(1) It occurs only when two immiscible fluids are in contact. The contact of any two phases will give rise to the free surface energy, but for two fluids that dissolve each other incompletely and thereby form no interface (e.g., water and ethanol, kerosene, and crude oil), the surface energy does not exist.

(2) The larger the area of the interface is, the greater the free surface energy will be. According to the second law of thermodynamics, the free surface energy tends to its minimum. Since among all geometric shapes of a certain volume sphere has the least surface area, sphere is also the shape that has the least surface energy. This is also the reason why a drop of mercury dropped on the floor takes the shape of sphere, which spreads over minimum surface area while occupies minimum space when placed over anything.

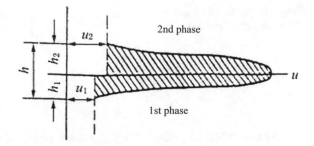

Fig. 8.2 Thickness and thermodynamic property, of interface, distribution of interfacial energy. h, h_1, h_2—respectively, the total thickness and the individual thickness of each phase; u_1, u_2—volume energy of each phase at the interface

(3) Figure 8.2 shows that the interface, or surface, is an interfacial layer with a certain thickness. As a transition molecular layer, the interfacial layer, changing gradually from one phase to the other, does not have the same property or structure with either of the two phases, and all of the molecules contained in this layer possess free surface energy to varying degrees. Besides, as shown in Fig. 8.2, the molecules in the transition layer exhibit a gradual and continuous change in the thermodynamic property. Taking the interfacial layer between water and gas as example, it has a thickness of at least several molecular layers.

(4) The magnitude of free surface energy is related to the molecular properties of the two phases in contact. A greater disparity in polarity between them leads to a greater surface energy of their interfacial layer. Since water has the strongest polarity among all liquids while clean air has very weak polarity, the surface energy of the water–air interface heads the list of all the two-phase systems. In contrast to this, as the polarity disparity between crude oil and carbon tetrachloride is so small, the interface between them disappears and a state of miscibility is finally reached. This is why carbon tetrachloride is usually used in laboratory for extracting oil from core samples.

(5) The magnitude of free surface energy is also related to the phase states. Generally speaking, the free surface energy of the interface between liquid and gas is greater than that between liquids, and the free surface energy of the interface between liquid and solid is greater than that between liquid and gas.

8.1.2 Specific Surface Energy and Surface Tension

Sustaining a set of asymmetric forces, the surface molecules have higher energy than the interior ones. Therefore, work is done against intermolecular forces in creating new surface or, equivalently, raising more molecules up to the surface from the interior of the phase, and it is obvious that the work done during this process is converted into inherent energy of the new surface. Figure 8.3 shows a frame out of metal wires, and the piece of wire on the right can be moved. By dipping it in soap solution, a soap film can be formed within the frame. If we want to increase the area

Fig. 8.3 Work and formation
of a gas–liquid interface

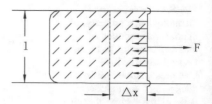

of the soap film, an external force is needed to pull the removable piece of wire, suggesting that equivalent work must be done for creating new surface.

In the case that a work has been done reversibly to add the surface by an area of ΔA, under the conditions of constant temperature, pressure, and composition, the work (W) done by the external force has the following relationship with the corresponding increment of free surface energy contained in this system:

$$\Delta U = -W \tag{8.1}$$

The work needed to create a new unit area of surface is:

$$\frac{\Delta U}{\Delta A} = \frac{-W}{\Delta A} = \sigma \tag{8.2}$$

And if expressed in differential form, it is:

$$\frac{\mathrm{d}U}{\mathrm{d}A} = \sigma \tag{8.3}$$

So, the specific surface energy can be defined as:

$$\sigma = \left(\frac{\partial U}{\partial A}\right)_{T,P,n} \tag{8.4}$$

where:

ΔU Increment of the free surface energy of the system, N.m;
ΔA Increment of the surface area, m²;
U Free surface energy of the system, N/m;
T, P, n Temperature, pressure, and composition of the system;
σ Specific surface energy, N/m.

The specific surface energy is the free surface energy possessed by per unit area of the surface and has the dimension of joule per square meter (J/m²), and, obviously, 1 J/m² equals 1 Newton per meter (N/m). In fact, the common unit used in engineering is milli-Newton per meter (mN/m).

In the centimeter–gram–second (CGS) system, the unit for specific surface energy is erg per square centimeter and 1 erg per square centimeter (erg/cm²) = 1 dyne per centimeter (dyn/cm), 1 milli-Newton per meter (mN/m) equals 1 dyne per centimeter (dyne/cm). From the perspective of dimension, the specific surface energy is equal to the force per unit length, so, it is also conventionally called "surface tension," symbolized as σ.

What should be pointed out is that the "specific surface energy" and the "surface tension," although equivalent numerically, are two concepts meaning different things. Generally speaking, the term "specific surface energy" is mostly used in thermodynamics, while the term "surface tension" is mostly used in mechanics and practical applications.

Strictly speaking, no tension force really exists at the interface of a two-phase system, and the term "surface tension" here is just used as one of the expression forms of specific surface energy. But the surface tensions really exist along with the three-phase perimeter, where the interfacial tensions between every two of phases vie with each other for contact points by virtue of their free surface energy. As shown in Fig. 8.4, as a drop of oil floats on the surface of water, three interfaces come into being: the oil–gas (2–3) interface, the oil–water (2–1) interface, and the water–gas (1–3) interface, and the three corresponding vying but balanced surface tensions are σ_{2-3}, σ_{1-2}, and σ_{1-3}. The correlation among them can be expressed as:

$$\overrightarrow{\sigma_{1-3}} = \overrightarrow{\sigma_{1-2}} + \overrightarrow{\sigma_{2-3}}$$

Note that the tensions only occur when the system has reached a state of equilibrium, being as the final result of the contest between the three sets of surface energy. As a vector, each of the surface tensions is numerically equal to the specific interface/surface energy between the corresponding phases and has a direction parallel to the plane or along with the tangential line of the curved plane, depending on the shape of the interface; and the points of application are the tricritical points of the system, or in other words, the contact perimeter of the three phases.

What should be noticed is that every time the term "surface tension" is mentioned, the corresponding two phases should be explicated definitely; and if no specification is given, as established by usage, one of the phases is air. For example, when we say that the surface tension of water is 72.75 mN/m, a hidden point is that it refers in particular to a water–air system. Table 8.1 gives the surface tensions of some common substances in contact with air or water.

Fig. 8.4 Diagram of the interfacial tensions at the three-phase perimeter

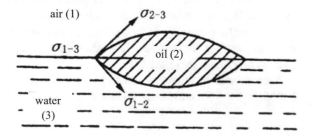

Table 8.1 Interfacial tension data

Substance	Interfacial tension in contact with air mN/m (at 20 °C)	Interfacial tension in contact with water mN/m (at 20 °C)
Mercury	484.0	375
Water	72.8	–
Transformer oil	39.1	45.1
Druzhnoye, Russian	27.2	30.3
Normal hexane	18.4	51.1
Normal octane	21.8	50.8
Benzene	28.9	35.0
Methylbenzene	28.4	/
Ethylether	17.0	10.7
Carbon tetrachloride	26.9	45.0
Carbon bisulfide	33.5	/
Methylene dichloride	28.5	/
Dichloroethane	32.5	/
Ethanol	22.3	/

8.1.3   Influencing Factors on Surface Tension

1. Composition

As shown in Table 8.1, the surface tension between mercury and air is greater than that between mercury and water. This is because that the larger the polarity disparity between the two contacting phases is, the greater the surface energy of them will be. Since water has the strongest polarity among all liquids while clean air has very weak polarity, the surface energy of the water–air interface heads the list of all the two-phase systems. While the methylbenzene and crude oil, both as organic substances, have such a small polarity, the interface between them disappears and a state of miscibility is finally reached.

2. Phase State

For instance, the surface tension between gas and liquids is usually greater than that between liquids. Anton off Rule states that the surface tension at the interface between two saturated liquid layers in equilibrium is equal to the difference between the individual surface tensions of similar layers when exposed to air. So, numerically the surface tension of a liquid–liquid interface lies between the two individual air–liquid surface tensions and is smaller than the larger one of the two.

3. Temperature and Pressure

Such factors as temperature and pressure directly affect the surface tension by changing the intermolecular distances by which the intermolecular forces and thereby the force field at the interface are decided by. For any liquid exposed to gas, an increase in temperature and/or pressure results in decrease of the surface tension. An increase in temperature will, on one hand, increase the distance between the liquid molecules, leading to weaker intermolecular attraction and, on the other hand, allow more liquid undergoing evaporation to reduce the disparity between the force fields of the liquid and the vapor, which decreases the surface tension. An increase in pressure tends to entail a progressive solution of the gas in the liquid with a concomitant smaller disparity between the densities of the two phases, a lessened bias of the force field and consequently a decreased surface tension.

8.1.4 Interfacial Tension Between Reservoir Fluids

As previously mentioned, there are a variety of interfaces coexisting in oil and gas reservoirs. But our discussion is generally restricted to the surface tension between fluids and usually excludes the solid phase from consideration, whose surface tension is rather difficult to determine.

As we all know, it is almost impossible for the temperature and pressure conditions to be homogenous throughout a reservoir, and the reservoir fluids are rather complex and variable in composition due to different T, P conditions. All these factors lead to the complexity of the interfacial tension between reservoir fluids. It is proven that the interfacial tension cannot be a constant even in a single formation, and the difference in different formations is bigger. The magnitude and variation regularity of interfacial tension will be discussed on the basis of its definition and characteristics.

1. Oil–gas Interface

Figure 8.5 shows several curves that describe the relationship between pressure and the surface tension of different oil–gas systems to demonstrate the effect of solution gas on the surface tension. As shown by the curves, the surface tension of the oil–air system is reduced slightly when the pressure keeps on increasing, due to the extremely low solubility of nitrogen (the major component of air) in oil; by contrast, the surface tension of the oil–CO_2 system suffers a quick descent as the pressure is increased, because CO_2 has a very low saturation vapor pressure and therefore a great solubility in oil; the oil–natural gas system, where the contained methane and small amounts of ethane, propane, and butane has much greater solubility in oil in comparison with nitrogen, also has a surface tension decreasing quickly with the increase in pressure.

With regard to the gasoline–CO_2 system, it has a surface tension even lower than that of the oil–CO_2 system. This is because that the gasoline, composed of lighter hydrocarbon components, has the capacity to dissolve more carbon dioxide than crude oil.

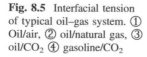

Fig. 8.5 Interfacial tension
of typical oil–gas system. ①
Oil/air, ② oil/natural gas, ③
oil/CO$_2$ ④ gasoline/CO$_2$

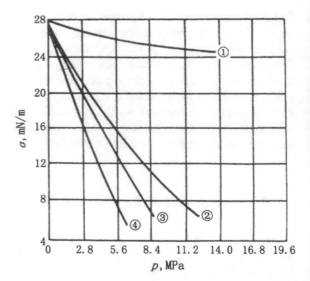

Therefore, lower surface tension of a crude oil–natural gas system can be caused
by a lower saturation vapor pressure of the gas phase, a greater solubility of the gas
in the liquid, a smaller relative density of the crude oil, a larger share of the heavy
hydrocarbons in the gas phase, or an increase in the temperature and/or pressure.

All in all, the oil–gas surface tension decreases with the increase in the solubility
of gas in oil. And, responding to decreasing pressure, a natural gas with more heavy
hydrocarbons in composition, allowing more of it dissolved in the oil, undergoes
larger amplitude of drop in surface tension.

Figure 8.6 shows the effect of temperature and pressure on surface tension. Seen
from the curves, an oil–gas system subjected to conditions of high temperature and
high pressure, containing large quantities of dissolved gas in the oil phase, has a
much lower surface tension than that between the gas phase and the tank-oil on the
land-surface. Likewise, the different parts within the same reservoir also vary in the
surface tension of oil. For example, the oil near gas-cap has a much lower surface
tension than the oil remote from the gas.

2. Oil–Water Interface

On the basis of the researches of the oil–water surface tension, we can arrive at
the following reviews:

(1) With regard to an oil–water system without solution gas, some scholars
believe that the surface tension is little affected by changes in temperature or
pressure as shown in Fig. 8.7. The reason is that an increase in temperature corresponds to the coinstantaneous expansion of both the water and the oil, and an
increase in pressure compresses the oil and the water simultaneously, too. That is to
say, the molecular thermodynamic properties of the oil and water phases are always
in substantially the same tendency of variation when exposed to temperature or

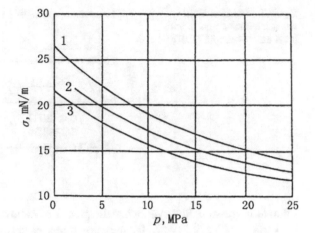

Fig. 8.6 Impact of temperature and pressure on interfacial tension 1 − *t* = 20 °C, 2 − *t* = 60 ° C, 3 − *t* = 80 °C

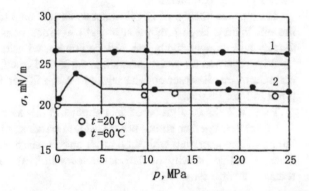

Fig. 8.7 Variation in oil–water interfacial tension with temperature and pressure (1956)

pressure changes, with the force field between the two phases maintained being constant and consequently leading to the resultant unaltered surface tension.

There are also some other scholars who think that the oil–water surface tension will decrease significantly with the increase in temperature, while pressure has little effect on it.

(2) With regard to an oil–water system with solution gas, the quantity of the gas dissolved in the oil is of decisive importance for the surface tension between the two phases. The effect of pressure on the surface tension in the presence of solution gas is shown in Fig. 8.8, where the curves ①, ②, and ③ represent three cases with different relative densities of the oil phase or different quantities of solution gas. As illustrated by the figures, the surface tension increases when the pressure below the saturation pressure P_b is increased, because of the enlarged polarity disparity between the water and the oil, caused by the entrance of more gas into the oil phase which has much higher capacity than water to dissolve gas when being below the

Fig. 8.8 Variation in oil–
water interfacial tension with
pressure by Cartmill (1956)

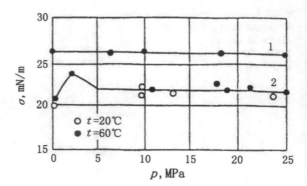

saturation pressure; however, when the pressure is above the saturation pressure P_b, the surface varies little when the pressure increases because that no more gas will be dissolved into the oil phase at a pressure higher than P_b, and what happens to the liquids is only compression.

The surface tension of an oil–water system is also related to the composition of the oil. With a large proportion of light hydrocarbons, the oil tends to have low density and comparatively low surface tension with its contacting water phase.

With regard to the active substances contained in oil, they also directly affect the magnitude of oil–water surface tension. We will soon discuss the details of this point in later lessons.

Table 8.2 shows a group of data about the oil–water surface tension.

No matter whether an oil–water system contains solution gas or not, in fact, an increase in temperature will cause faster and fiercer motion of the molecules, which leads to worse polarity of the molecules within the interface layer and thereby reduced surface tension.

Table 8.2 Data of oil–water interfacial tension from some oil fields at home and abroad by Gengsheng He (1994)

Oil field	Oil–water interfacial tension (mN/m)	Test conditions
Druzhnoye, Russian	30.2	Surface temperature
Romashkino, Russian	25.6	Surface temperature
Glozini, USA	26.0	Surface temperature
Texas, USA	13.6–34.3	Surface temperature
Hengli	23–31	70 °C
Liaohe	9–24	45–85 °C
Daqing	30–36	Formation temperature
Changqing	28.6	51 °C
Renqiu	40	Surface temperature

Table 8.3   Scope of application of different methods

Method	Scope of application	Interfacial tension (mN/m)	Notes
Pendant-plate	Comparatively high interfacial tension	1 to 10^2	
Pendant-drop	Moderate interfacial tension	10^{-1} to 10^2	
Rotating-drop	Low or ultra low interfacial tension	10^{-1} to 10^{-3}	e.g., Microemulsion and oil (or water)

8.1.5   Measurement of Surface Tension

Many a method can be employed to measure the surface tension of liquids. Table 8.3 shows the measuring ranges of them.

1. Pendant-Drop Method

A pendant-drop interfacial tensiometer consists of injector, needle, and optical image pickup device. Different sized needles should be chosen according to the viscosity of the liquid required to be measured: The 0.7-mm needle is for normal liquids, and the 1.5–2.0-mm needles are for viscous liquids. Note that this method requires precision machining for the needle head, which needs to be smooth and even-molded.

When measuring, the liquid of larger density (Liquid 1) is put into the container and the other liquid (Liquid 2) is put into the injector, the needle head of which is submerged into the Liquid 1. Slightly push the piston of the injector to make a bit of Liquid 2 flow out of the needle and form into a droplet under the action of gravity and surface tension. As the surface tension tries to pull the droplet upward while the gravity tries to drag the droplet downward, the magnitude of the surface tension is proportional to shape of the liquid droplet being ready to drop. Then, as shown in Fig. 8.9, the image of the droplet is picked up by the optical system, and its maximum diameter d and the particular diameter d_i which is at a distance d above the top of the droplet are measured. The following equation is used to calculate the surface tension:

$$\sigma = \frac{\Delta \rho g d^2}{H} \qquad (8.5)$$

where:

σ Surface tension, mN/m;

g Acceleration of gravity, 980 cm/s^2;

d Maximum diameter of the droplet, cm;

$\Delta \rho$ Density difference between the two phases awaiting measurement, $\Delta \rho = \rho_1 - \rho_2$, g/cm^3;

Fig. 8.9 Pendant-drop method for the measurement of surface tension.
a Overview diagram, **b** partial enlarged detail

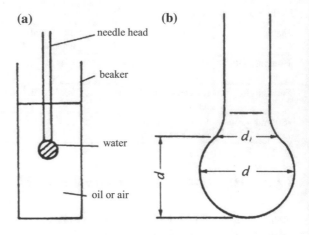

H   Correction value for the shape of the droplet; a function of d and d_i, can be read in related charts.

If the measured liquid has a density lower than its ambient medium, it is better to use a L-type needle for the experiment. The pendant-drop method is also applicable for the measurement of liquid–gas surface tension.

2. Pendant-Plate Method

The basic working principle of the pendant-plate interfacial tensiometer is shown in Fig. 8.10. When measuring, the vitreous plate is accommodated by the lifter to a proper position where the bottom of it is rightly in contact with the oil–water interface in the vessel, so that a liquid–solid interfacial tension and a liquid–liquid interfacial tension simultaneously occur to jointly create a resultant force dragging the vitreous plate downwards. Thus, with the recording and readout instruments to obtain the upward pulling force R that balances the mentioned resultant downward force and the size of the weights, the interfacial tension can be calculated from the equation below:

$$\sigma = R + \frac{Y \times 10^{-3} \times 980}{L} = R + 0.49Y \qquad (8.6)$$

where:

R   Recorded value, mN/m (=dyn/cm);

Y   Weight value, mg;

L   Perimeter length of the bottom of the tension plate; for the matching tension plate of the instrument, $L = 2$ cm.

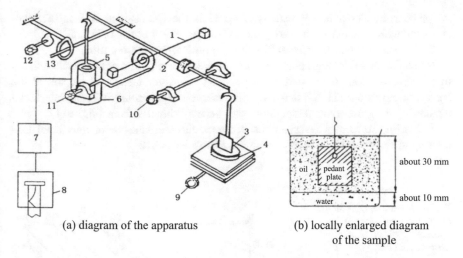

(a) diagram of the apparatus

(b) locally enlarged diagram
of the sample

Fig. 8.10 Pendant-plate method for the measurement of surface tension. *1*—torsion tungsten wire, *2*—beam, *3*—pendant plate, *4*—sample, *5*—iron core, *6*—differential transducer, *7*—amplifier, *8*—recording instrument, *9*—vernier adjustment knob, *10*—brake knob, *11*—torsion knob, *12*—knob controlling the weights, *13*—weights

This instrument for the measurement of surface tension has the characteristics listed below:

(1) The instrument is semiautomatic. Both the plus of weights and the adjustment of the contact between the liquid and the plate bottom are operated by controlling the rotary knob of the instrument, and the weight value Y can be read directly from the graduation of the scale. Actually, the measurement of normal oil–water surface tension needs no plus of weights, and the value of the surface tension R can be read directly from the recording paper.

(2) As the output of data is recorded automatically, this method permits a continuous record of the variation in the tension with time, whereupon the variation regularity of the interfacial tension with the addition of active substances can be traced.

(3) In addition to the gas–liquid and liquid–liquid interfacial tensions, this instrument is also able to measure the selective wettability of solid surface for liquid. This will be discussed in Sect. 8.3.

It should also be noted that in accordance with related industry standards must be ensured during the operating procedure, and the operators must have enough practical experience.

The pendant-plate method applies to the cases limited to density difference not larger than 0.4 g/cm^3 and surface tension ranging between 5 and 100 mN/m.

At present, in China what's usually adopted is the CBVP-type A-3 surface ten-siometer with a matched vitreous plate of 2.4 cm × 2.4 cm × 0.03 cm.

3. Measurement of Interfacial Tension Upon Reservoir Conditions

Related charts (Nomographs) are used for the determination of interfacial tension upon reservoir conditions. If it is known that oil–water interfacial tension at the surface temperature (21 °C) is σ_0 and the reservoir temperature is T, the interfacial tension σ_t at the reservoir temperature can be read directly from Fig. 8.11.

According to known reservoir temperature, the methane–water interfacial ten-sion upon reservoir conditions can be read from Fig. 8.12.

Fig. 8.11 Nomograph for the oil–water interfacial tension upon reservoir conditions by Schowalter (1970)

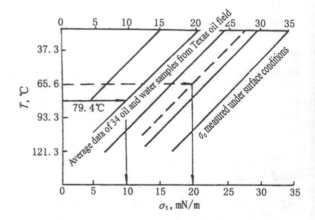

Fig. 8.12 Nomograph for the methane–water interfacial tension upon reservoir conditions by Schowalter (1970)

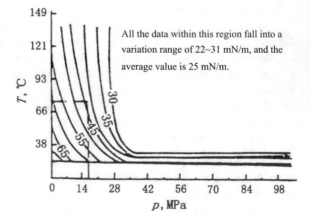

All the data within this region fall into a variation range of 22~31 mN/m, and the average value is 25 mN/m.

8.2 Interfacial Adsorption

8.2.1 Concept of Adsorption

As we all know, the facial layer possess free surface energy due to the disequilibrium and asymmetry of its molecular force field. According to the second law of thermodynamics, the free surface energy of any substance tends to its minimum spontaneously. The ways to reduce the free surface energy of a system, besides changing in shape, include adsorption and wetting. Here, we will focus on the problems associated with adsorption.

Adsorption occurs at the interface between two phases. As there are a variety and vast area of interfaces present within reservoir rocks, the phenomenon of adsorption is ubiquitous in underground formations.

According to the nature of the cohesive force between the adsorbate molecules and the adsorbent solid surface, there are two types of adsorption, chemisorption, and physisorption, which may take place under either static or dynamic conditions. Physisorption occurs when adsorbate molecules physically interact with a surface through van der Waals forces, which are of long range, weak attractive forces. In fact, the process of adsorption is accompanied by the desorption. At the beginning, the adsorption rate greatly exceeds the desorption rate; subsequently, the disparity between the rates of the two reverse processes will gradually go smaller and smaller until they become equal with each other numerically and reach a state of adsorption. The layer formed by the adsorbate molecules that attach to the surface of the adsorbent material is called "interfacial layer" or "absorbed layer." Chemisorption occurs when an adsorbate molecule chemically reacts with the adsorbent and consequently bonds to the surface as chemical compound, usually through a covalent bond. In most cases what happens to the interface is a mixed adsorption, a combination of physisorption and chemisorption.

8.2.2 Adsorption at the Gas–Liquid Interface (Surface Adsorption)

As an extremely complex mixture, the crude oil contains, as far as the property of polarity is concerned, the nonpolar substances and the polar active substances. The former part includes alkanes, cycloalkanes, and aromatic hydrocarbons; the latter part mainly includes compounds consisting of one hydrocarbon combined with elements such as oxygen, sulfur, or nitrogen, for example, naphthenic acid, colloid, and asphalt. Actually, crude oil can be regarded as a solution in which some surface active substances are dissolved in nonpolar hydrocarbons. And formation water is also a solution, too. So, further study about the interfacial tension between two immiscible solutions will help us get better understanding of the oil–water interface present in real reservoirs.

The adsorption between liquids will be explicated through the following examples.

The pure water has a much lower surface tension than a soap aqueous solution where the soap plays a key role as surfactant. Seen from the example, the surface tension between two phases is so sensitive to some extraneous substances that it can be decreased sharply even by a little amount of surfactants such as soap. Both of the characteristics of the surface tension and the extraneous substances should be taken into account when we give an explanation of this phenomenon.

Soap, a particular type of active agents performing as a surfactant, structurally consists of certain salts of higher fatty acids and of organic compounds that are amphiphilic, meaning they contain both hydrophobic and hydrophilic groups. Figure 8.13 gives the molecular structure of the sodium soap, $C_nH_{2n+1}COONa$ (e.g., $C_{16}H_{33}COONa$). Seen from this chemical expression, which can be figuratively illustrated as "—O," its "tail," conventionally called hydrocarbon chain (e.g., $C_{16}H_{33}$-), is a symmetrical, nonpolar, and hydrophobic group formed from the elements hydrogen and carbons; while its "head" is an asymmetrical, polar, and hydrophilic group. The amphiphilicity makes it soluble in both organic solvents and water.

All of the compounds that have structures like this are able to reduce the surface tension of water by absorbing at the interfaces. Here, we will take about the mechanism by which the surfactant molecules lower the surface tension. As shown in Fig. 8.14, if a soupcon of soap surfactant is put into pure water, the surfactant molecules will assemble spontaneously onto the interface, the water surface. The polar ends of these molecules tend to face towards the polar water phase while the nonpolar ends of them are attracted by the nonpolar air at the same time, trying to reduce the polarity disparity over the water surface. As a result, the free surface energy and the surface tension are decreased accordingly.

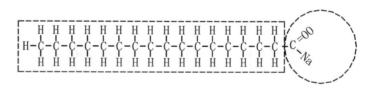

Fig. 8.13 Chemical structural formula of sodium soap molecule

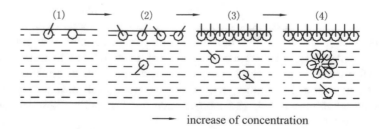

— increase of concentration

Fig. 8.14 Distribution of the adsorbed substances at the interface between two phases

In a word, it is the assemblage of the surfactant molecules at the interface that lower the polarity disparity and thereby reduce the free surface energy (i.e., the surface tension). This is a process during which the free energy tries to descend to its minimum, and so it occurs spontaneously. The soap surfactant can greatly reduce the surface tension of pure water (72.8 mN/m) when even used in very low concentrations.

The term adsorption refers to the process in which the molecules of a substance that is soluble in a two-phase system move from the bulk phase and collect at the interface where they reduce the surface tension considerably. And the substance adsorbed and located at the interface and reducing the surface tension between phases is called surfactant or surface-active substance or surface-active agent. The surface excess, or the concentration of solute at the surface above the concentration in the bulk phase, is referred to as specific adsorption, symbolized as G.

The Gibbs adsorption isotherm written below is an equation used to relate the specific adsorption with the solute concentration and the surface activity:

$$G = -\frac{1}{RT} C \left(\frac{\partial \sigma}{\partial C} \right)_T \tag{8.7}$$

where:

G Gibbs surface excess;
C Solute concentration;
$\left(\frac{\partial \sigma}{\partial C} \right)_T$ Surface activity, meaning the change rate of the surface tension with the solute concentration at a certain temperature;
T, R Absolute temperature and universal gas constant.

According to the sign of $\frac{\partial \sigma}{\partial C}$, adsorption can be classified into to two types:

(1) Positive adsorption: When $\frac{\partial \sigma}{\partial C} < 0$, the specific adsorption G is positive, indicating that the surface tension tends to decrease when the solute concentration is increased, and the solute exists as surface-active substance;

(2) Negative adsorption: When $\frac{\partial \sigma}{\partial C} > 0$, the specific adsorption G is negative, indicating that the surface tension tends to increase when the solute concentration is increased, and the solute exists as surface-inactive substance. Most inorganic salts such as NaCl, $MgCl_2$, and $CaCl_2$ usually act as surface-inactive substances in aqueous solution, increasing the cohesive forces among the water molecules and thence increasing the water–oil and water–gas surface tension.

Obviously, in dealing with interfacial problems, on the basis of Eq. (8.7), we can get the relation, as shown in Fig. 8.15, between the surfactant concentrations in solution with the specific adsorption G and with the surface tension σ.

At low surfactant concentration, as demonstrated by the two curves, an increase in the concentration of the surface active solute will result in a quick increase in the specific adsorption and a quick decrease in the surface tension. However, the

Fig. 8.15 Specific adsorption and interfacial tension at constant temperature

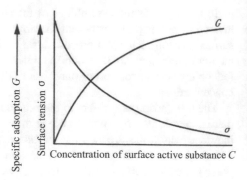

specific adsorption will stop rising and reach its maximum once the surfactant concentration increases a definite value, which means that the interface is already saturated with the adsorbed surfactants. Evidently, at this moment, the surface tension stops increasing with the increase in the surfactant concentration in the solution, either. The right side of Fig. 8.14 shows how the surfactant molecules form aggregates in the water bulk: When surfactants assemble, the nonpolar, hydrophobic tails of them form a core that can encapsulate an oil droplet, and their polar, hydrophilic heads form an outer shell that maintains favorable contact with water—an aggregate of surfactant molecules like this is named "micelle." As the nonpolar ends of the surfactants can be encompassed completely inside the outer shell formed by the polar ends, without contact with water, the surfactant molecules can be dissolved stably in water and thence have little effect on the surface tension of it.

If C is the substance adsorbed at the interface between the phase A and the phase B, the three substances follow the principle of "polarity equilibrium," which is essentially the polarity sequence of them: $A > C > B$. This principle applies to all adsorptions.

Figure 8.16 shows the variation in the oil–water surface tension with the change in pressure in the case where both the phases, the oil and the water, are saturated

Fig. 8.16 Variation in oil–water interfacial tension with pressure

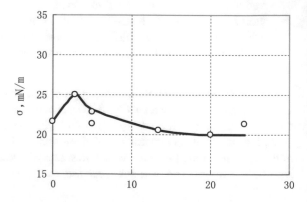

with a mixture of ethane and propane. At low pressure, the surface tension is increased when the pressure is raised; but once the pressure surpasses a definite value, the surface tension is decreased when the pressure is raised; finally, as the pressure continues going higher, the surface tension is tending towards a constant. This process is due to not only the solution of ethane and propane in the two phases but also the concentration variation in the polar constituents contained in the oil phase and collected at the interface.

8.2.3   Adsorption at the Gas–Solid Interface

A few theories and calculation equations concerning the adsorption of gas at solid surface, such as the Langmuir monolayer adsorption theory, the BET (Brunauer–Emmett–Teller) multilayer adsorption theory, and so on, have already been developed by people and can explain well some related phenomena under certain limited applying conditions.

The monolayer theory proposed by Languir in 1916 considered that molecular force field at the solid surface is in a state of undersaturation, and the surplus force yielded from the undersaturation enables the solid surface capable of capturing and adsorbing s gas molecules hitting it. Thus, adsorption is the net result of the two reverse processes concurring at the solid surface and maintaining in dynamic equilibrium: the aggregation and escape of the gas molecules.

Langmuir presented an equation (the Langmuir isotherm) describing the correlation between the adsorbance and the pressure of gas:

$$V = V_\infty \frac{bP}{1 + bP} \tag{8.8}$$

where:

V Amount of the gas adsorbate adsorbed on a certain deal of adsorbent, expressed as number of moles;

P Gas pressure, MPa;

V_∞ Maximum amount of the gas adsorbate adsorbed on a certain deal of adsorbent, expressed as number of moles;

b Langmuir adsorption coefficient or characteristic constant.

It should be noted that when the pressure (P) is quite low, the value of bP is far smaller than 1; thus $V \approx V_\infty bP$, and V is be inversely proportional to P; when pressure (P) is quite high, the value of bP is far larger than 1; thus $V=V_\infty$, meaning that under high pressure the surface of the adsorbent will be covered by the monolayer film formed by the adsorbate and that an further increase in pressure will fail to increase the adsorbance.

The solid surface, where the atoms and molecules are sustaining an asymmetrical cooperation of forces, also possesses surface energy, and it is always trying to adsorb gas molecules or other substances hitting onto it to lower the surface energy. In fact, although people's scientific researches and industrial applications of adsorption in history started with the solid surface, there are still some adsorption mechanisms that we fail to clarify. Generally speaking, the adsorption at solid surface is characteristic of:

(1) The adsorbance (usually expressed as the quantity adsorbed onto per unit of mass) onto the solid surface increases with the increase of the interface;
(2) A real solid surface does not have perfect smoothness, and it is usually rough and anisotropic in mineral composition across the surface. As different mineral components vary widely in adsorption, the solid surface conducts its adsorption capacity with selectivity.
(3) Adsorption is exothermic process, so the adsorption capacity will be reduced if the temperature is raised. However, sometimes, the solubility of the solute in system is decreased with the raised temperature, and the precipitation of some solutes from the solution tends to enhance the adsorption on the solid. That is to say, in the case that the decrease in solubility exerts such a great effect on the adsorption that it overwhelms the diminution of the adsorption capacity due to the thermal effect, the adsorbance increases with the increase in temperature.
(4) The adsorbance held onto the solid surface is proportional to the concentration of the adsorbates in the phase contacting with the solid: The higher the concentration, the greater the adsorbance. Besides, for gas has great compressibility, more gas molecules will be adsorbed onto the solid surface if the pressure is increased.

8.2.4 Adsorption at the Liquid–Solid Interface

When liquid and solid come into contact, a boundary layer of liquid is usually formed at the solid surface, partly induced by the force field at the interface and partly affected by the molecules involved in the boundary layer, for example, some active agents. Real solid is usually rough and anisotropic in mineral composition across the surface. Therefore, the solid surface, allowing different parts of it have different adsorbabilities, exhibits great selectivity in conducting adsorption: The polar parts tend to adsorb polar adsorbates, while the nonpolar parts tend to adsorb nonpolar adsorbates.

Among the reservoir fluids, water has strong polarity and the various non-hydrocarbon compounds containing oil, such as naphthenic acids, asphalts, also have polar structures, so they could easily been adsorbed onto the rock surface composed by polar mineral grains. It is considered that the degree of adsorption of oil on rock surfaces depends on the amount of polar substances contained in oil.

Studies have shown that the boundary layer, or adsorption layer, is firmly attached at the solid surface and has abnormal mechanical property as well as high resistance to shearing, which is the reason why it is very hard to get rid of this film coating the rock. In the rock pore system where the films of water, the boundary layers, are formed and firmly attached on the inner walls of the pores, those pores with radius equal to or smaller than the thickness of the adsorbed boundary layer have no capacity to store reservoir oil. So, from the perspective of adsorption, effective pores are those who have radius larger than the thickness of the water film adsorbed on the pore walls.

The thickness of the adsorption layer is dependent of many factors: the surface characteristics of rock, the pore structure, the surface roughness, the liquid property, and the temperature. All of these factors contribute to the solid surface's characteristic of selective absorption.

The adsorption of liquid at solid surface is much more complex than that of gas. In dealing with the adsorption occurring at the interface between a solid and a solution, more than one empirical equation are put forward by people to correlate the amount of the solute adsorbed on the solid surface with the solute concentration in the liquid phase where the adsorbates come form. Taking the Langmuir adsorption isotherm as example:

$$\Gamma = \Gamma_\infty \frac{bc}{1 + bc} \tag{8.9}$$

where:

Γ Adsorbance, mol/g;
Γ_∞ Saturation adsorbance (the amount of the adsorbed solute that coats 1 g absorbent by one single layer), mol/g;
c Solute concentration, mol/L;
b Adsorption constant, depending on the temperature, as well as the properties of the solute and the solvents, L/mol.

8.2.5   Wetting Phenomena and Capillary Force

As we all know, adsorption refers to the process in which the surface of a material spontaneously adsorb something from the neighboring medium to lower its free surface energy. Wetting is another process spontaneously occurring in nature. When two immiscible fluids (e.g., water and oil) come into contact with the solid phase in conjunction, one of them will spread across the surface where the free surface energy of the system can be thus lowered. This phenomenon is called wetting, and the phase that spread across the rock surface is called the wetting phase. As the wetting effect dominates the microdistribution of the oil, gas, and water in the underground pore system, it is a problem deserving of deepened study. It will be discussed in detail in our next lesson.

Supposing that the wetting phenomenon occurs on capillary pores of reservoir rocks, in the capillary tubes there would appear a meniscus (bended liquid interface/surface), in virtue of which a capillary pressure also occurs. The capillary force, playing a key role in the seepage flowing mechanism of the reservoir fluids, leads to many additional resistances with which the engineers are constantly concerned. Relevant contents are included in the next chapter.

8.2.6 Interfacial Viscosity

Because of the mentioned wetting, adsorption phenomena occurring in a liquid–solid system, the fluid that moves past an object exhibits at the interface a viscosity (the so-called interfacial viscosity) different from its bulk viscosity.

The determination of the interfacial viscosity is challenging. In dealing with this problem, a lot of scholars have measured the interfacial viscosity in different ways, but the results from their measurements suffer great discrepancy between each other. This is understandable in the presence of the difference among the measuring instruments which decide the experimental result a lot, and the discrepancy may also grow out of the great errors due to the sensibility of the interfacial viscosity to the impurities; what's more, we can not ascertain whether some other factors such as the "edge effect" have made an impact significant enough to disturb the measurements and to make great errors.

Currently, concerning the interfacial viscosity, the following conclusions are realized and accepted by people:

Within the interfacial region, it is the structured polar groups (e.g., groups structured by hydrogen bonds) that restrain the molecules from performing relative movements and thus render the fluid near the rock surface exhibit abnormally high viscosity. Some people consider that the interfacial viscosity is 10^6 times greater than the viscosity of the bulk liquid.

8.3 Wettability of Reservoir Rocks

Wettability of rocks is a combined character reflecting the interaction between the rock minerals and the reservoir fluids, is fully as important as other parameters such as the porosity, the permeability, the saturation, and the pore structure. In fact, it plays a key role in deciding the microdistribution of the reservoir fluids in the rock pore space, and therefore deciding the level of difficulty will be encountered by the injected fluids.

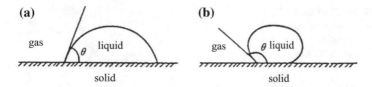

Fig. 8.17 Selective wetting of liquid on solid surface

8.3.1 Basic Concepts About Wettability of Reservoir Rocks

Wetting refers to the phenomenon of how a liquid deposited on a solid substrate spreads out under the force of surface tension. Understanding wetting enables us to explain why water spreads readily on clean glass, where a droplet of mercury cannot spread at all. Figure 8.17 shows the phenomenon.

When the term "wetting" is mentioned, three phases, solid, liquid, and gas, must be involved in the studied system. And, moreover, whether a liquid can wet up a solid or not is always suggesting that it is a comparison relative to the third phase (gas or another liquid). That is to say, if one of the non-solid phases can wet up the solid, the other cannot.

1. Parameters Describing the Degree of Wetting

The degree of wetting can be described by the contact angle or the adhesion work.

(1) Contact Angle (Also Called Wetting Angle)

At any point located on the liquid–liquid–solid (or gas–liquid–solid) triple line, as shown in Fig. 8.18, where each sketch illustrates a small liquid droplet is resting on a flat horizontal solid surface, the tangent line drawn tangential to the liquid–liquid (or gas–solid) interface forms an angle with the solid–liquid interface. This angle, symbolized as θ, is rightly called the contact angle. It is stipulated that θ should be reckoned from the side presented by the liquid phase with higher polarity.

For a system composed of oil, water, and rock, wetting can be characterized into the following types

Fig. 8.18 Selective wetting of oil and water on solid surface. *1*—Water; *2*—oil; *3*—solid

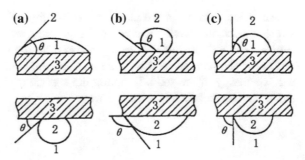

Fig. 8.19 Equilibrium of interfacial tensions at the three-phase perimeter. *1*— Liquid (L), *2*—gas (g), *3*— solid (s)

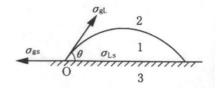

(1) Water-wetting: $\theta < 90°$—in this case, the rock can be wetted up by water, meaning that the rock has good hydrophilicity;

(2) Neutral-wetting: $\theta = 90°$—in this case, the rock has equivalent capacities to wet up oil and water, meaning that the rock has neither hydrophilicity nor lipophilicity;

(3) Oil-wetting: $\theta > 90°$—in this case, the rock can be wetted up by oil, meaning that the rock has good lipophilicity.

The spreading of a fluid covering the solid surface results from the interactions among the surface tensions occurring along the line of contact (the triple line) where three phases meet. Figure 8.19, taking the contact point O as an example, gives the three surface tensions at each point of the contact triple line: the gas–liquid surface tension σ_{gL}, the gas–solid surface tension σ_{gs}, and the liquid–solid surface tension σ_{Ls}. In equilibrium, the three surface tensions satisfy such a correlation equation:

$$\sigma_{gs} = \sigma_{Ls} + \sigma_{gL} \cos \theta \qquad (8.10)$$

The above equation is rightly the famous Young's Equation.

By rearranging Eq. (8.10), we can get:

$$\cos \theta = \frac{\sigma_{gs} - \sigma_{Ls}}{\sigma_{gL}} \qquad (8.11)$$

Or,

$$\theta = \arccos \frac{\sigma_{gs} - \sigma_{Ls}}{\sigma_{gL}} \qquad (8.12)$$

(2) Adhesion Work

Another indicator for the magnitude of wetting of a rock is the adhesion work, which means the work required in the environment of a non-wetting phase to separate per unit area of the wetting phase from the solid surface.

During the process, as shown in Fig. 8.20, the work done is converted into new surface energy of the solid. Denoting the surface energy increment as the symbol ΔU_s, we can calculate it in such a way:

Fig. 8.20 Diagram of
adhesion work

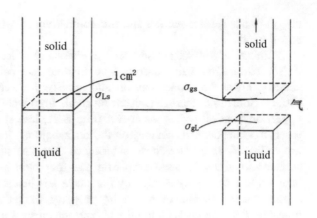

$$\Delta U_s = U_2 - U_1 = \left(\sigma_{gL} + \sigma_{gs}\right) - \sigma_{Ls} \qquad (8.13)$$

where:

U_1, U_2 The specific surface energy before and after the leaving of the wetting
            phase from the solid surface;

$\sigma_{gL}, \sigma_{gs}, \sigma_{Ls}$ The gas–liquid, gas–solid, and liquid–solid surface tension.

In terms of the concept of surface tension, we can know that $(\sigma_{gL} + \sigma_{gs}) > \sigma_{Ls}$.
So, $\Delta U_s > 0$, meaning that the surface energy of the system has increased. This
increment of surface energy is equal to the adhesion work W:

$$W = \Delta U_s = (\sigma_{gL} + \sigma_{gs}) - \sigma_{Ls} = (\sigma_{gs} - \sigma_{Ls}) + \sigma_{gL} \qquad (8.14)$$

Then, by bringing in the Young's Equation, we obtain:

$$\sigma_{gs} - \sigma_{Ls} = \sigma_{gL} \cos \theta \qquad (8.15)$$

Therefore, the correlation between the contact angle and the adhesion work is:

$$W = \sigma_{gL}(1 + \cos \theta) \qquad (8.16)$$

As seen from Eq. (8.16), a smaller contact angle θ is indicative of a greater
adhesion work W and thereby a better spreading of the wetting phase on the solid
surface. Therefore, the adhesion work can be used to tell the level of rock wettability: for a oil–gas–water three-phase system, the rock is hydrophilic (water-wet) if
the adhesion work is greater than the oil–water surface tension; and the rock is
lipophilic (oil-wet) if the adhesion work is smaller than the oil–water surface tension; and the rock is neutral-wet if the adhesion work is equal to the oil–water
surface tension.

Despite no direct measurements available for the oil–solid or water–solid surface
tension, the adhesion work can be calculated through equations correlating it with

the oil–water surface tension and the contact angle, which can be obtained from experiments.

2. Effects of Surfactants on Wetting—Wettability Reversal

The surfactants, being adsorbed at the interface between two phases spontaneously, lower the surface tension and modify the wetting behavior of the solid surface. Sometimes there may occur reversals from water-wet to oil-wet (Fig. 8.21, upper) or from oil-wet to water-wet (Fig. 8.21, lower). The degree to which the reversal occurs depends not only on the surface properties of the solid as well as the properties of the surface-active substances, but also on the concentration of the surface-active substances. Comparing the four curves given in Fig. 8.22, what happens to the case represented by the curve 1 is just a modification in its wetting degree, but no transition of the wetting property, while the curves 2 and 3 undergo a transition from water-wet into oil-wet, and the curve 4 undergoes a transition from oil-wet into water-wet.

The term "wettability reversal" refers to the reversal of the wettability of a solid surface where surface-active substances are adsorbed. Sand grains, for example, are water-wet originally, but they usually undergo a wettability reversal in the presence

Fig. 8.21 Wettablility reversal of solid surface. *1*—water, *2*—oil; *3*—solid

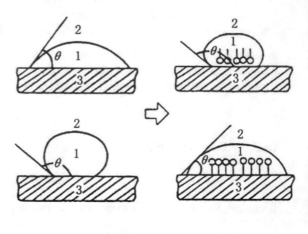

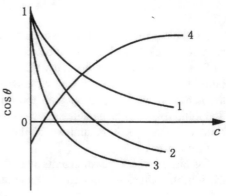

Fig. 8.22 Relationship between selective wetting and concentration of active agent

of some surface-active substances and thus become oil-wet. Similarly, in terms of the principle of wettability, we can also perform some technological processes reversal so as to enhance the oil recovery ratio. For example, if we inject a certain deal of active water into the formation, the surface-active agents in it will form a second adsorption layer coating the rock surface, acting on the principle of "polarity affinity" and consequently counteracting the effect of the original, or the first, active substances. As shown in Fig. 8.21, the originally oil-wet solid surface is finally reversed to a water-wet one, which makes it easier for the oil to be driven away and ensures a higher recovery.

8.3.2   Wetting Hysteresis

As shown in Fig. 8.23, if we tilt a horizontally placed solid surface to an angle α, it can be observed that what firstly happens to liquid drop resting on the surface is the deformation of the oil–water surface, which changes the corresponding original contact angle; then, the perimeter of the triple line where the oil, water, and solid meet each other follows to advance downwards. At the point A, where the oil is driven by the water, the newly formed contact angle θ_1 is named the advancing contact angle. Obviously, $\theta_1 > \theta$ (θ is the original contact angle). At the point B, where the water is driven by the oil, the newly formed contact angle θ_2 is named the receding contact angle, and $\theta_2 < \theta$.

When a liquid is going to move on a solid surface by external forces, the term "wetting hysteresis" refers to the change in the wetting contact angles caused by the sluggish movement of the triple line across the surface. The wetting hysteresis is considered to be related to the flowing four factors:

(1) The direction in which the triple line is advancing—the static wetting hysteresis.

The static wetting hysteresis occurs because that the oil and the water do not get into contact with the rock surface simultaneously, but in specified order—when water displacing oil, $\theta_1 > \theta$; and when oil displacing water, $\theta_2 < \theta$. In other words, the static wetting hysteresis is due to the wetting sequence, respectively, before and after the arrival of water; the point A is in contact firstly with the oil phase and then with the water phase, representing the wetting hysteresis during the process of water

Fig. 8.23 Wetting hysteresis

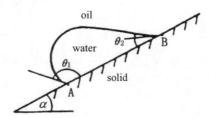

Fig. 8.24 Dynamic wetting hysteresis

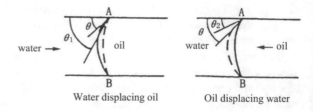

displacing oil, and vice versa. The cosine difference between the advancing angle and the receding angle is usually used to indicate the degree of wetting hysteresis.

Researches have proved that the hysteresis of the contact angles is one of the main causes of the hysteresis of the capillary force.

(2) The velocity at which the triple line is advancing—the dynamic wetting hysteresis.

No matter during the process of water displacing oil or the process of oil displacing water, changes in the contact angles must occur along with the triple line because that the different segments of the oil–water interface advance at different velocities. This kind of changes of the contact angles are referred to as the dynamic wetting hysteresis. As shown in Fig. 8.24, when the oil–water interface stays in a stationary equilibrium, the meniscus exhibits a contact angle θ smaller than 90°; when water displacing oil, the interface begins to move but the triple line does not advance immediately, and what happens before that is the deformation of the oil–water interface and the thence the increase in the contact angle (enlarged to θ_1), which represents the so-called dynamic wetting hysteresis; when oil displacing water, the contact angle decrease to θ_2. With regard to the magnitude of the three angles, including the advancing contact angle θ_1 (the enlarged angle) at the water-displacing-oil interface, the receding contact angle θ_2 (the lessened angle) at the oil-displacing-water interface and the wetting angle θ measured when the triple line stops moving and stays in stationary equilibrium can be arranged in such a sequence: $\theta_1 > \theta > \theta_2$.

Figure 8.25 shows that both of the advancing angle and the receding angle are related to the velocity at which the wetting perimeter is moving. Generally

Fig. 8.25 Relationship between the contact angle and the movement of three-phase perimeter

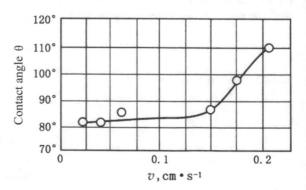

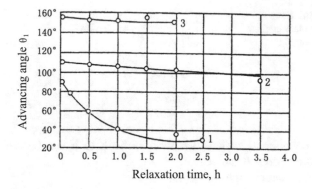

Fig. 8.26 Relationship between the Relaxation time and advancing angle

speaking, the higher the velocity, the more serious the effect of dynamic wetting hysteresis will be. Once the velocity exceeds a specified critical value, there would occur a wettability reversal as shown in Fig. 8.26. That is to say, in the case that the water remains at rest or moves at a low velocity, it can wet the formation rocks quite well; however, once the injected water drives the oil with such a rapid speed that the meniscus moves at a velocity exceeding the critical velocity at which the water can wet the solid surface, leading to enlarged wetting angle and consequently the wettability reversal. In fact, this effect is everything but a circumstance favoring the extraction of oil from the reservoir rocks—after the water following through the pores and channels, a film of oil will be left coating the pore walls, which impedes the further displacement of oil considerably.

(3) The adsorption of those surface-active substances on the rock surface.

As shown in Fig. 8.27, the wetting of liquid experimented on a smooth, clean marble is quick to reach an equilibrium, while the relaxation time is longer for the experiment performed on a marble that has been immersed in crude oil for 59 days; the relaxation time increases again when the experiment is performed on a marble that has undergone a pretreatment in oleic acid. Obviously, the shape of the curves 2 and 3 is associated with the adsorption of the surface-active substances. Oil acid is an active substance itself, and there are also a certain amount of surface-active substances contained in the oil, which can be adsorbed onto the rock surface. The effect of static wetting hysteresis tends to be aggravated by the adsorption layer attached to the solid surface. The more firmly the oil film is attached to the inner

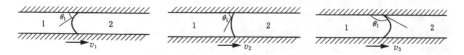

Fig. 8.27 Dynamic wetting hysteresis and wettability reversal

Fig. 8.28 Relationship
between the contact angle θ
and the shape angle τ

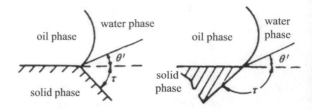

wall of the pores and channels, the more difficult to it can be removed, which
implies a more serious hysteresis.

(4) The roughness of the rock surface.

Generally speaking, the more rough and anisotropic the solid surface, the bigger
difficulty for the triple wetting line to move, and therefore the more serious the
effect of wetting hysteresis.

Wetting hysteresis is strongly affected by the edges and corners and the sharply
pinnacled angles of the solid. As proven by experiments, the sharp parts distributed
across the solid surface put up profound resistances against the movement of the
triple line. Figure 8.28 shows that the degree of wetting hysteresis at a rough
surface must be expressed as $\theta = \theta' + \tau$, where θ is the real contact angle that
reflects the wetting hysteresis, θ' is the contact angle measured by definition or, in
another way to illustrate, the contact angle if hypothetically not encountering the
emergent change in the surface, and τ is the "shape angle." The bigger the shape
angle is, the more serious the hysteresis is.

The phenomenon of wetting hysteresis has considerable impact on the shape and
position of the capillary pressure curves as well as the relative permeability curves
measured from different types of displacing processes. A full explanation about this
point will be given in the following chapters.

8.3.3 Influencing Factors for the Wettability of Reservoir
Rocks

The wettability of reservoir rocks is a long-lasting problem being concerned,
argued, and researched by petroleum engineers. The original viewpoint put forward
by some scholars, especially the geologists, is that all the reservoir rocks must be
water-wet because that almost all of them are formed in water environments
through sedimentation and diagenesis, and this point was supported by the fact that
the fresh surfaces of the minerals composed in reservoir rocks are hydrophilic.
Then, another viewpoint was proposed: All the reservoir rocks must be oil-wet
because that the rock surfaces in a formed reservoir have been in touch with oil
during such a long period that the wettability there is already reversed to oil-wet due
to the adsorption of the active substances contained in oil.

As researches progressed, however, people realize that:

(1) Considering the diverse conditions of sedimentation and formation for real reservoirs, the rocks as well as the reservoir fluids display a wide range of variation in physical property. So, it is hard to give a sweeping generalization of the wettability of reservoir rocks—some are water-wet and some are oil-wet. Statistics given in the early 1970s showed that among the 50 plus investigated reservoirs being under development, 27 % of them are water-wet ($\theta = 0$–$75°$), 66 % of them are oil-wet ($\theta = 105$–$180°$), and the rest ones are neutral-wet ($\theta = 75$–$105°$).

Another report of the same class said that water-wet rocks account for 26 %, oil-wet rocks account for 27 %, and neutral-wet rocks accounts for 47 %. From the two reports, we can know that water-wet rocks, with similar percentage in the two given reports, are in the minority; and the oil-wet rocks, although taking shares utterly different percentages in the two reports, can be believed to be more than a few.

(2) Even for a single reservoir, it is also possible for the rocks to possess microanisotropy. Because of the roughness and variation in mineral composition of the rock surface, plus the complexity of the oil complexity, it is normal for the different parts of the surface of a real rock to be different in wettability, which is referred to as anisotropic wettability—parts of a surface are water-wet, and the other parts are oil-wet.

Regarding the anisotropic wettability, it falls into two categories: speckled wettability and mixed wettability. The speckled wettability, also called partial wettability, indicates a rock with surface exhibiting various local wettabilities according to the different mineral compositions in different parts. Besides, the distribution of the water-wet or the oil-wet does follow a random rather than regular way. As far as a single pore is concerned, part of its surface can be strongly water-wet while the other part of it is strongly oil-wet, and, moreover, maybe the oil-wet part is discontinuous in distribution which is shown in Fig. 8.29.

Mixed wettability is the wettability that varies with the size of the pore space—small pores maintain to be water-wet and contain no oil at all, while big pores, because of the sand-grains' long time of exposure to crude oil, exhibit to be oil-wet

Fig. 8.29 Speckled wettability

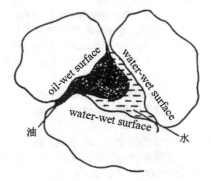

Fig. 8.30 Mixed wettability

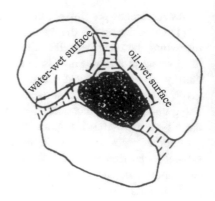

and allow the contained oil to form a continuous channel flow as shown in Fig. 8.30.

It will be seen that the wettability of reservoir rocks is an extraordinarily complex problem—some rocks are water-wet (hydrophilic) and some others are oil-wet (oleophylic) or even mixed-wet. The wettability of reservoir rocks is mainly dependent on:

1. The mineral composition of rock

Sandstone and carbonate rock are the two major types of reservoir rocks. The carbonate rocks, being comparatively simple in composition, are mainly composed of calcite and dolomite; sandstones are usually composed of silicate minerals of different properties and crystal structures, for example, feldspar, quartz, mica, as well as clay minerals and sulphates (gypsum). So sandstone is much more complex than carbonate rock in properties such as surface characteristics and wettability, since the diversity of its mineral composition.

According to the definition of wettability, the minerals contained in rock can be classified into two types: one is the water-wet minerals, such as quartz, feldspar, mica, silicates, glass, carbonates, and aluminosilicates, and on the surface of these minerals, water drops have a wetting contact angle smaller than $90°$; the other one is the oil-wet minerals, mainly consisting of talcum, graphite, solid organic hydrocarbons, and metal sulfides.

Clay minerals have great impact on the wettability of reservoir rocks. Because of the swelling behavior of some clay minerals in water, for example, the montmorillonite, the mud cements increases the hydrophilicity of rock. Apart from this, those clay minerals containing iron exert another impact on the rock: for example, the element iron present in the oolith chlorite ($Fe_3AL_2S_4O_{10}·3H_2O$) can adsorb surface-active substances from the crude oil, leading to local wettability reversals into lipophilicity across the rock surface.

2. Composition of the Reservoir Fluids

In consideration of their impacts on rock wettability, the complex constituents comprising crude oil can be classified into three categories: (1) the nonpolar hydrocarbons (the main part of oil), (2) the compounds containing polar oxygen,

Table 8.4 Advancing angle of different hydrocarbon components in contact with the smooth surface of polyfluortetraethylene

Hydrocarbon	Pentane (C$_5$H$_{12}$)	Hexane (C$_6$H$_{14}$)	Octane (C$_8$H$_{18}$)	Dodecane (C$_{12}$H$_{26}$)
Advancing angle (degree)	0	8	26	42

sulfur, or nitrogen, and (3) the polar substances, also called active substances, contained in crude oil.

The wettability degree of the nonpolar hydrocarbons varies with their number of carbon atoms. Table 8.4 shows the contact angles, respectively, of pentane, hexane, octane, and dodecane in contact with the smooth surface of polyfluortetraethylene. Obviously, more the number of carbon atoms contained in hydrocarbon is, the larger the contact angle is.

3. The Surface-Active Substances

The surface-active substances may alter, or even reverse, the rock wettability by adsorbing onto the surfaces, so they exert even greater impact on rock wettability than the polar substances mentioned above. At the present time, it is common in oil fields to add certain amounts of surface-active agents into the injected water, so as to reduce the oil–water interfacial tension and alter the rock wettability and to ultimately improve the displacement efficiency.

The surface-active substances can also be adsorbed onto the rock surface, and their adsorbance tends to decrease with the increase in the electrolyte amount present in water. Besides, certain metal ions contained in water are also able to alter the wettability of rock surface.

4. The Roughness of Rock Surface

Due to the roughness of rock surface, the surface energy across its appearance is not uniform, and therefore its surface energy varies from here to there, too. So, reservoir rocks may exhibit speckled wettability and mixed wettability at their surfaces. As proven by experiments, the sharp parts distributed across the solid surface put up profound resistances against the movement of the triple line. In Fig. 8.28, we can see that to reasonably represent the real degree of hysteresis at a sharp part of a rough rock surface, the shape angle τ must be added to the contact angle. Obviously, the larger the shape angle is, the more serious the hysteresis effect is.

Besides, the wettability is also related to some other factors such as pore structure, temperature, and pressure. But researches show that temperature and pressure affect little on the wettability of the oil-bearing formations. In short, wettability is a manifestation of rock–fluid interactions associated with fluid distribution in porous media.

8.3.4 Distribution of Oil and Water in Rock Pore Space

The wetting characteristics of the reservoir rocks, being rather complex under reservoir conditions, play a key role in deciding the distribution of irreducible oil and the ultimate recovery ratio by affecting microdistribution of the reservoir fluids in rock and controlling the magnitude and direction of the capillary forces in the pore system, which impact the process of water displacing oil remarkably. Conversely, the injected water into the reservoir may work a significant impact on the rock wetting characteristics, too. Each of the two aspects of the problem will be discussed in the following section.

1. Wettability Affects the Oil and Water Microdistribution

The distribution of the liquids in a pore system is dependent upon the wetting characteristics. So, because of the variation in wettability across the surfaces the rock grains, the oil–gas distribution in the pores and channels within reservoir rocks differ in thousands of ways. In Fig. 8.31, the diagram (a–c), respectively, illustrate the intergranular oil–water distribution in the case of different saturations within a water-wet rock; while the diagram (d–f), respectively, illustrate the intergranular oil–water distribution in the case of different saturations within an oil-wet rock.

Resulting from the interactions between the surface tensions of the present phases, the wetting phase is always attached to the grain surfaces and trying to occupy the narrow corners and interstices, impelling the non-wetting phase to the mid-intergranular space which is comparatively open and unobstructed.

Within packings of water-wet grains, the wetting phase, water, is attached to the grain surfaces. As shown in Fig. 8.31a, the pendular-ring distribution is a state in which the water collects around the intergrain contacts in forming the pendular-ring

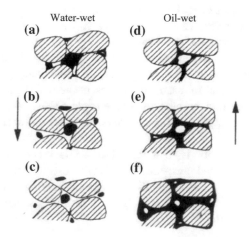

Fig. 8.31 Distribution of oil and water in rock pore system. **a** Water: pendular ring; oil: circuitousness, **b** water, oil: circuitousness, **c** water: circuitousness; oil: isolated droplet, **d** water: circuitousness, oil: pendular ring, **e** water, oil:circuitousness, **f** water: isolated droplet; oil: circuitousness

distribution. Since the pendular saturation is too low to allow the water rings to contact with each other and to form a continuous body, it is possible for the water in such a distribution to flow. The water pendular rings, as the non-continuous and non-flowing phase in the pore space, exist as connate water. Correspondingly, the oil, the non-wetting phase sharing a higher saturation in the pore space, is distributed continuously and tortuously in the mid of the interconnected intergranular pores, and, what's more, it will migrate under pressure differential to form a channel flowing.

As the water saturation progresses, the water pendular rings grow in size until the rings encounter each other and connect to form a whole body which is one of the forms of coexisting water. In such a case, whether the water is able to flow depends on the magnitude of the pressure differential. As long as the amount of the water exceeds the coexisting water saturation, the water will form a tortuous distribution and join in the flowing, as shown in Fig. 8.31b.

Responding to further increase of the water saturation in the intergranular space, the oil, as shown in Fig. 8.31c, loses its continuity and break up into drops, which exist as isolated oil droplets distributed in the mid-pore space. The some big oil droplets, driven away by comparatively great pressure differential, may be stuck at the narrow pore channels and put up great resistance against the flow through the pore system.

Within packings of oil-wet grains, the oil–water distribution corresponding to different saturations in the pore system is the inverse of the case described above, and the case of water displacing oil is shown in Fig. 8.31d–f.

Observation of the flowing process of the oil and water through artificial porous mediums show that the oil and water flow along with their, respectively, own paths in the form of "channel flowing," which is shown in Fig. 8.32. The flowing channels vary greatly in diameter: somewhere it can be as narrow as a sand-grain-diameter minus, while somewhere it can be as wide as several times a sand-grain-diameter. The water- or oil-flowing channels are usually sandwiched between the oil–water or solid–liquid interfaces, being winding and tortuous. With

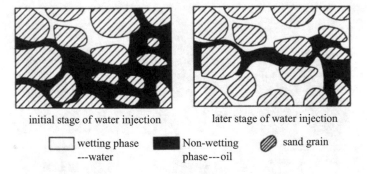

initial stage of water injection   later stage of water injection

☐ wetting phase   ■ Non-wetting   ⬭ sand grain
---water   phase---oil

Fig. 8.32 Diagram of channel flowing—wetting phase displacing non-wetting phase

the variation in the saturations, the geometric configuration of the flowing-channel network also changes. Generally speaking, when the water saturation is raised, the number of the water flowing channels increases, while that of the oil-flowing channels decreases. Additionally, through the numerous interconnected pores impenetrating the porous medium, the flowing channels are always able to find their way out.

As concerns the flow pattern, it is commonly considered that almost all the observed cases of liquid channel flowing is in the pattern of laminar flow—although so irregular and tortuous the rock pores are, no vortex flow occurs. But during the water-displacing-oil experiments conducted on some photoetching-glass micro-models, the author once did observe the phenomenon of vortex flow, in which the oil droplets, being scoured by the flowing water, eddied at the throats.

Figures 8.33 and 8.34 give the water-injection process and the oil–water distribution in the case of water-wet reservoir rocks. Within the region that has not been swept by the injected water, the coexisting water saturation is very low and the water is distributed, coating the grain surfaces as water films or stagnating at corners shaped by the sharply pinnacled angles of the solid, while the other pore

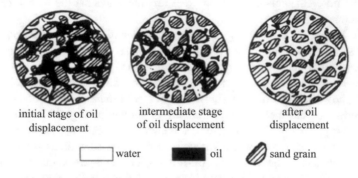

initial stage of oil   intermediate stage   after oil
displacement   of oil displacement   displacement

☐ water   ■ oil   ⬭ sand grain

Fig. 8.33 Oil–water distribution in water-wet rock during water injection

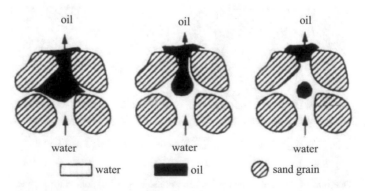

oil   oil   oil

water   water   water

☐ water   ■ oil   ⬭ sand grain

Fig. 8.34 Water-displacing-oil process in water-wet rock (imbibition)

space is occupied by oil. Within the section where the simultaneous flow of oil and water occurs, most of the oil flows in the pattern of channel following although some of it is trapped in blind alleys, and a small proportion of the oil is divided into small oil droplets by the intruding water. Within the water-flooded region, there are only a few oil droplets left in the pore space.

As shown in Figs. 8.35 and 8.36, the non-wetting water entering the oil-wet reservoir rocks firstly forms into tortuous, continuous flowing channels along with the pores that are comparatively large in size and good in interconnectivity. With the water injection going on, more and more comparatively small pores turn to permit the water in and are gradually connected in series, whereupon some new flowing channels of water are formed. Consequently, so many flowing channels have been tunneled through the rocks that the water can easily conduct seepage flowing in the pore system, where the flowing channels of oil have almost disappeared. But there is still some residual oil, some of which is trapped in small channels, and the left exists as oil film coating the solid surface within those big flowing channels of water. It is difficult for the injected water to displace the oil in the form of thin film. Besides, with regard to the residual oil distribution within neutral-wet rocks, there is not only some stagnant oil occupying the "dead alley"

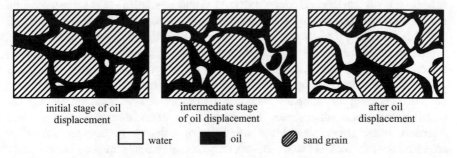

initial stage of oil
displacement

intermediate stage
of oil displacement

after oil
displacement

☐ water   ■ oil   ⬮ sand grain

Fig. 8.35 Oil–water distribution in oil-wet rock during water injection

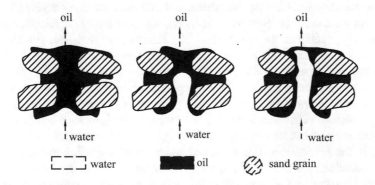

oil   oil   oil

water   water   water

⌐ ⌐ water   ■ oil   ⬮ sand grain

Fig. 8.36 Water-displacing-oil process in oil-wet rock (drainage)

pores, but also a certain amount of oil drops adhering to the rock surfaces enclosing the pores.

Not only is the oil–water distribution within a rock pore system a function of saturation, but is also a function of the saturation history of the particular porous medium that is being investigated—that is to say, it depends on whether the displacement process is a "wetting phase displacing non-wetting phase" one or a "non-wetting phase displacing wetting phase" one.

The process of forcing the entry of a non-wetting phase into a system initially saturated with a wetting phase is called the "drainage process," during which the non-wetting-phase saturation increases by reducing the wetting-phase saturation. For instance, driving the oil out of an oil-wet reservoir by the method of water injection is a drainage process.

Conversely, the process of forcing the entry of a wetting phase into a system initially saturated with a non-wetting phase is called the "imbibition process," during which the wetting-phase saturation increases by reducing the non-wetting-phase saturation. For instance, the process of water displacing oil in a water-wet reservoir belongs to imbibition.

Figures 8.34 and 8.36 show, respectively, two oil-water distribution variations during the imbibition and the drainage processes. Comparing the rock with different saturation histories, their oil–water distribution characteristics at the same saturation are not identical. The pore system shown in Fig. 8.36 was initially saturated with oil and then invaded by water that caused the oil to be displaced—that is to say, the saturation sequence of this rock is "oil first and water later." In contrast, the rock shown in Fig. 8.33 was initially saturated with water before the entry of oil, and only a film of water coating the rock grains is left, which can be considered as a process occurring in company with the hydrocarbon accumulation during the reservoir formation; then, when the reservoir is under development by water injection, water comes and drives away the oil—that is to say, the saturation sequence for this rock is "water–oil–water." The saturation sequence is essentially the so-called saturation history, which also implies the effect of static hysteresis. Static hysteresis is a hysteresis caused by the sequence in which the fluids get to saturate the solid.

2. Wettability Decides the Magnitude and Direction of CapillaryForce

As introduced in the previous discussion, the distribution regularities of fluids are not only dependent upon the wetting characteristics of rock, but also affected by the presence of the capillary force in the rock pores and channels. In fact, the magnitude and direction of the capillary force are also decided by whether the capillary tubes formed by the rock pore space are water-wet or oil-wet. When the capillary tube is water-wet, the capillary pressure is in the same direction as the driving pressure difference formed due to water injection, so it exists as driving force; conversely, because that the capillary pressure present in an oil-wet capillary tube is in the reverse direction to that of water displacing oil, it exists as resisting force. Therefore, in the development practice of oilfields, the influence of the capillary pressure on the displacement should be sufficiently considered, and more details about this point will be given in the next chapter.

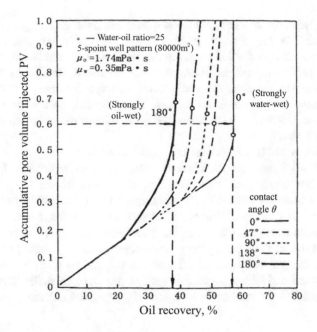

Fig. 8.37 Effect of rock wettability on the recovery of water-injection reservoir

3. Wettability Affects the Recovery Ratio

In the case of water displacing oil, the oil recovery has much to do with the wetting characteristics of reservoir rocks. From above paragraphs, we can infer that the oil recovery of a water-wet reservoir should be higher than that of an oil-wet reservoir.

As far as normal isotropic wetting systems are concerned, water-wet formations have higher recovery than the oil-wet ones. The reason for this is that the water injected into a water-wet formation prefers to be imbibed into the pores and channels spontaneously, reducing the velocity differences of flow between the pores of different sizes and overcoming the disadvantageous viscous fingering, with the result of a larger swept region by water and thereby a higher oil recovery.

As shown in Fig. 8.37, if the same volume of water is injected into reservoirs with different rock wetting characteristics, the oil recoveries obtained are not identical. For example, when 0.7 pore volume of water is injected into a strongly oil-wet reservoir, the recovery is 37 %, and when the same volume of water is injected into a strongly water-wet reservoir, the recovery is 57 %.

But different conclusions are drawn by Warren and Calboun from their experiments in which they inject specified volume of water (20PV) into artificial cores with various wetting characteristics and obtain a collection of recovery data for comparison and analysis. They propose that the highest recovery is reached in the approximately neutral-wettability cores, and this is because that under such a condition the oil exists as an uncontinuous phase and the interfacial tension remains minimized. However, in a strongly water-wet system, the smaller pore openings are so preferential for the injected water to enter that the larger pore openings are skirted and the oil contained in them are upswept. Besides, too strong an interfacial tension is likely to nip off, or disconnect, the oil stream flowing in the pores. In strongly water-wet systems, the water tends to finger into the comparatively big pores and thus leave some oil untouched. These are rare such unfavorable occurrences in neutral-wet reservoirs where the oil is seldom disconnected or skirted by the injected water.

3. Injected Water Has Effect on the Rock Wettability

Long period of water injection causes changes in rock wettability. For example, sealed corings targeting at some water-flooded areas of the Daqing Oil Field manifest that when the water saturation exceeds 40 %, the wettability of most of the reservoir rocks is converted from weak oil-wet into weak water-wet, and when the water saturation exceeds 60 %, all of the reservoir rocks become water-wet. This is because that the oil film coating the rocks, subjected to the washing of the injected water, has been gradually peeled off and the original rock surface composed of water-wet quartz and feldspar is consequently exposed to the fluids. The conversion of rock wettability favors the displacement efficiency and enhances the tertiary oil recovery.

8.3.5   Measurement of Wettability of Reservoir Rocks

Because of the complexities of rock wettability under reservoir conditions, no methods of measurement for it have so far succeeded in yielding satisfactory results. In fact, the wettability of rocks under reservoir conditions is usually estimated from the in-laboratory experiments, which can be generally classified into two types: (1) direct measurements, for example, contact-angle method, liquid-drop method, and hanger-plate method; (2) indirect measurements, for example, Aamodt method, imbibition method, and centrifuge method. The following paragraphs will briefly introduce several most commonly used methods.

1. The Contact-Angle Method

As shown in Fig. 8.38, a drop of liquid (oil or water) of about $1 \sim 2$ in diameter is placed upon the smooth surface of the specially prepared plate of substratum, a core sample (mineral) which has already been processed to the shape of flat slab and polished on the surface. Contact angles of the liquid drop can be directly determined according to its images photographed and magnified via a special optical system or camera. What's more, the advancing angle and receding angle can also be determined by tilting the flat slab to some degree.

During the measurement, it is better to use fresh oil or water samples obtained from wellbore bottom, and oxidation of the samples should be avoided. But simulated reservoir oil and simulated reservoir water can also be employed when no fresh oil or water samples are available. Usually, for convenience, we use the major mineral crystals to stand in for the rock which needs to be measured. For example,

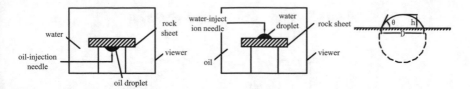

Fig. 8.38 Measurement of contact angle

in China, the surfaces of carbonate rock and sandstone are usually studied with reference, respectively, to a crystal of calcite and Iceland spar, whose surfaces are polished before measurement.

To determine the contact angle in the case of reservoir temperature and pressure, the measuring installation introduced above should be put in a high-pressure chamber where the reservoir conditions can be simulated. Generally speaking, the characteristic of wettability of a water–oil–rock system is greatly affected by temperature but insignificantly by pressure.

This method features as follows:

(1) The measuring principle is simple, the operating procedure is straightforward, and the result is visualized.
(2) It strictly requires specified conditions of measurement. For example, the minerals for test must be smooth, clean, and free from contamination on the surface, and the temperature must be held a constant.
(3) It requires a great deal of time on account of the stabilization of the liquid drop, which might otherwise cause great errors if the waiting time is not long enough. Sometimes it consumes a period of time as long as hundreds or even thousands of hours.
(4) Strictly speaking, the results yielded from this method are not exactly the real contact angles of reservoir rocks of which the experimented "major mineral crystals" are representative.

According to the SY/T 5153-2007 petroleum industry standard, a contact angle of 0°–75° indicates a wettability of water-wet, a contact angle of 75°–105° indicates a wettability of neutral-wet, and a contact angle of 105°–180° indicates a wettability of oil-wet.

2. The Hanger-Plate Method

The measuring instruments for the hanger-rod method to determine rock wettability are approximately the same as that used in the measurement to determine surface tension, and what needs to do for change is just to replace the hanger one in the latter by the target rock plate. Firstly, measure the oil–water interfacial tension σ_{ow}; secondly, pull out the rock plate and measure the variation in the force f during the process in which the plate goes through the air–oil–saltwater interface; finally, calculate the values of $\cos\theta$ and θ according to related correlation formulas.

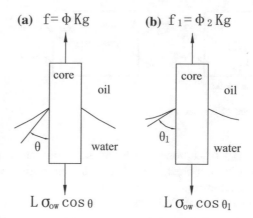

Fig. 8.39 Measurement of contact angle with hanger-plate

The initial state of the hanger plate is shown in Fig. 8.39a. As the oil–water interface moves upwards, the contact angle θ grows to the advancing contact angle θ_1 ($\theta_1 > \theta$) due to the wetting hysteresis effect shown in Fig. 8.39b. At this moment, the force dragging the plate downwards is:

$$L\sigma_{ow} \cos \theta_1 \tag{8.16}$$

where:

L The length of the triple line around the hanger plate immersed in the liquid, cm;
σ_{ow} The oil–water interfacial tension, dyn/cm.

For force equilibrium, the upward force applied to the other end of the torsion balance sustaining the rock place amounts to:

$$f_1 = \varphi_1 Kg \tag{8.17}$$

where:

φ_1 The number of intervals or divisions shown on the scale graduated on the torque-force knob of the torsion balance;
K The torsional constant, which is equal to the mass of the counterpoising weights corresponding to a single interval on the scale graduated on the torque-force knob of the torsion-balance, g/interval;
g Acceleration of gravity.

Equating the two quantities would yield a force balance:

$$f_1 = \varphi_1 Kg = L\sigma_{ow} \cos \theta_1 \tag{8.18}$$

By transformation, the equation above can be written as:

$$\cos \theta_1 = \frac{\phi_1 Kg}{\sigma_{ow} L} \tag{8.19}$$

All of the parameters on the right side of Eq. (8.19) can be obtained with the aid of instruments, so:

$$\theta_1 = \arccos \frac{\phi_1 Kg}{\sigma_{ow} L} \tag{8.20}$$

As the oil–water interface moves downwards, conversely, the receding contact angle resulting from the static wetting hysteresis effect is θ_2, and it meets the equation as follows:

$$\theta_2 = \arccos \frac{\phi_2 Kg}{\sigma_{ow} L} \tag{8.21}$$

where:

φ_2 The number of intervals or divisions shown on the scale graduated on the torque-force knob of the torsion balance in the case of downward movement of the oil–water interface.

Because of the macro- and microanisotropy, it is also probable for a single rock to present itself with a variety of wettability, which the direct measurements fail to reflect. In such cases, e.g., sparkled wetting and mixed wetting, indirect methods, especially the imbibition method, should be resorted to for the measurement of rock wettability.

3. The Imbibition Method

As a conventional wettability evaluation approach, the imbibition method is performed by a sequence of spontaneous and forced displacements of different fluids into a porous sample, whereupon the rock wettability can be determined by the rate of the spontaneous imbibition of the wetting phase, as well as the quantity of the non-wetting phase displaced out of the experimented core.

In order to yield wettability data reflecting the real reservoir rocks, it is required to use fresh cores and to protect the cores against contamination during the experiments. What's more, the experimental fluids should be prepared elaborately so as to simulate the oil and water under reservoir conditions. Usually, the reservoir oil with specified viscosity is often represented by crude oil that is diluted by kerosene to some degree.

The imbibition methods include the spontaneous imbibition method, the imbibition and displacement method, and the imbibition centrifugation method. The first two will be introduced here.

Fig. 8.40 Measurement of
wettability by spontaneous
imbibition method.
a Imbibition of oil to displace
water, **b** imbibition of water
to displace oil

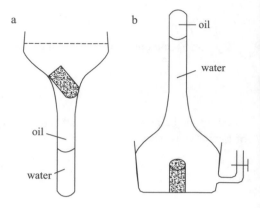

(1) The Spontaneous Imbibition Method

The experiment procedure of the spontaneous imbibition method:

In the case of water-wet rock core as shown in Fig. 8.40b, immerse the core initially saturated with oil into the specially shaped container of water (water absorption meter). In this way, the water would spontaneously enter the rock pore system by capillary force and displace the oil that originally occupies the pore openings. Consequently, with buoyancy, the oil forced out rise to the top of water and collect in the slim graded tube where its volume can be read. The imbibition of water into the core indicates that the core possesses a certain degree of hydrophilicity.

In the case of oil-wet rock core as shown in Fig. 8.40a, immerse the core initially saturated with water into the specially shaped container of oil (oil absorption meter). In a similar manner, the oil would spontaneously enter the core and displace the water. The water driven out would sink to the bottom of the container where its volume can be read.

In practice, taking account of the wettability anisotropy of rock, the experiment of "imbibition of water to displace oil" and the experiment of "imbibition of oil to displace water" are performed in parallel, respectively, on a couple of identical rock cores. Then, compare the experimental results yielded from them to get conclusions: If the imbibed amount of water is greater than that of oil, the core is water-wet, whereas contrariwise the core is oil-wet; if the imbibed amount of water is close to that of oil, the rock is neutral-wet.

The virtue of this method resides in its simplicity of experimental apparatus, the convenience of operation, and, although just qualitatively, it is well able to reflect the actual situation in real reservoirs.

(2) The Imbibition and Displacement Method

In order to evaluate the relative wettability of rock quantitatively semiquantitatively, Amott put forward the imbibition and drainage method, which is essentially an improved spontaneous imbibition method. It is the recommend method in China for the determination of rock wettability.

The experiment procedure of the imbibition and displacement method:

Step 1 Imbibition of oil to spontaneously displace water: Immerse the core initially saturated with water into the oil absorption meter where the oil will enter the core to drive the water out. Measure the volume of the water discharged in this step, which is expressed as the spontaneously imbibed water volume.

Step 2 Displacement of water by oil in core holder: Then, put the core in a core holder and, by applying pressure to it, force more oil into the core to further displace the water that is still contained in the rock pore space. Measure the volume of the water discharged in this step, which is expressed as the drainage water volume due to oil displacement.

Step 3 Imbibition of water to spontaneously displace oil: Conduct the previously introduced water-imbibition experiment on the core, which is now saturated with oil but still contains connate water. Measure the volume of the oil discharged in this step, which is expressed as the spontaneously imbibed oil volume.

Step 4 Displacement of oil by water in core holder: Put the core in a core holder and, by applying pressure to it, force more water into the core to further displace the oil that is still contained in the rock pore space. Measure the volume of the oil discharged in this step, which is expressed as the drainage oil volume due to water displacement.

Figure 8.41 shows the measurement.

Finally, calculate the water-wettability index and the oil-wettability index:

$$\text{water - wettability index}: I_w = \frac{\text{wolume of the discharged oil in water imbibition}}{\text{volume of the discharged oil in water imbibition} + \text{volume of the discharged oil in water driving}}$$

$$\text{oil - wettability index}: I_o = \frac{\text{volume of the discharged water in oil imbibition}}{\text{volume of the discharged water in oil imbibition} + \text{volume of the discharged water in oil driving}}$$

$$(8.23)$$

The Amott wettability index is defined as:

$$I_A = I_w - I_o \tag{8.24}$$

On the basis of the Amott wettability index, it is suggested by the IFP that the rock wettability should be determined according to Table 8.5.

No matter which method is used, to maintain the original rock formation is the premise for a correct evaluation of the wettability of reservoir rock. Improper preservation methods, such as too long exposure time of the core in the air and sedimentation of asphalt or other heavy fractions on the surface of rock pores, will cause the reservoir core to be more and more oil-wet. It should be noted that changes in rock wettability occur when cores containing natural clay are cleaned with solvents, leading to expansion and dispersion of clay, ion exchanges, as well as changes in effective porosity. Therefore, cores obtained from coring wells with ordinary water-based mud are often used in laboratory, and a processing of manual restoration must be given to the core before measuring its wettability. Manual restoration refers to the process in which certain artificial methods are used to

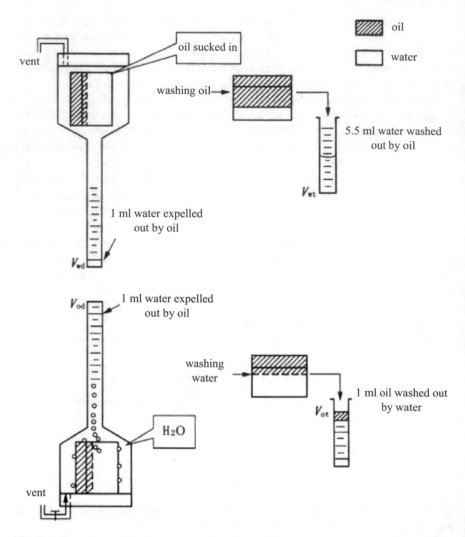

Fig. 8.41 Procedure of the measurement of rock wettability by spontaneous imbibition method

Table 8.5 Evaluation of rock wettability by Wettability Index

Wettability index	−1 to −0.3	−0.3 to −0.1	−0.1 to 0.1	0.1 to 0.3	0.3 to 1
Wettability	Oil-wet	Weakly oil-wet	Neutral-wet	Weakly water-wet	Water-wet
Wettability (by rough and ready classification)	Oil-wet	Neutral-wet			Water-wet

restore the wettability of the rock, but it can be performed only when the wettability of rock is already known.

The determination of rock wettability, being the same as the measurement of the other parameters, should follow a standard procedure and be carried out upon experimental conditions in accordance with industry standards, so that more accurate and comparable data can be obtained.

In addition to the most commonly used methods mentioned above, it is also feasible to make use of capillary pressure curve to determine the wettability of the rock. More details will be discussed in the following chapters.

Chapter 9
Capillary Pressure and Capillary Pressure Curve

Reservoir rocks contain huge amounts of tortuous, rambling, and interconnected pores of all sizes and shapes. So, this complexity rock pore space entails a simplification of it for the study of the multiphase flowing within reservoir rocks. Usually, we simplify the miscellaneous pores and channels into a capillary model, regarding the pore-channels as capillary tubes wherein the internal diameter varies along the tube and the internal wall is coarse. In this chapter, we will deal with capillary pressure characteristics of the multiphase fluids in porous mediums.

9.1 Concept of Capillary Pressure

9.1.1 Rise of Fluids in Capillary Tube

Consider the case in which we place a clean capillary tube in a large open vessel containing wetting liquid (e.g., water). As shown in Fig. 9.1a, the air–water interface would shape into a concave due to the upward additional force applied to the water, and, consequently, the water, which is the wetting phase with respect to the inner surface of the tube, would rise in the capillary tube to some height above the liquid surface in the large vessel. This rise in height results from a set of balanced forces: the upward adhesion tension between the tube and the liquid and the downward gravitational pull on the column of liquid in the tube. Conversely, in the case that the capillary tube is immersed into a non-wetting phase (e.g., mercury) rather than water, the air–mercury interface, as shown in Fig. 9.1c, would shape into a convex due to the downward additional force applied to the mercury, whose surface would consequently fall to a level below the liquid surface in the large vessel. The mentioned additional forces applied at the curved interfaces are usually referred as capillary forces, and they cause the rise or fall of liquid in the capillary tubes.

© Petroleum Industry Press and Springer-Verlag GmbH Germany 2017
S. Yang, *Fundamentals of Petrophysics*, Springer Geophysics,
DOI 10.1007/978-3-662-55029-8_9

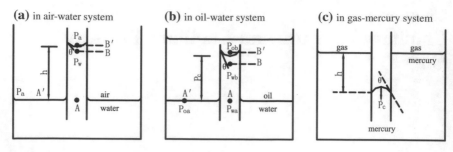

Fig. 9.1 Rise and fall of the liquid level in capillary tube

1. Capillary force in gas–liquid system

Figure 9.1a shows the gas–liquid system. By equating the upward adhesion tension and the downward gravitational force of the liquid column, we can establish the force equilibrium equation for this system:

$$A2\pi r = \pi r^2 h \rho_w g$$

$$h = \frac{2A}{r\rho_w g} = \frac{2\sigma\cos\theta}{r\rho_w g} \tag{9.1}$$

where

A adhesion tension, $A = \sigma\cos\theta$;
r radius of tube;
σ surface tension of water;
θ contact angle;
ρ_w density of water.

It should be noted that in the equations above the weight of the air column is not taken into account. In fact, the weight of the air column is customarily ignored in the force equilibrium calculation of air–liquid systems.

The pressure in the air immediately above the curved surface, represented by the pressure P_a at the point B', is essentially equal to the pressure at the point A', which represents the pressure at the whole free water level in the large open vessel. Furthermore, as the point A' stays at the same horizontal level with the point A, the pressure at the point A must equal to that at the point A' and therefore also equal P_a. The value of P_a has a correlation as follows with the pressure immediately below the curved surface, represented by the pressure P_w at the point B:

$$P_a = \rho_w g h + P_w \tag{9.2}$$

$$P_a - P_w = \rho_w g h$$

The above equation indicates that the pressure existing in the wetting water phase beneath the air–liquid interface is less than the pressure which exists in the non-wetting gaseous phase above the interface. This difference in pressure existing across the interface is referred to as the capillary pressure or capillary force of the system. The capillary force is the force to balance the pressure difference across the curved interface in the tube, resultantly pointing toward the direction to which the concave interface opens. Expressed mathematically:

$$P_c = P_a - P_w = \rho_w g h \tag{9.3}$$

By substituting Eq. (9.1) into Eq. (9.3), we can get the following:

$$P_c = \frac{2\sigma \cos \theta}{r} \tag{9.4}$$

2. Capillary force in oil–water system

Consider the capillary tube immersed in a beaker of water upon which the other phase is oil rather than air. As shown in Fig. 9.1b, the wetting-phase water would rise in the tube to a height of h. By denoting σ as the oil–water interfacial tension, θ as the contact angle, and ρ_o, ρ_w as the oil density and the water density, the equation yielded from balancing the upward and downward forces can be written as follows:

$$A \cdot 2\pi r = \pi r^2 h(\rho_w - \rho_o)g$$

$$h = \frac{2A}{r(\rho_w - \rho_o)g} = \frac{2\sigma \cos \theta}{r(\rho_w - \rho_o)g} = \frac{2\sigma \cos \theta}{r \Delta \rho g} \tag{9.5}$$

Assume the pressures at the point B' and B are, respectively, P_{ob} and P_{wb}, which, respectively, represent the pressure in the oil and water phases immediately nearby the interface; assume the pressures at the point A' and A nearby the interface in the big open beaker are, respectively, P_{oa} and P_{wa}. The four pressures satisfy the equations below:

In the oil phase: $P_{ob} = P_{oa} - \rho_o g h \tag{9.6}$

In the water phase: $P_{wb} = P_{wa} - \rho_w g h \tag{9.7}$

Since the beaker is large compared with the capillary tube, the oil–water interface in the beaker is essentially horizontal. The capillary pressure is zero in a horizontal or plane interface. As we all know, in terms of communicating vessels containing liquid the pressures at the same level are equivalent; therefore, the pressure in the water at the bottom of the column is equal to the pressure in the oil at the surface of water in the large vessel:

$$P_{oa} = P_{wa} \tag{9.8}$$

By definition, the capillary pressure is the pressure difference across an interface, pointing from the wetting phase toward the non-wetting phase. It is therefore equal to the non-wetting-phase pressure minus the wetting-phase pressure:

$$P_c = P_{ob} - P_{wb} = (\rho_w - \rho_o)gh = \Delta\rho gh \tag{9.9}$$

where

P_c capillary force, Pa;
$\Delta\rho$ density difference between the two phases of fluids, kg/m^3;
h height of the column of wetting phase in the tube above that in the large vessel, m;
g acceleration of gravity, m/s^2.

Equation (9.9) is the fundamental formula for the capillary equilibrium in oil formations. It indicates that the greater the capillary force, the higher the liquid column will rise before an equilibrium system is obtained. By substituting Eq. (9.5) into it, we can get:

$$\Delta\rho gh = \frac{2\sigma \cos\theta}{r}$$

or

$$P_c = \frac{2\sigma_{wo} \cos\theta_{wo}}{r} \tag{9.10}$$

where

P_c capillary force, Pa;
r radius of tube, mm;
σ interfacial tension, mN/m.

The characteristics of capillary force can be summarized as follows:

(1) Capillary force does exist in capillary tubes at various positions. Providing that the other parameters such as σ, r, $\cos\theta$, and $\Delta\rho$ are all held constant, the height supported in the capillary tubes at various positions above the free water level in the container is equivalent with that supported in the capillary tube suspended vertically. This phenomenon shown in Fig. 9.2 can be explained by principle of communicating vessels. It deserves to be specially noted that, in the case of horizontally placed capillary tube, the wetting water would climb into

Fig. 9.2 Water displacing oil
in a horizontally placed
capillary tube

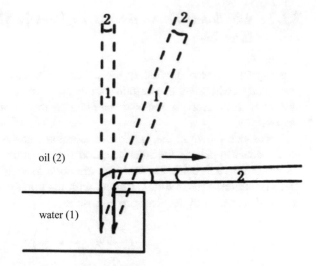

oil (2)

water (1)

the tube and try to exit from the other opening of it, which simply illustrates
how the capillary force in a water-wet formation contributes to the displacement of oil as a driving force.

(2) Capillary force P_c is directly proportional to $\cos \theta$. For a water-wet reservoir,
the contact angle with regard to a water–oil system is smaller than 90°
($\cos \theta > 0$) and therefore, the capillary force is positive, causing the water to be
spontaneously imbibed into the rock. Conversely, for an oil-wet reservoir, the
contact angle with regard to a water–oil system is larger than 90° and therefore,
the capillary is negative, and in such a case the capillary force is a resisting
force against which an external force is needed to force the water into the core.

(3) Capillary force P_c is inversely proportional to tube radius r. As noted in
Eq. (9.10), the height of the water column will be increased with the decrease in
the tube radius because the magnitude of the capillary force increases in inverse
proportional relation to the radius. Therefore, the height of water surface varies
from each other with different diameters, leading the interfaces of oil–water or
gas–oil systems to be a transition zone with a certain height, instead of a
distinctively separated zone. Generally, as $(\rho_w - \rho_o) < (\rho_o - \rho_g)$, the thickness
of an oil–water transition zone is greater than that of an oil–gas transition zone.

(4) In the case of water-wet reservoir rock, the water can be spontaneously imbibed
into the rock and the initially contained oil would be drive out—this process is
classified as "imbibition"; in the case of oil-wet reservoir rock, conversely, the
rock fails to spontaneously imbibe the water and an external force is needed to
overcome the capillary force before flowing the water into the core—this
process is termed "drainage."

9.1.2 Additional Force Across Differently Shaped Curved Interfaces

The curved interface previously discussed in the case of capillary tubes can be particularly considered as a certain part of spherical surface. However, in the pore system of real reservoir rocks there are numerous curved interfaces of various shapes.

It was also shown in the previous discussion that the greater pressure is always on the concave side of the interface, and the direction of the additional force, as shown in Fig. 9.3, coincides with the direction to which the concave side opens. The magnitude of the additional force, with the gravity effect deliberately ignored, can be determined by the Laplace equation:

$$P_c = \sigma\left(\frac{1}{R_1} + \frac{1}{R_2}\right) \tag{9.11}$$

where

P_c additional force across the curved interface;
σ interfacial tension between the two fluids;
R_1, R_2 the principal radii of curvature of the interface.

Equation (9.11) is the fundamental formula for capillary pressure calculations. Obviously, the additional force P_c is inversely proportional to the radii of curvature R of the interface. It is also able to explain why the static liquid surface in a large container is a horizontal or plane one—this is because of the radii of curvature of the large liquid surface $R \to \infty$ and thus, the capillary force equals zero. However, the curved interface in a capillary tube usually has small radii of curvature and therefore considerable additional force. Customarily, the additional force occurring in a capillary tube is particularly called capillary pressure or capillary force.

Fig. 9.3 Additional force across curved interface

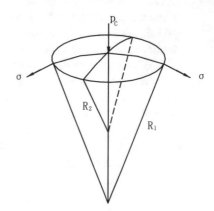

The following paragraphs will present the capillary pressure calculations for several specially shaped interfaces that are commonly encountered in reservoir rocks.

1. Capillary-force calculation for spherical surface

In the case that the interface formed in a capillary tube takes the shape of a spherical surface, the radii of curvature are $R_1 = R_2 = R$. So, Eq. (9.11) becomes the following:

$$P_c = \sigma\left(\frac{1}{R_1} + \frac{1}{R_2}\right) = \frac{2\sigma}{R} \tag{9.12}$$

Figure 9.4 shows that: $R = \frac{r}{\cos\theta}$
By substituting it into Eq. (9.12), we obtain:

$$P_c = \frac{2\sigma}{R} = \frac{2\sigma\cos\theta}{r} \tag{9.13}$$

where

θ wetting contact angle;
r radius of tube.

The capillary pressure has a direction pointing to the inner side of the curved interface, or in other words, pointing to the non-wetting phase.

Fig. 9.4 Relationship between capillary radius and interface curvature

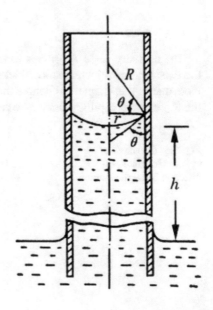

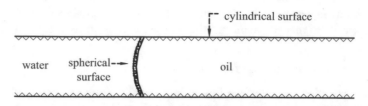

Fig. 9.5 Meniscus in water-wet channel—spherical interface and cylinder interface

Equation (9.13) indicates that the capillary pressure is inversely proportional to the tube radius—the smaller the tube radius, the greater the capillary pressure, and that the capillary pressure will increase with the increase in the interfacial tension or the decrease in the contact angle.

2. Capillary-force calculation for cylindrical surface

The case of cylindrical surface is also encountered in the rock pores and channels. Figure 9.5 shows a water-wet channel wherein a film of connate water is attached to the rock wall, and when a stream of oil (or gas) flows through the central space of the channel, an oil–water (or a gas–water) cylinder interface, in addition to the spherical interface between the two fluid phase, will be formed between the oil (or gas) and the connate water.

As for a cylinder interface, it is noted in Fig. 9.6 that $R_1 = r$ and $R_2 = \infty$. Substitute the two quantities into Eq. (9.11) to obtain the calculation equation for cylinder interface:

$$P_{cz} = \sigma\left(\frac{1}{\infty} + \frac{1}{r}\right) = \frac{\sigma}{r} \tag{9.14}$$

The capillary force P_{cz} points to the geometric central axis of the cylinder and the existence of it increases the thickness of the water film coating the solid wall. Note that in this chapter we denote the capillary force across a cylinder interface as the P_{cz} and the capillary force across a spherical interface as P_c.

Fig. 9.6 Curvature of cylinder interface and direction of capillary force

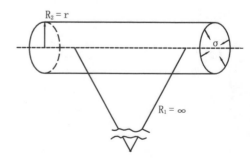

Fig. 9.7 Meniscus of conical capillary tube

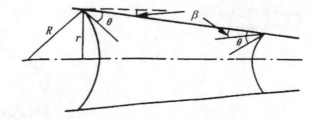

3. Capillary-force calculation for conical surface

There also exist channels with gradually varying radius in the pore system of real reservoir rock. In the case of such nonconstant-diameter curved interfaces as the conical one shown in Fig. 9.7, it is better to consider its capillary pressure problem in light of the way in which the inclination of the solid surface affects the contact angle. As illustrated in the figure, the radii of curvature at the two specified section are, respectively, $R_1 = r_1 / \cos(\theta + \beta)$ and $R_2 = r_2 / \cos(\theta - \beta)$; thus, the capillary pressure across the conical interface is:

$$P_{ci} = \frac{2\sigma \cos(\theta \pm \beta)}{r_i} \qquad (9.15)$$

where

β the angle between the wall and the central line of the conical capillary tube, i.e., half of the cone angle;

θ the equilibrium contact angle when the system is at rest.

Proven by Eq. (9.15), in a capillary tube which changes in diameter form small to large the magnitude of the capillary force depends on where the curved interface between the two fluids is—the strongest capillary force takes place in the thinnest cross section, and the weakest capillary force takes place in the largest cross section.

4. Capillary-force calculation for two phases within fracture

The fracture where the two phases of fluids coexist can be seen as a pair of parallel plates as shown in Fig. 9.8, and the distance between the two plates is W. In the terms of shape, the interface between the two fluid phases is a certain part of a cylinder surface—one of the radii of curvature of which (R_1) is usually greater than $W/2$ and the other one is infinitely great ($R_2 = \infty$).

Known from Fig. 9.8, $\cos\theta = \frac{W/2}{R_1}$, that is to say, $\frac{W}{2} = R_1 \cos\theta$. If we substitute this correlation into Eq. (9.14), we can get the calculation equation for the capillary force between the parallel plates:

$$P_{cz} = \frac{\sigma}{R_1} = \frac{2\sigma \cos\theta}{W} \qquad (9.16)$$

Fig. 9.8 Capillary force
between parallel plates

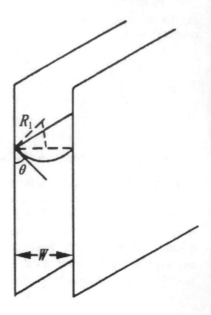

Comparing it with Eq. (9.14), the tube radius in the latter one is replaced by the distance between the two parallel plates. Equation (9.16) can be used for the capillary-force calculations of fractured reservoir rocks. It is shown that the narrower the fracture, the greater the capillary force.

5. Capillary-force calculation for the pendular rings at the contact points of ideal sand grains

Figure 9.9 shows two sand grains with the same radii and a pendular ring composed of the wetting phase around their contact point. As the non-wetting-phase fluid exists in the central space of the pores and channels, a curved interface between the two fluid phases is formed as illustrated in the diagram. By denoting the radius of curvature in the vertical sectional plane is R_1 and the radius of curvature in the horizontal sectional plane is R_2, the capillary force can be expressed by the equation below:

$$P_c = \sigma \left(\frac{1}{R_1} + \frac{1}{R_2} \right) \tag{9.17}$$

It is practically impossible to measure the values of R_1 and R_2, so another term named "the mean radius of curvature" is generally used instead:

$$\frac{1}{R_m} = \frac{1}{R_1} + \frac{1}{R_2} \tag{9.18}$$

Fig. 9.9 Pendular-ring
distribution and meniscus of
the wetting phase between
two sand grains (by Leverett
1941)

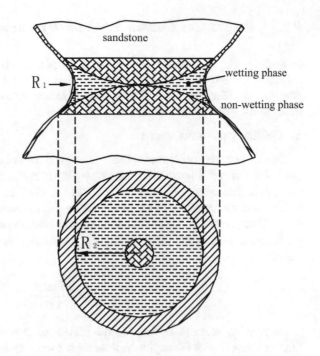

so

$$P_c = \frac{\sigma}{R_m} \qquad (9.19)$$

In the equations, the symbol R_m is the mean radius of curvature, which can be obtained through the formula $P_c = \Delta\rho g h = \sigma/R_m$.

Considering the wetting characteristics of real reservoir rocks, the capillary force present in the pore system always renders the wetting phase to firstly occupy the small pore spaces, leaving the large open channels to the non-wetting phase. Then, as the wetting phase increases in amount or, in other words, the saturation of it increases, the wetting phase will go on to enter the capillary tubes with larger and larger diameter. Let us restate what's mentioned above: The higher the saturation, the bigger will be the mean radius of curvature and the smaller will be the capillary force P_c, which means that the wetting-phase material will enter the bigger crevices and openings of the pore system. This behavior indicates that there is an inverse functional relationship between the capillary pressure and the wetting-phase saturation in a rock:

$$P_c = f(S_w) \qquad (9.20)$$

The relation curve between the capillary pressure and the wetting-phase saturation within a rock is called the capillary pressure curve of rock.

9.1.3   Resistances Caused by Capillary Force

In dealing with multiphase systems in the capillary tubes involved in a reservoir rock, it is necessary to consider the resistances against the liquid flow which result from the capillary pressure. Three most common cases encountered in reservoir rocks will be discussed here.

1. The first type of resistance

The first type of resistance refers to the resistance occurs to an oil column or a gas bubble at rest. Figure 9.10 shows a cylinder capillary tube where a column of oil inhibits and applies to an extrusion force to the tube wall.

Consider the capillary effect exhibited by this oil column. Based on Eq. (9.13), the capillary forces P'_c, respectively, induced at the two ends of the column, which are spherical interfaces in shape, have the same magnitude decided by the equation below:

$$P'_c = \frac{2\sigma}{R} = \frac{2\sigma \cos \theta}{r} \qquad (9.21)$$

As their components of force in the horizontal direction cancel out each other, the resultant effect of the capillary pressure is a force applied to the tube wall, which can be explained by the Pascal's principle.

At the same time, the capillary force P''_{CZ} induced at the cylinder interface across the flank of the oil column, pointing to the central axis of the column, can be determined according to Eq. (9.14):

$$P''_{CZ} = \frac{\sigma}{r} \qquad (9.22)$$

It becomes evident that, in the case of a liquid column at rest, the spherical-interface capillary force P'_c and the cylindrical-interface capillary force P''_{CZ} act in the opposite directions, respectively, upon the tube wall and the central axis of the tube. As P'_c causes thinning of the liquid film while P''_{CZ} tends to thicken it, the liquid film will finally reach a certain equilibrium thickness. If we specify the resultant capillary force P_I pointing to the tube wall, the magnitude of it can be obtained by this equation:

Fig. 9.10 Liquid globule or gas bubble in cylinder capillary tube

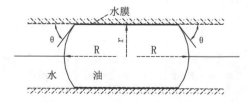

$$P_{\mathrm{I}} = \frac{2\sigma \cos\theta}{r} - \frac{\sigma}{r} = \frac{2\sigma}{r}(\cos\theta - 0.5) \tag{9.23}$$

As shown in the equation above, greater resultant capillary force in a capillary tube can be caused by greater oil–water (or oil–gas) interfacial tension or smaller tube radius. Note that when the resultant capillary force is increased, the liquid film will become thinner and the resistance against the horizontal flow through the rock grows stronger.

Furthermore, it is noteworthy that the liquid film coating the solid surface possesses abnormal properties such as high viscosity and high strength; it therefore requires substantial pressure difference to overcome P_{I} as well as the frictional drag caused by the thin liquid film before successfully forcing the oil column to move in the capillary tube.

2. The second type of resistance

Due to wetting hysteresis, when an oil column is about to move, each of the spherical interfaces of it will undergo a deformation as shown by the dotted lines in Fig. 9.11, wherein is also given the two radii of curvature that consequently no longer equal. The capillary forces at the two ends of the oil column are as follows:

$$P' = \frac{2\sigma}{R'}$$
$$P'' = \frac{2\sigma}{R''} \tag{9.24}$$

Hence, the second type of resistance, symbolized as P_{II}, is caused by the deformation of the curved liquid interfaces and equals the difference between the two spherical capillary forces:

$$P_{\mathrm{II}} = P'' - P' = 2\sigma\left(\frac{1}{R''} - \frac{1}{R'}\right) = \frac{2\sigma}{r}(\cos\theta'' - \cos\theta') \tag{9.25}$$

As P_{II} is opposite to the direction of flow, it exists as a resisting force. Therefore, the pressure difference subjected to which the oil column begins to move must be greater than the sum of P_{I}, P_{II}, and the fractional drag. Note that P_{I} and P_{II} are not allowed to be simply added algebraically to obtain the resultant force, because in

Fig. 9.11 Transform of meniscus under external pressure

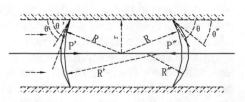

Fig. 9.12 Transform of
globule or bubble at
constriction

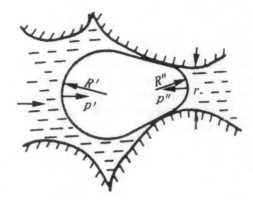

dealing with the resistance due to P_I, which is actually a vector pointing to the tube wall, we must take into some more factors into account before determining the magnitude of its component force acting against the liquid flow, for example, the frictional resistance coefficient caused by the water film coating the tube wall.

3. The third type of resistance

Figure 9.12 shows a single globule of oil or gas bubble of gas entering a pore-throat constriction smaller than itself. It proves a radius of curvature R' in the leading portion smaller than the radius R'' of the portion of the drop still in the pore. Thus, the distorted oil globule, giving the third type of resistance P_{III} an opportunity to develop, has different capillary pressures at its front and back menisci, and therefore, pressure is required to force the droplet through. Obviously, the third additional resistance due to capillary force P_{III} is determined by the equation below:

$$P_{III} = 2\sigma\left(\frac{1}{R''} - \frac{1}{R'}\right) \tag{9.26}$$

This effect is referred to as "Jamin's effect."

The globule is not allowed to flow into the throat until its front radius of curvature is distorted to be equal to the radius of the narrowest part of the throat, that is, $R'' = r$. In such a case, the resistance produced by the front meniscus is $P_I = \frac{2\sigma}{r}$ and therefore, a pressure difference greater than P_I is required to applied over the globule to force it through.

The resistances discussed above are commonly met by multiphase flowing through reservoir rocks. Although the globules and bubbles cause quite strong resistances, it does not indicate that they would completely shut up the channels for oil flow. This is because that the pore openings involved in reservoir rocks are irregular in shape and comprise a spatially interconnected network in which an oil drop will make a detour and try find another way forward when encountered by a blockage in a channel as shown in Fig. 9.13. However, it is still possible for all the channels for oil flow to be blocked if there are huge amounts of globules and/or

Fig. 9.13 Oil droplet making
a detour to flow

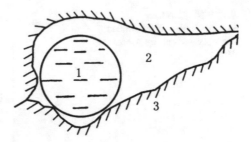

bubbles are present in the pore system, and under such a case, even if not all the channels are blocked, the flow of oil toward the bottom of well will be significantly affected, accompanied by a reduction of the wells' natural capability of production.

With pore openings of a range of sizes and numerous throats, the reservoir rocks exhibit obvious effect due to these mentioned resistances. In the case of two-phase flow, the total resistance can be quite considerable as the capillary forces caused by all the droplets are superimposed.

The capillary resisting effect does harm to the extraction of oil and gas, so measurements should be adopted to avoid two-phase flow during drilling, well completion, downhole operation, and oil extraction.

9.1.4   Capillary Hysteresis

Holding the other conditions as constant, it is found that the height of the liquid column raised up in the capillary tube during the imbibition process is much larger than that during the drainage process, and this difference due to different saturation histories is called "capillary hysteresis." There is an alliance between the wetting hysteresis and the capillary hysteresis, which is a particular phenomenon occurring in capillary tubes.

Consider a hollow capillary tube stuck into a container wherein the fluid is the wetting phase (e.g., water), while the air is the non-wetting phase. In such a case, the wetting phase (water) will displace the non-wetting phase (air) and finally, under the action of the capillary pressure, be raised up to a specified height in the tube. This is the process of imbibition, as shown in Fig. 9.14a.

Consider another capillary tube prefilled with water and then stick it into the container with water. In such a case, the water in the tube will be displaced by the air and finally descend to a certain height under the action of gravity. This is the process of drainage, as shown in Fig. 9.14b.

Although the two processes are both at the atmospheric pressure, the height of water produced in the imbibition process is shorter than that produced in the drainage process, as a result of the difference in saturation history.

Fig. 9.14 Capillary
hysteresis caused by contact
angle hysteresis (by Morrow
1976)

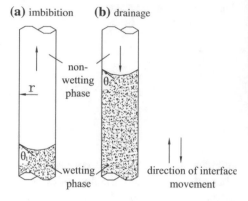

The phenomenon of capillary hysteresis is mainly caused by wetting hysteresis and the geometrical shape of the pore-channels in the rock. There are several types of capillary hysteresis as follows.

1. Capillary hysteresis caused by contact angle hysteresis

Figure 9.14 shows that both the fluid interfaces produced in the imbibition process and the drainage process are, respectively, given in a capillary tube, and the advancing angle θ_1 is larger than the receding angle θ_2 as a result of the different wetting histories in the two tubes. Furthermore, as $\theta_1 > \theta_2$, the capillary pressure produced in the process of imbibition (P_{im}) is smaller than that in the drainage process (P_{dr}).

So, the liquid column raised up during the process of imbibition, in comparison with the process of drainage under the same non-wetting-phase pressure, is balanced at a lower height. In other words, when subjected to the same displacement pressure the drainage process exhibits higher saturation of the wetting-phase fluid than the imbibition process. This is the theoretical basis for the study of capillary pressure and pore structure. In the case of water-wet rocks, the drainage method, which is a process of oil-displacing-water, should be adopted to determine the connate-water saturation; and the imbibition method, which is a process of water displacing oil, should be adopted to determine the residual oil saturation.

2. Capillary hysteresis caused by abrupt change in capillary radius

Another cause of the phenomenon of capillary hysteresis is the abrupt change in capillary radius. Consider a capillary tube which is thick in the center and slim at each end, as shown in Fig. 9.15, the pressure P_t in the upper section of tube, which can be considered as a throat, is larger than the pressure P_p in the middle thick section of the tube. If it is assumed that process of imbibition in the diagram (a) and the process of drainage in the diagram (b) have the same pressure of non-wetting phase, and that contact angle $\theta = 0°$, then it is easy to know that $P_t = \frac{2\sigma}{r_t}, P_p = \frac{2\sigma}{r_p}$.

So, during the process of imbibition, the uplifted meniscus which has crossed the

Fig. 9.15 Capillary
hysteresis caused by abrupt
change in capillary radius (by
Morrow 1976)

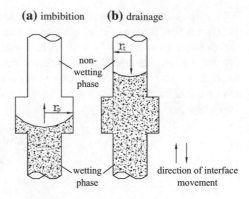

(a) imbibition   **(b)** drainage

constriction will cease to stay firmly in the middle thick section of the tube; conversely, during the process of drainage, the fall off meniscus, pulled up by the force of P_p which is comparatively large, will stay firmly in the upper slim section in the section. Obviously, the saturation of the wetting phase in the imbibition process is lower than that in the drainage process.

If the wetting contact angle $\theta = 0°$, the capillary hysteresis is only associated with the change in capillary radius; but if the wetting contact angle $\theta \neq 0°$, the capillary hysteresis is associated not only with the change in capillary radius, but also with the wetting hysteresis.

3. Capillary hysteresis caused by continuous change of capillary radius

For conical capillary tubes that change in diameter from small to large to small, as shown in Fig. 9.16, the capillary hysteresis is caused by the variation of the contact angle and the capillary diameter. The tilt of the inner wall of such a capillary

Fig. 9.16 Capillary
hysteresis caused by
continuous change of
capillary radius (by Morrow
1976)

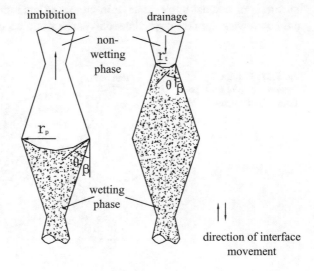

tube causes the variation of the contact angle, and, as a result, it is needed to subtract a half apex angle of cone from the contact angle when using Eq. (9.15), which states $P_c = 2\sigma \cos{(\beta \pm \theta)}/r$, to calculate the displacement capillary pressure P_t; when using the equation to calculate the imbibition capillary pressure P_p, it is needed to add a half apex angle of cone to the contact angle. Expressed mathematically:

Imbibition:

$$P_p = \frac{2\sigma \cos(\theta + \beta)}{r_p} \tag{9.27}$$

Drainage:

$$P_t = \frac{2\sigma \cos(\theta - \beta)}{r_t} \tag{9.28}$$

If merely consider the change in capillary radius and leave the change in contact angle out, then the situation is similar to the second case discussed above:

$$P_t = \frac{2\sigma}{r_t}, \quad P_p = \frac{2\sigma}{r_p}$$

As shown in Fig. 9.16, the change in the capillary radius and the change in the contact angle result in the same consequence: $P_p < P_t$, and the saturation of the wetting phase in the imbibition process is lower than that in the drainage process.

4. Capillary hysteresis in real rock pore system

As shown in Fig. 9.17, the real rock usually has coarse pore walls and a pore-throat system varying in cross-sectional area. All of the related factors such as hysteresis of contact angle, change in cross-sectional area, change in pore radius, and coarseness should be simultaneously taken into account when considering the

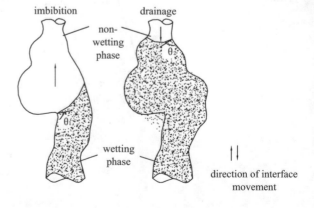

Fig. 9.17 Capillary hysteresis in real rock pore system (by Morrow 1976)

phenomenon of capillary hysteresis occurring in it. However, due to the difficulty in experimental measurement, the impact of the surface coarseness on the advancing angle and the receding angle is usually determined on the basis of the capillary pressure data obtained from the imbibition and drainage tests. By defining the apparent advancing contact angle as $\theta_A = \theta_1 + \beta$ and the as $\theta_R = \theta_2 - \beta$ (θ_1 is the advancing contact angle and θ_2 is the receding contact angle), the capillary pressure is given as:

Drainage:

$$P_{dr} = \frac{2\sigma \cos \theta_R}{r_t} = \frac{2\sigma \cos(\theta_2 - \beta)}{r_t} \tag{9.29}$$

Imbibition:

$$P_{im} = \frac{2\sigma \cos \theta_A}{r_P} = \frac{2\sigma \cos(\theta_1 + \beta)}{r_P} \tag{9.30}$$

where

P_{dr}, P_{im} ,respectively, the displacement pressure and the imbibition pressure;
θ_R, θ_A ,respectively, the apparent advancing contact angle and the apparent receding contact angle.

Seen from the two equations given above, the imbibition test provides capillary pressure data reflecting the pore radius (large capillary tubes), and the drainage process provides capillary pressure data reflecting the throat radius (small capillary tubes). Thus, the size and distribution of the pore-throat openings involved in a rock core can be determined through a combination of imbibition and drainage tests, and the two sets of capillary pressure curves are obtained from them.

9.2 Measurement and Calculation of Capillary Pressure Curves of Rock

Known from the discussion given above, there is a functional relationship of some kind between the capillary pressure and the saturation of the wetting phase. Because of the sheer complexity of the pore structure, it is hard to directly deduce the mathematic model for the reservoir rocks. So, the most commonly used approach is to experimentally measure the capillary pressure data at different saturations of the wetting phase. As shown in Fig. 9.18, the relation curve between the capillary pressure and the saturation of the wetting phase (or the non-wetting phase) is called the capillary pressure curve.

As one of the dominant contents of the course of petrophysics, the capillary pressure curve is used as an important basis for the study of the pore structure of rock and the seepage-flow performance of multiphase fluid. There are various

Fig. 9.18 Typical capillary
pressure curve

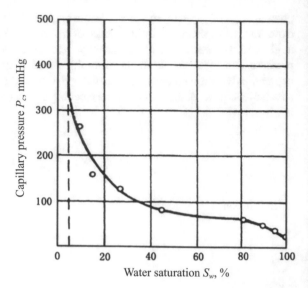

methods of obtaining the capillary curves. These methods, although being different in the fluid media used for test, the pattern of pressurization, and the interval of time required for test, share the same basic principles. Three commonly used measuring methods will be introduced here: the porous diaphragm method, the mercury injection method, and the centrifuge method.

9.2.1 Porous Diaphragm Plate Method

Figure 9.19 shows a porous diaphragm plate device for the measurement of capillary pressure. By means of evacuation as sketched in the figure, a pressure difference can be established between the two ends of the test core, whereupon the water initially saturated in the rock core would flow through the semipermeable diaphragm and into the subjacent graduated tube. A range of pressure differences can be created by creating different degrees of vacuum, but it is impossible for the pressure difference to exceed 0.1 MPa in the case of atmospheric pressure tests. As a drainage process under the pressure difference, the non-wetting phase (air) expels the wetting phase (water) out of the core and results in lower and lower wetting-phase saturation. At each equilibrium test point where the established displacement pressure is balanced by the corresponding capillary pressure, the pressure difference is equal to the capillary pressure, and the saturation of the wetting phase can be calculated according to the initial water saturation and the total volume driven out. With incrementally larger and larger pressure difference for the drainage, the relation curve between the capillary pressure and the wetting-phase saturation can be obtained by a series of point-by-point tests.

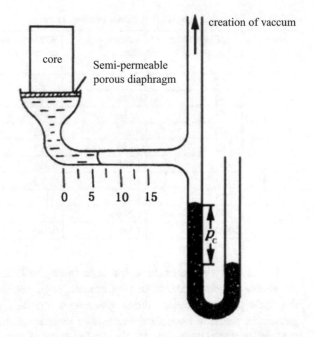

Fig. 9.19 Room pressure porous diaphragm plate device

In experiments, the pressure difference is increased stepwise from the smallest one. With the enlargement of the pressure difference for drainage, the saturation of the non-wetting phase (oil or gas) increases while the saturation of the wetting phase (water) decreases.

At each stop, equilibrium state, at which the pressure of the system is stabilized and no movement of the liquid level in the graduated tube can be observed any longer, has to be reached before a record is made of the produced volume of the wetting phase. Make a note of the equilibrium pressure and corresponding cumulative volume of the produced water, and then raise the pressure difference to conduct the next test, which can be done in the same manner.

A dataset involving a series of capillary pressures and their corresponding cumulative volumes of produced water is obtained from experiment and shown in Table 9.1. The water saturation can be calculated through the following equation:

$$S_w = \frac{V_p - \sum V_w}{V_p} \tag{9.31}$$

where

S_w water saturation;
$\sum V_w$ cumulative volume of the produced water;
V_p pore volume of the core, i.e., volume of the water initially saturating in the rock.

Table 9.1 Measured results for capillary pressure curve

Number	Capillary force (P_c), mmHg	Value on the tube, cm^3	Water volume contained in core, cm^3	Water saturation (S_w), %
1	20	0	1.365	100
2	40	0.075	1.290	94.6
3	50	0.150	1.215	89.0
4	60	0.250	1.115	81.8
5	80	0.750	0.615	45.0
6	120	1.000	0.365	27.0
7	160	1.125	0.240	17.9
8	260	1.225	0.140	10.7
9	390	1.285	0.080	6.3
10	>390	1.285	0.080	6.3

The capillary pressure curve drawn according to Table 9.1 is shown in Fig. 9.18. As shown in the curve, there is a breakthrough pressure, which is the so-called threshold pressure, before the displacement succeeds to take place. As the displacement pressure exceeds the threshold pressure and increases, the saturation of the wetting phase decreases deeply and the curve slopes gently with a certain range. Then, after the saturation of the wetting phase has been reduced to a certain point, the curve exhibits a steep rise and get into a section of vertical straight line.

The major component of the setup for the porous diaphragm plate method is a core chamber composed of a semipermeable diaphragm plate and a glass funnel. The semipermeable diaphragm plate is a porous medium made of ceramic or glass disk, powdered-metal sintered plate, polycarbonate rubber membrane, or biofilm. Whether a water-wet plate or an oil-wet plate should be employed depends on the purposes of respective experiments. Because that the holes of the semipermeable diaphragm, which has already been saturated with the wetting phase before the experiment, are smaller than the pores of the core in size, the plate conducts flow of the wetting phase only while the non-wetting phase is repelled as long as the pressure difference is below the capillary pressure of the largest pressure. This is the reason why the method got its name.

The setup for the high-pressure porous diaphragm plate method, which works on the same principle with the porous diaphragm plate method but provides displacement pressures larger than 0.1 MPa, is shown in Fig. 9.20. The setup is composed of: (1) high-pressure vessel, (2) replaceable diaphragm plate, (3) graduated tube, and (4) pressure source.

When using a water-wet plate to conduct measurement, as shown in Fig. 9.20a, the experimental procedure is as follows: Firstly, the diaphragm plate is saturated with water and the room below the plate is also filled with the wetting-phase water. Secondly, the core that has already been saturated with the wetting phase is mounted on the plate and then compressed by the spring which presses it to be closely against the plate. Note that a filter paper is usually placed between the core

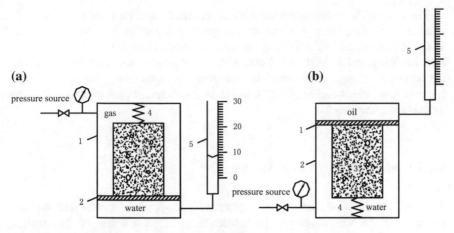

Fig. 9.20 High-pressure porous diaphragm plate device

and the plate in order to make them closely contacted. Thirdly, as the valve controlling the pressure source (e.g., a nitrogen cylinder) is opened, the non-wetting-phase fluid (oil or gas) enters and pervades the chamber and pressure is applied upon the core. Hereupon, the non-wetting phase, overcoming the capillary pressure under the displacement pressure, permeates into the core and drives the wetting-phase water into the U-tube through the diaphragm plate. The volume of the produced water can be read from the graduation of the tube. When the displacement pressure is lower than the lowest capillary pressure of the diaphragm plate, it is impossible for the non-wetting phase to enter the pores of the plate (Fig. 9.21).

When using an oil-wet plate to conduct measurement, as shown in Fig. 9.20b, the diaphragm plate must be saturated with oil beforehand.

Fig. 9.21 Capillary pressure curve measured from porous diaphragm plate device

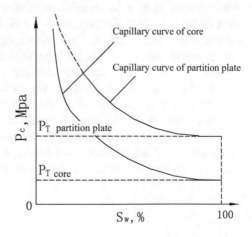

The porous diaphragm plate method, as a classical and standard method for the measurement of capillary pressure data, employs comparatively simple instruments and provides results well reflecting the real saturations in reservoir. But it often fails to satisfy the practical field tests due to its fatal defect of time-consuming. However, the porous diaphragm plate method is more nearly approaching the actual wetting reservoir conditions and therefore utilized as the standard against which all other methods are compared.

9.2.2 Mercury Injection Method

The mercury injection method, as a conventional approach to measure capillary pressure curves, was developed to accelerate the determination of the capillary pressure–saturation relationship. As compared to the porous diaphragm plate method, it consumes much shorter time (about 30–120 for a single test) and permits higher applying pressure. Because the actuating medium, mercury, features steady chemical property, substantial interfacial tension, and small compressibility, its volume still can be accurately metered when subjected to high pressure.

The mercury injection apparatus used at home and abroad work on the same principle but differ from each other in the way of pressurization or metering modes. The pressurization work can be done by metering pump, oil pump, or high-pressure gas cylinder. Among them, the metering pump directly exerts pressure on the mercury, while the other two, the oil pump and the high-pressure gas cylinder, apply pressure on the mercury via hydraulic oil or nitrogen. The volume of the mercury injected into the core can be metered by metering pump or dilatometer.

Figure 9.22 shows a mercury injection apparatus which is pressurized by high-pressure nitrogen cylinder and metered by metering pump. The main body of such an apparatus is composed of a mercury-metering pump and a core chamber. The mercury-metering pump is constructed of the screw 5, the plunger case 1, and the measuring scale 4. In experiment, the core 6 is put into the core chamber 6 and then sealed in it. The metering pump contains mercury and communicates with the core chamber. The principle of this measuring method is as follows: A liquid cannot spontaneously enter a small pore which has a wetting angle of more than 90° because of the surface tension; however, this resistance may be overcome by exerting a certain external pressure. The mercury is a non-wetting fluid for rock, so high pressure must be applied to force the mercury into the core to overcome the capillary pressure. The volume of the mercury injected at each pressure determines the non-wetting-phase saturation. With the capillary pressure data balancing the applied pressure and the saturation data according to the injected mercury volume, the capillary pressure–saturation relationship is easy to get.

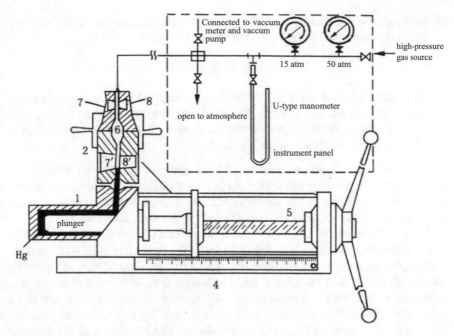

Fig. 9.22 Capillary pressure cell for mercury injection (by Purcell 1949)

The measuring procedure is listed below:

(1) The core that has already been cleaned and dried is put into the core chamber; the chamber is then sealed with the gland.

(2) With the liquid level of mercury being controlled slightly below the lower marked line (8'), the system is evacuated; the liquid level can be seen from the lower window (7').

(3) The mercury is introduced into the chamber and reaches the lower marked line (8'). At this moment, the measuring scale of the metering pump reads zero.

(4) The liquid level of mercury is raised to the upper marked line (8) of the upper window (7). At this moment, the measuring scale reads V, so the volume of the hollow space outside the core is: $V_f = V_E$ (volume of the chamber cavity)—V.

(5) Close the vacuum system and introduce the high-pressure gas into it. As the injection of the mercury into the rock pore system causes the diminution of the mercury in the chamber, the liquid level of mercury drops below the upper marked line (8). Balance is accomplished when the predetermined pressure applied upon the system is counteracted by the corresponding capillary pressure, which is denoted as P_c. Introduce more mercury into the chamber to raise the liquid level back to the upper marked line (8), and the variation of the reading on the scale corresponds to the volume of mercury V_{Hg} injected into the core during this process.

(6) According to the core's porosity φ and bulk volume, the saturation of mercury under the pressure V_f can be determined by the equation below:

$$S_{Hg} = V_{Hg}/\varphi V_f$$

(7) By conducting more experiments with incrementally higher pressures of the gas source and repeating the steps (5) and (6), a dataset about the P_c–S_{Hg} relationship can be obtained, whereupon the mercury injection curve, also called the drainage curve is shown in Fig. 9.23 ①.

After the injection process has reached the highest experimental pressure, in similar manner, a mercury withdrawal curve shown in Fig. 9.23 ②, also called the withdrawal curve or the imbibition curve, can be obtained by releasing the pressure in a stepwise fashion to eject the mercury held in the rock pore system.

Abroad dilatometer is also used to meter the volume of the mercury injected into core. Dilatometer, as shown in Fig. 9.24, is a stainless steel tube that contains platinum filament and is able to withstand high pressures. Considering the mercury and the platinum filament as a single conductor, the electric resistance of this conductor varies with the volume of the mercury. Therefore, the calibrated dilatometer, conducting measurement by correlating the volume of the mercury filling the rock pores and the electric resistance, yields volume of the injected mercury with higher accuracy than the metering pump.

Two advantages are gained by the mercury injection method: The time for determination is reduced and the range of pressure for investigation is increased as the limitation of the properties of the diaphragm is removed. Besides, it can also be used to test samples with irregular shapes, for example, rock debris. Disadvantages are: (1) the difference in such aspects as wetting properties between the

Fig. 9.23 Capillary pressure curves measured form mercury injection. ① Mercury injection curve (drainage), ② mercury ejection curve (imbibition)

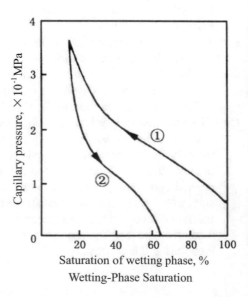

Fig. 9.24 Mercury
dilatometer

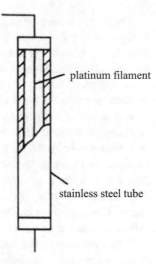

platinum filament

stainless steel tube

experimental conditions and the reservoir realities, (2) the permanent loss of the
core sample, and (3) the toxicity of mercury. Mercury is extremely toxic and should
be generally handled with care to avoid spills and evaporation of it. Besides, it is
stipulated that the laboratory in the presence of mercury must be kept in conditions
of good ventilation and low temperature (<15 °C).

9.2.3   Centrifuge Method

The centrifuge method as shown in Fig. 9.25 is an indirect approach for the
determination of capillary-force curves of reservoir rocks. In this method, the
centrifugal force, which builds up the displacement pressure, is produced by the
high-speed centrifuge to drive the non-wetting phase to displace the wetting phase.
The working principle is presented below.

Fig. 9.25 Principle of
centrifuge method

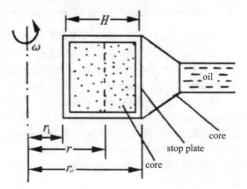

Firstly put a core initially saturated with the wetting-phase water into the centrifugal tube (also called core box) filled with the non-wetting-phase oil. Then, embed the centrifugal tube into the centrifuge which can rotate with specified angular velocities. Due to the density difference, the oil and the water even at the same radius of rotation in the rotated tube are subjected to centrifugal forces of different magnitude. By the capillary force, the water initially contained in the core is thrown out of the pore system, and the oil in the tube is forced into the core and occupies pore space of it. Finally, the difference of centrifugal force between the two phases is balanced by the capillary force between the two phases in the porous medium, and the system reaches its equilibrium at the give rotating speed. That is to say, the difference of centrifugal force between the two phases is equal to the displacement pressure. Besides, as the volume of the water thrown out of the core can be observed through the view window and with the aid of the stroboscope, the saturation data at equilibrium are easy to obtain.

When the sample is rotated at a variety of constant speeds, a complete capillary pressure curve can be obtained.

According to the Petroleum and Natural Gas Industry Standard, the correlation between the capillary force and the wetting-phase saturation is:

$$P_c = 0.05656 \, \Delta\rho n^2 H \left(r_e - \frac{H}{4} \right) \times 10^{-8} \tag{9.32}$$

$$\overline{S_w} = (V_{wi} - V_{wc})/V_p \tag{9.33}$$

where

P_c capillary force, MPa;
$\Delta\rho$ density difference between water and water, g/mL;
n rotating speed of the centrifuge, r/min;
H length of the core, cm;
r_e the external radius of rotation of the core, cm;
$\overline{S_w}$ average water saturation of the core, %;
V_{wi} water volume initially saturated in the core, cm^3;
V_{wc} water volume cumulatively driven out of the core, cm^3;
V_p effective pore volume of the core, cm^3.

The wetting and non-wetting fluids shall be selected in accordance with the specific reservoir conditions and the different purposes of the respective experiments, so the experiments may be gas displacing water, gas displacing oil, water displacing oil, oil displacing water, and so on.

In the case of water displacing oil, the core boxes should be fixed on the centrifuge in the pattern shown in Fig. 9.26a, wherein the oil forced out is near to the central axis of the centrifuge; conversely, in the case of oil displacing water the core boxes should be fixed in the pattern shown in Fig. 9.26b, which is quite the reverse of that for the water displacing oil experiment.

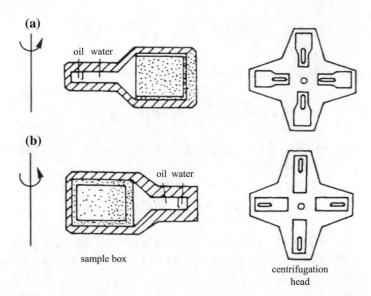

Fig. 9.26 Placement of the core box in the water-displacing-oil test and the oil-displacing-oil test

The centrifuge method for the determination of capillary-force curves combines the advantages of the porous diaphragm plate method and the mercury injection method. Firstly, it yields the curves directly and needs no conversions (the mercury–gas curve obtained by the mercury injection method must be converted into an oil–gas curve); secondly, it allows convenient measurements in both the imbibition and the drainage ways; besides, it requires much less time for determining a complete curve. However, this method involves a more complex calculation process and requires a more complex set of instruments.

9.2.4 Conversion Between the Capillary Forces Obtained Through Different Methods of Measurement

As we all know, different approaches usually use different fluids to measure capillary pressure curves and the in-laboratory experimental conditions are almost impossible to be all the same with the reservoir realities. This problem necessitates conversions among the different methods. For example, in dealing with the difference in experimental fluid medium, conversion equation bridging different methods can be deduced as follows:

Given that the same cores will be used for the experiments, the capillary pressure on laboratory conditions is:

$$P_{cL} = \frac{2\sigma_L \cos \theta_L}{r} \quad \text{or} \quad r = \frac{2\sigma_L \cos \theta_L}{P_{cL}} \tag{9.34}$$

And the capillary pressure on reservoir conditions is:

$$P_{cR} = \frac{2\sigma_R \cos \theta_R}{r} \quad \text{or} \quad r = \frac{2\sigma_R \cos \theta_R}{P_{cR}} \tag{9.35}$$

For the same cores have equivalent pore radius r, the conversion formula between the two situations is:

$$P_{cR} = \frac{\sigma_R \cos \theta_R}{\sigma_L \cos \theta_L} P_{cL} \tag{9.36}$$

This formula applies to the conversion between the capillary pressure data obtained from different methods and experimental conditions. Detailed calculations are as follows.

1. Conversion between the P_{Hg} obtained from the mercury injection method and the water–gas P_{wg} obtained from the porous diaphragm plate method.

It is known that the surface tension of mercury is $\sigma_{Hg} = 480$ mN/m and $\theta_{Hg} = 140°$; the water–air interfacial tension is $\sigma_{wg} = 72$ mN/m and $\theta_{wg} = 0°$. Thus:

$$P_{wg} = \frac{\sigma_{wg} \cos \theta_{wg}}{\sigma_{Hg} \cos \theta_{Hg}} P_{Hg} = \frac{72 \times \cos 0°}{480 \times \cos 140°} P_{Hg} \approx \frac{1}{5} P_{Hg} \tag{9.37}$$

Indicated by Eq. (9.37), the capillary pressure scale for the curves determined by mercury injection is approximately 5 times of that for the curves determined by the porous diaphragm plate method shown in Fig. 9.27. It is generally considered that the porous diaphragm plate method is more nearly approaching the actual wetting reservoir conditions and yields results with higher accuracy, and therefore, it is utilized as the standard against which all other methods are compared. Proven by practice, the results yielded from the other methods are in substantial agreement with that from the porous diaphragm method.

2. Conversion between the P_{Hg} obtained from the mercury injection method and the oil–water P_{ow} obtained on reservoir conditions.

It is known that the surface tension of mercury is $\sigma_{Hg} = 480$ mN/m and $\theta_{Hg} = 140°$; the oil–water interfacial tension is $\sigma_{ow} = 25$ mN/m and $\theta_{ow} = 0°$. Thus:

$$P_{ow} = \frac{\sigma_{ow} \cos \theta_{ow}}{\sigma_{Hg} \cos \theta_{Hg}} P_{Hg} = \frac{25 \times \cos 0°}{480 \times \cos 140°} P_{Hg} \approx \frac{1}{15} P_{Hg} \tag{9.38}$$

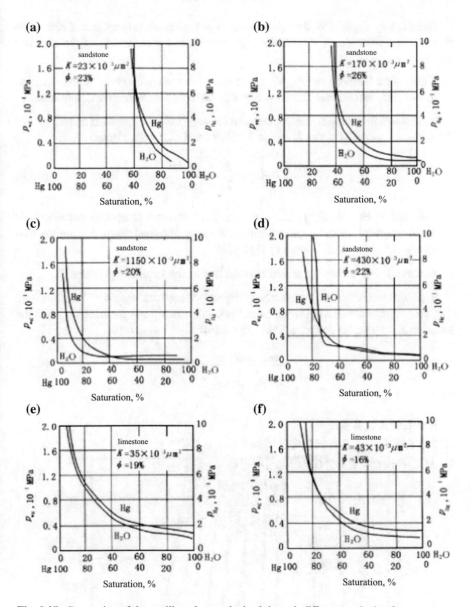

Fig. 9.27 Conversion of the capillary forces obtained through different methods of measurement (mercury injection method and porous diaphragm plate method. **a** Sandstone, $K = 23$ mdarcy, $\varphi = 23$ %; **b** sandstone, $K = 170$ mdarcy, $\varphi = 26$ %; **c** sandstone, $K = 1150$ mdarcy, $\varphi = 20$ %; **d** sandstone, $K = 430$ mdarcy, $\varphi = 22$ %; **e** sandstone, $K = 35$ mdarcy, $\varphi = 19$ %; **f** sandstone, $K = 43$ mdarcy, $\varphi = 16$ %

Indicated by Eq. (9.38), the oil–water capillary pressure present in real reservoirs is about one-fifteenth of the capillary pressure value obtained from the mercury injection method.

3. Conversion between the P_{wg} obtained from the porous diaphragm plate method (air–water system) and the oil–water P_{ow} obtained on reservoir conditions.

It is known that the oil–water interfacial tension is $\sigma_{ow} = 25\text{mN/m}$ and $\theta_{ow} = 0°$; the surface tension of water is $\sigma_{wg} = 72\text{mN/m}$ and $\theta_{wg} = 0$. Thus:

$$P_{ow} = \frac{\sigma_{ow}\cos\theta_{ow}}{\sigma_{wg}\cos\theta_{wg}}P_{wg} = \frac{25 \times \cos 0°}{72 \times \cos 0°}P_{wg} \approx \frac{1}{3}P_{wg} \qquad (9.39)$$

Indicated by Eq. (9.39), the oil–water capillary pressure present in real reservoirs is about one-third of the capillary pressure value obtained from the porous diaphragm plate method, as shown in Fig. 9.28.

4. Calculation of the throat radius r according to the capillary pressure P_c.

It is known that in the mercury injection method $\sigma_{Hg} = 480$ mN/m and $\theta = 140°$. The conversion formula correlating the capillary pressure P_{Hg} and the throat radius r can be deduced from Eq. (9.34):

$$P_{Hg} = \frac{2\sigma_{Hg}\cos\theta_{Hg}}{r} = \frac{0.75}{r} \ \text{或} \ r = \frac{0.75}{P_{Hg}} \qquad (9.40)$$

where

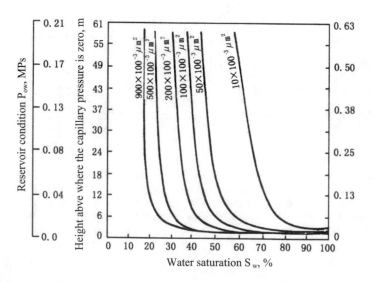

Fig. 9.28 Conversion between the capillary-force curves under laboratory conditions and under reservoir conditions (by Frick)

P_{Hg} capillary pressure measured from the mercury injection method, MPa;
r throat radius at the capillary pressure P_{Hg}, μm.

Using the equation above, the capillary pressure curve could be converted into the relation curve between the throat radius and the saturation of the wetting phase. As shown in Fig. 9.29, the throat radius corresponding to the variation of saturation can also be directly given as another vertical coordinate in the coordinate system of the capillary pressure curve.

5. Calculation of the height of liquid column h according to the capillary pressure P_{c}.

The equation below is obtained from Eq. (9.9):

$$h = \frac{P_{\mathrm{cR}}}{(\rho_{\mathrm{w}} - \rho_{\mathrm{o}})9.81} \tag{9.41}$$

where

h height of the wetting-phase (water) column above oil–water interface, m;
P_{cR} oil–water capillary pressure upon reservoir conditions, Pa;
$\rho_{\mathrm{w}}, \rho_{\mathrm{o}}$ water density and oil density upon reservoir conditions, kg/m^3.

The height of the liquid column can also be directly given as another vertical coordinate in the coordinate system of the capillary pressure curve, as shown in Figs. 9.28 and 9.29.

(**a**) capillary pressure with the diameter of pore throat

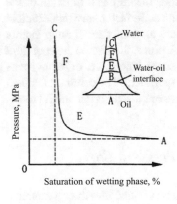
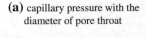

(**b**) The height of liquid column with saturation

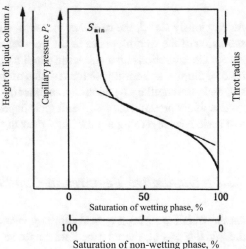

Fig. 9.29 Qualitative features of capillary pressure curve

Example 9.1 Find the in-laboratory capillary pressure curve for the core of K = 200 md in Fig. 9.28. Convert this curve into a curve reflecting the distribution of the oil–water saturations corresponding to the different heights above free water surface. It is assumed that, upon reservoir conditions, σ_{wo} = 24 mN/m, ρ_w = 1088 kg/m³, and ρ_o = 848 kg/m³; and that the surface tension of water at room pressure is σ_{wg} = 72 mN/m.

Solution: Arbitrarily select a water saturation point on the curve, e.g., S_w = 40 %, and read the in-laboratory capillary pressure corresponding to it: P_{cL} = 0.065 MPa. According to Eq. (9.36), convert P_{cL} into the capillary pressure P_{cR} on reservoir conditions:

$$P_{cR} = \frac{\sigma_{wo}}{\sigma_{wg}} P_{cL} = \frac{24}{72} \times 0.065 = 0.022 \text{ MPa} = 22{,}000 \text{ Pa}$$

Calculate the height h above the free water surface according to Eq. (9.41):

$$h = \frac{P_c}{(\rho_w - \rho_o)9.81} = \frac{22000}{(1088 - 848)9.81} = 9.34(\text{m})$$

Find the point of P_{cL} = 0.065 MPa on the right vertical coordinate and draw a horizontal line from this point, which intersects the left vertical coordinate at the point a. Mark the point a as 9.34 m.

In a similar way, the height corresponding to the other water saturations can be marked on the left vertical coordinate.

9.3   Essential Features of Capillary Pressure Curve

As previously stated, the capillary pressure of a porous medium is a function of the saturation of the wetting phase or the non-wetting phase. In fact, many other factors except the saturations also affect the capillary pressure to some extent. Such factors involve the size and sorting characteristics of the rock pores, the composition of fluid and rock, the capillary hysteresis phenomenon, and so on. Therefore, different rocks give capillary pressure curves of different shapes in experiment. The common features and basic laws regarding the capillary pressure curves are discussed in detail below.

9.3.1   Qualitative Features of Capillary Pressure Curve

As shown in Fig. 9.29, a typical capillary cure could be divided into three segments: the initial segment, the intermediate gentle segment, and the end upward segment. The curve exhibits a shape of being steep in both ends and gentle in middle.

During the initial period, the saturation of the wetting phase decreases slowly with the increasing capillary pressure, while the saturation of the non-wetting phase increases slowly at the same time.

Actually, what have been entered by the non-wetting phase in this period are not the inner pore system of the rock, but merely the small pits on the core surface and the dissected large pore openings.

The intermediate segment of the capillary pressure curve slopes gently. This feature indicates that the non-wetting phase is being gradually pushed into the smaller and smaller pores of the rock under the pressure within this variation interval. With the rapid increase of the saturation of the non-wetting phase, the corresponding pressure varies little. The longer the intermediate segment, the more concentratedly distributed and better sorted the pore-throat sizes. The lower the gentle segment is located, the larger the pore-throat radius will be.

The upward segment at the end of the curve indicates that, as the capillary pressure increases dramatically, less and less non-wetting phase is being forced into the pore system. Finally, only quite a small portion of pore openings still hold wetting phase that the non-wetting phase fails to expel. As a result, the saturation of non-wetting phase ceases increasing although the pressure is still being raised.

9.3.2   Quantitative Features of Capillary Pressure Curve

Many a quantitative index are used to describe the capillary pressure curve: the threshold pressure P_T, the pressure of median saturation P_{c50}, and the minimum wetting-phase saturation S_{min}. They are marked in Fig. 9.30.

1. Threshold pressure P_T

The threshold pressure is the minimum pressure for the non-wetting phase to be forced into the core. It corresponds to the capillary pressure at the largest pore. The

Fig. 9.30 Quantitative features of capillary pressure curve

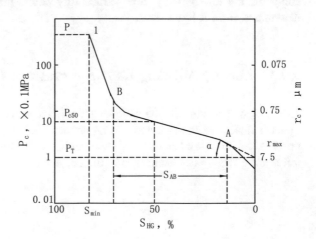

threshold pressure is also called the entrance pressure or the displacement pressure. Generally speaking, the threshold pressure is the pressure at the point where the elongation line of the intermediate gentle segment of the capillary pressure curve intersects with the vertical coordinate representing the zero saturation of non-wetting phase.

Rocks with better permeability and larger pore openings exhibit lower threshold pressure P_T, which is indicative of a more favorable physical property, and vice versa. Therefore, the threshold pressure can be used for the evaluation of rock permeability. Besides, the threshold pressure also provides a way to judge the rock wettability and to determine the largest pore radius present in the rock.

2. Pressure of median saturation P_{c50}

The pressure of median saturation P_{c50} refers to the capillary pressure at the point of 50 % saturation. The pore radius corresponding to this pressure is called the radius of median saturation and denoted as r_{50}.

Obviously, lower P_{c50} reflects larger r_{50} and more favorable porosity-permeability property of the studied rock. Provided that the pore-size distribution of a rock is nearly a normal distribution, the r_{50} is roughly considered to be equivalent to the average pore radius.

3. Minimum wetting-phase saturation S_{min}

The minimum wetting-phase saturation S_{min} is the volume percentage of the pores that have not been intruded by the non-wetting phase when the displacement pressure reaches its highest value. In the case of water-wet rocks, S_{min} represents the connate-water saturation, and in the case of oil-wet rocks, S_{min} represents the residual oil saturation. The minimum wetting-phase saturation is an index reflecting the pore structure of rock—the better the rock's physical property, the lower the minimum wetting-phase saturation of it. In addition, the valve of S_{min} is also limited by the highest pressure that could be afforded by the apparatus. If the steep segment of the capillary pressure curve is not parallel to the coordinate of pressure, especially for the low-porosity, low-permeability cores, the connate-water saturation determined from it bears error.

9.3.3 Factors Affecting Capillary Pressure Curve

As mentioned previously, the features of capillary pressure curve are also impacted by such factors as the size and sorting characteristics of the rock pores, the composition of fluid and rock, and the capillary hysteresis phenomenon. These influencing factors cause changes in the curve.

1. Pore structure and physical property of rock

As we all know, capillary pressure is correlated with pore radius through the equation $r = 2\sigma \cos \theta / P_c$. Thus, the pore-throat distribution can be gained from the capillary pressure curve, and the vertical coordinate shown in Fig. 9.29 is convertible into a coordinate representing the pore radius.

In order to quantitatively study the pore-throat distribution of rock, a frequency distribution diagram (Fig. 5.13) and a cumulative frequency distribution diagram (Fig. 5.14) are usually drawn on the basis of capillary pressure curve. The two distribution diagrams allow determination of such parameters as the largest pore-throat radius, the mean pore-throat radius, and the smallest pore-throat radius accessible for the wetting-phase flow.

More concentrated distribution as well as better sorting of the pore-throat sizes lead to longer intermediate gentle segment of the capillary pressure curve, which is simultaneously closer to being horizontal. Larger pore radius leads the intermediate gentle segment of the curve to be closer to the horizontal coordinate and thereby the capillary pressure to be smaller. Regarding the curvilinear characters, what's significantly affected by the size and the concentration of the pore openings involved in a rock is the skewness of the curve, which is a measurement to characterize the shape of the curve and to indicate the concentration of the pore openings—whether the pore size is concentrated on the side of large pore-throats or the side of slim pore-throats. The more the large pores and throats, the closer to the lower left the curve will be, which is a case named coarse skewness; the more the slim pores and throats, the closer to the upper right the curve will be, which is a case named "thin skewness."

Thus, the shape of capillary pressure curve depends on the pore-structure characteristics such as the sorting and the size of the pores and channels. Based on this conclusion, the storage and collection property of reservoir rocks can be evaluated from the capillary pressure curves. Figure 9.31 gives 6 idealized capillary pressure curves. Obviously, the curve (a) represents rocks with very favorable physical properties while the curve (b) represents rocks with very unfavorable physical properties.

The geological classification and development of carbonate reservoirs are also based on the capillary pressure curves. Due to the fractures involved in carbonate rocks, it is no longer enough to describe the storage and permeability characteristics of this kind of rocks only with the porosity and the permeability. Besides, the capillary pressure curve can also clearly reveals the double-porosity characteristics of carbonate reservoirs. Figure 9.32 shows a typical capillary pressure curve obtained from mercury injection method for a double-porosity rock.

2. Direction of the variation of the wetting-phase saturation

As shown in Fig. 9.33, the features of capillary pressure curve are also related to the direction of the variation of the wetting-phase saturation. Subjected to the same condition of saturation, the capillary pressure in the drainage process of non-wetting

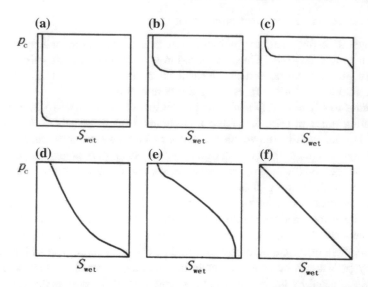

Fig. 9.31 Six idealized capillary pressure curves. **a** Good sorting, even fracture distribution, 粗歪度; **b** good sorting, even fracture distribution; **c** good sorting, even fracture distribution, "thin skewness"; **d** bad sorting, even fracture distribution, 略粗歪度; **e** bad sorting, uneven fracture distribution, 略 "thin skewness"; **f** unsorted, extremely uneven

Fig. 9.32 Qualitative features of the capillary pressure curve of double-porosity medium

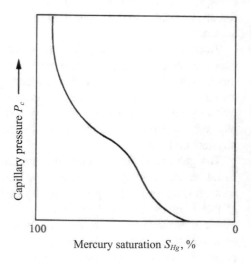

phase displacing wetting phase is always higher than that in the imbibition process of wetting phase displacing non-wetting phase. This difference is caused by the wetting hysteresis, which renders the advancing angle to be always larger than the receding angle.

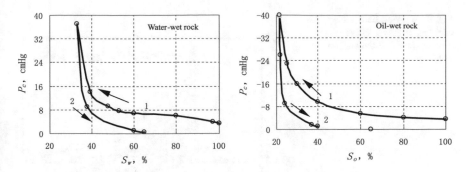

Fig. 9.33 Capillary pressure curves highlighting the change direction of saturation. *1* Drainage curve, *2* imbibition curve

The cores with similar drainage capillary pressure curves may exhibit unlike imbibition capillary pressure curves. This proves the complexity of the pore structure of reservoir rocks. Therefore, the combination of the mercury-injection curve and the mercury-withdrawal curve should be used for the study of the pore structure of reservoir rocks, especially the carbonate rocks.

In the in-laboratory simulation experiments for secondary and tertiary recovery, the difference between the drainage curve and the imbibition curve should be paid special attention.

3. Wettability of reservoir rocks

As shown in Fig. 9.33, different wettability characteristics of reservoir rock yield capillary pressure curves of different shapes. Taking the wettability of rock into consideration, assign the vertical coordinate in the right-hand diagram with negative sign.

Figures 9.34 and 9.35 show the capillary pressure curves measured from centrifuge method. The dashed line I represents the capillary pressure curve obtained by displacing a initially water-saturated with oil, aiming at simulating the connate-water saturation of the reservoir; the capillary pressure curve II is gained by displacing the core with water after the process of curve I; the curve is measured subsequent to the process of curve II, which is again an oil-displacing-water test.

Putting the water-displacing-oil capillary pressure curve underneath, the horizontal coordinate is for the convenience of comparison and statistics. A_{od} is the area under the curve III, and A_{wd} is the area under the curve II. The rock wettability can be determined by comparing A_{od} and A_{wd}: if $A_{od} > A_{wd}$, it is indicated that the work done by the oil displacing the water is larger than the work one by the water displacing the oil, or in other words, the rock is water-wet shown in Fig. 9.35; conversely, if $A_{od} < A_{wd}$, it is indicated that the rock is oil-wet shown in Fig. 9.34; if $A_{od} = A_{wd}$, it is indicated that the rock is intermediate-wet.

Fig. 9.34 Area below the capillary pressure curve of oil-wet rock (by Donaldson 1969)

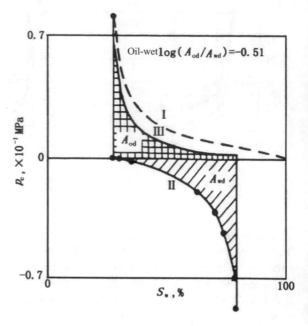

Fig. 9.35 Area below the capillary pressure curve of water-wet rock (by Donaldson 1969)

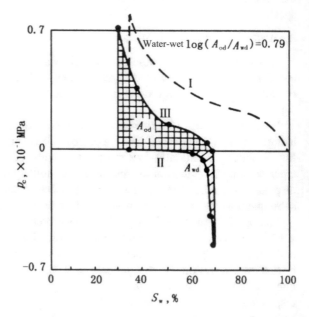

In order to give a quantitative description for rock wettability, the logarithm of the area ratio $\log \frac{A_{od}}{A_{wd}}$ is usually taken as a quantitative index for the determination of rock wettability based on capillary pressure curves. If $\log \frac{A_{od}}{A_{wd}} > 0$, the rock is water-wet; if $\log \frac{A_{od}}{A_{wd}} < 0$, the rock is oil-wet; and if $\log \frac{A_{od}}{A_{wd}} = 0$, the rock is intermediate-wet.

9.4 Application of Capillary Pressure Curve

The capillary pressure data of oil-bearing reservoir rocks are commonly used to directly or indirectly determine some petrophysical parameters of formation, for example, porosity, absolute permeability, relative permeability, connate-water saturation, residual oil saturation, rock wettability, specific surface, pore-throat size distribution, and sweep efficiency. Applications of capillary pressure curves will be discussed in detail in this section.

In the Sect. 9.3.2, it is already been discussed the aids of capillary pressure curve in investigating pore structure and in evaluating physical property and storage and collection performance of reservoir rocks, so relevant contents to that point will not be repeated here.

9.4.1 Determine the Wettability of Reservoir Rocks

1. The area ratio method

This approach to determine the wettability of reservoir rocks is based on the centrifuge capillary pressure curve. Go back to the last section for more details.

2. The wettability index and apparent contact angle method

Taking the pore surface wetted by the oil of an air–oil system as standard, this method compares the pore surface wetted by the water of an oil–water system with the standard and thereby yield a judgment of water's ability to wet the rock.

Get two cores from the same rock and, respectively, saturate them with oil and water. Perform air-displacing-oil test and the oil-displacing-water test on the cores, and then draw the capillary pressure curves for them. Calculate the threshold pressure for each curve. The threshold pressures are symbolized as P_{Tog} and P_{Two}.

By definition, the wettability index is:

$$W = \frac{\cos \theta_{wo}}{\cos \theta_{og}} = \frac{P_{Two} \sigma_{og}}{P_{Tog} \sigma_{wo}} \tag{9.42}$$

where

θ_{wo}, θ_{og} the water–oil–rock contact angle, the oil–gas–rock contact angle;

P_{Two}, P_{Tog} the threshold pressure at which the oil enters the water-saturated core, and the threshold pressure at which the air enters the oil-saturated core;

σ_{og}, σ_{wo} the oil–air interfacial tension and the water–oil interfacial tension.

As liquid is much more capable of getting rock wetted than air, the oil is able to wet the rock, whereas the air is not. Expressed mathematically, $\theta_{og} \approx 0$ and $\cos \theta_{og} = 1$. Thus, the air requires quite sufficient threshold pressure P_{Tog} before entering the oil-saturated core.

When $W = 1$, it is indicated that the water of an oil–water system has equivalent capacity to wet the rock with the oil of an oil–gas system. In other words, the water rock is fully water-wet.

When $W = 0$, it is indicated that the water of an oil–water system does not wet the rock at all, whereas the oil is able to enter the rock pores easily. In other words, the rock is fully oil-wet.

Generally, the closer the W is to 0, the more oil-wet the rock will be; the closer the W is to 1, the more the more water-wet the rock will be.

Substituting $\theta_{og} = 0$ and $\cos \theta_{og} = 1$ into Eq. (9.42), we can get:

$$\cos \theta_{wo} = \frac{P_{Two}\sigma_{og}}{P_{Tog}\sigma_{wo}} \tag{9.43}$$

And thence the apparent contact angle is:

$$\theta_{wo} = \arccos \frac{P_{Two}\sigma_{og}}{P_{Tog}\sigma_{wo}} \tag{9.44}$$

As the parameters on the right of the equal sign are measurable, the apparent contact angle of oil–water system θ_{wo} can be obtained through Eq. (9.44). As a quantitative indicator to reflect the degree of wetting of rocks, the closer the θ_{wo} is to 0, the more water-wet the rock will be; the closer the θ_{wo} is to 90°, the more oil-wet the rock will be.

9.4.2   Averaging Capillary Pressure Data: $J(S_w)$ Function

As capillary pressure data are obtained from small cores (e.g., sidewall cores and drill cuttings) which represent extremely small and local parts of the reservoir, it therefore becomes necessary to combine all the capillary data in some way so as to manifest the entire reservoir. In dealing with the capillary data from oil-field practice, a method called J-function is commonly used to do the averaging work. J (S_w) characterizes the multiphase flow through porous medium by integrating such various relevant elements as the interfacial tensions present in formation, the rock wettability, the rock permeability, the porosity, and the capillary pressure curves Amounts of data and researches indicate that when all the capillary pressure data of a formation are converted into a universal curve by using the $J(S_w)$ function, the curve turns out to be a monotone one. Thus, the $J(S_w)$ function provides a good way to evaluate and compare formations. $J(S_w)$ is defined as:

$$J(S_w) = \frac{P_c}{\sigma \cos\theta} \left(\frac{K}{\phi}\right)^{\frac{1}{2}} \tag{9.45}$$

where

$J(S_w)$ J-function, dimensionless;
P_c capillary pressure, dyn/cm^2, $1 dyn = 0.1$ MPa;
K air permeability, cm^2;
σ interfacial tension, dyn/cm;
θ wetting contact angle, (°);
φ porosity, fraction.

Some authors alter the above expression by excluding the $\cos\theta$ because the value of θ is rather difficult to measure but reasonable to be considered as a constant for a reservoir. As the experimented cores and fluids in laboratory represent real situation of the reservoir, Eq. (9.45) can be reduced to a simpler form as follows:

$$J(S_w) = \frac{P_c}{\sigma} \left(\frac{K}{\phi}\right)^{\frac{1}{2}} \tag{9.46}$$

$J(S_w)$, as a semiempirical dimensionless correlation function, was firstly deduced from dimensional analysis. This correlation is then confirmed by numerous experiments and serves as a comprehensive treatment for capillary pressure curves. As shown in Fig. 9.36 wherein field data obtained by Brown et al. are given, the J (S_w) functions for the cores of a specific lithologic type from the same formation exhibit good conformity.

Figure 9.36a shows the $J(S_w)$ data converted from the capillary pressure curves for the cores taken from the same reservoir. Although the data points are rather scattered, general trend (the solid curve) of them is still obvious. The core materials were subdivided into limestone and dolomites, both materials occurring within the productive section of the same formation. The dolomite samples, as shown in Fig. 9.36c wherein the data points are comparatively concentrated, indicate a good correlation while the limestone samples exhibit a scattering of data in the range of low water saturations. In an attempt to obtain a better correlation, the limestone cores were further subdivided into microgranular (d) and coarse-grained samples

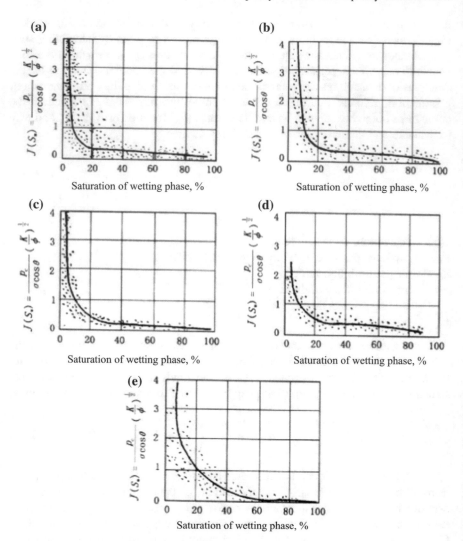

Fig. 9.36 *J*-function correlation of capillary pressure data (by Brown 1951). **a** All cores, **b** limestone cores, **c** dolomite cores, **d** fine-grained limestone cores, **e** coarse-grained limestone cores

(e). The dispersion of data points given in Fig. 9.36e is rather great. Further identification shows that this group of data pertains to vuggy-fractured limestone.

Through the treatment of $J(S_w)$, we can: (1) obtain the averaged capillary pressure data for the rocks of the same type; (2) determine the physical property characteristics of the rocks of different types.

Generally speaking, there are significant differences in correlation of the *J*-function with water saturation from formation to formation, and even the cores from the same formation fail to yield a universal curve if the difference in permeability between the cores is too great.

9.4.3 Determination of the Fluid Saturation in the Oil–Water Transition Zone

Although oil and water are immiscible, the contact between oil and water is commonly a transition zone rather than a distinct interface. Another application of capillary pressure curve is to determine the distribution of fluid saturation in the oil–water or oil–gas transition zone of petroleum reservoirs.

Figure 9.37 shows the distribution of fluid saturation in the oil–water transition zone of a petroleum reservoir. To convert capillary pressure–saturation ($P_c \sim S_w$) relation to height-saturation ($h \sim S_w$) relation, it is only necessary to rearrange the terms in Eq. (9.41) so as to solve for the height instead of the capillary pressure. That is:

$$h(S_w) = \frac{P_c(S_w)}{9.81(\rho_w - \rho_o)}$$

However, as the connate-water saturation directly corresponds to the height at which the water begins to be immovable and irreducible, the relative permeability curves are required before the determination of the span of oil–water transition zone is accomplished.

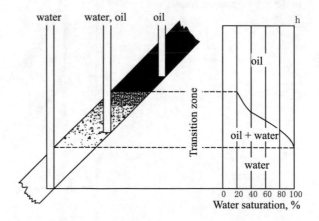

Fig. 9.37 Distribution of fluid saturation in a petroleum reservoir

9.4.4   Calculation of Recovery by Using Capillary Loop

When measuring capillary pressure curves by means of mercury injection method, the curve measured by pressurizing the core and gradually displacing the wetting phase with the non-wetting phase is called the drainage capillary pressure curve, also named drainage curve or injection curve for short, and the curve measured by depressurizing the core and displacing the non-wetting phase with the wetting phase is called the imbibition capillary pressure curve, also called imbibition curve or withdrawal curve.

That is to say, the drainage process is also called injection while the imbibition process is also called withdrawal. In test, firstly the core is pressurized and mercury is injected into it; then, the core is depressurized and the mercury in it is expelled out; and after that, the core is performed to undergo a secondary injection process and a subsequent secondary withdrawal process of the mercury by the same token. Thus, a family of capillary pressure curves can be obtained by repeating the process on the core. The measurement of capillary pressure loop, which is given in Fig. 9.38, involves at least three component processes: mercury injection, mercury withdrawal, and secondary mercury injection. Study of the drainage curve and the imbibition curve is of great importance in petroleum engineering because the curves offer helpful information about the rock pore structure, the fluid distribution, the mechanism of flow, and the impact of the non-wetting's capillary effect on recovery. The family of capillary pressure curves presented in Fig. 9.38 includes the primary injection curve (the curve I), the primary withdrawal curve (the curve W), and the secondary injection curve (R). During the process of primary mercury injection wherein the pressure is raised from 0 to the highest, the saturation of the

Fig. 9.38 Drainage and imbibition capillary pressure curves

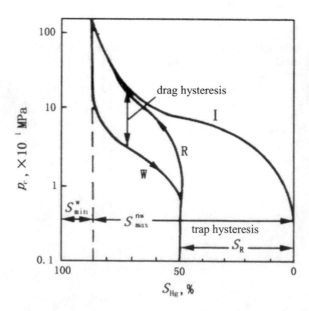

wetting phase decreases from 100 % to S_{min}^{w} (the minimum wetting-phase saturation), while the saturation of the non-wetting phase increases from 0 to S_{min}^{nw}. During the process of primary mercury withdrawal wherein the pressure is reduced from the highest to 0, it is impossible for all the non-wetting phase to be expelled out of the core in the presence of capillary pressure, which renders a portion of it to be left in the pore system and establish the residual-mercury saturation S_R. The difference between the primary mercury injection curve and the secondary mercury injection curve, as shown in the figure, is decided by the pore structure of rock and usually called trap hysteresis. In other words, rocks with different pore structures, giving rise to different degrees of trap hysteresis phenomenon when experimented by means of the mercury injection method, yield primary and secondary mercury injection curves with different in-between deviations. In comparison with the secondary injection curve and the primary withdrawal curve, the displacement pressure is higher than the imbibition pressure on the same condition of saturations. This phenomenon is called capillary hysteresis or drag hysteresis, which results in the change of the mercury–rock contact angle during the mercury withdrawal process and reduction of the surface tension of mercury due to contamination.

The minimum wetting-phase saturations yielded from the secondary injection curve and the withdrawal curve are equal to each other. Thus, a closed loop is formed by the secondary injection curve together with the withdrawal curve, and this loop is named capillary loop (the R–W loop in Fig. 9.38).

On the basis of the maximum mercury saturation and the residual-mercury saturation obtained from the capillary pressure curves, Wardlaw proposed the term of "mercury withdrawal efficiency" in 1976. The mercury withdrawal efficiency E_w is defined as the ratio of the volume of mercury that has withdrawn from the core after the depressurization to the volume of mercury that had been injected into the core before the depressurization:

$$E_w = \frac{S_{Hgmax} - S_R}{S_{Hgmax}} \times 100\% \qquad (9.47)$$

Essentially speaking, the mercury withdrawal efficiency is the recovery of the non-wetting phase. For those water-wet formations, the mercury withdrawal efficiency is rightly the oil recovery. So, this term provides a new path for the study of the water-displacing-oil recovery and the research on the relationship between pore structure and oil recovery.

9.4.5   Calculation of Absolute Permeability of Rock

Equation (9.48) is the formula for the calculation of rock permeability based on the mercury injection method. Related principles and derivations will be given in Sect. 10.4.

$$K = \frac{\lambda (\sigma \cos \theta)^2 \phi}{2 \times 10^4} \int_{S=0}^{S=100} \frac{\mathrm{d}S_{\mathrm{Hg}}}{P_{\mathrm{c}}^2} = 0.66\lambda\phi \int_{S=0}^{S=100} \frac{\mathrm{d}S_{\mathrm{Hg}}}{(P_{\mathrm{c}}^2)} \qquad (9.48)$$

where

K permeability, cm^2;
φ porosity;
S_{Hg} saturation of mercury, dyn/cm;
P_{c} capillary pressure, Pa;
λ lithology factor, dimensionless.

The key to the application of the equation above is the determination of the lithology factor. This computing method gives a good reference to the calculation of rock permeability if the lithology factor of formation can be assigned dependable value. In fact, because it is difficult to measure the permeability of rock debris by conventional means, this method is of more special importance for the determination of the permeability of rock debris.

9.4.6   Estimation of the Degree of Damage on Formation

In 1984, Amaefule et al. proposed a set of capillary pressure curves measured from a high-speed centrifuge method and a fast in-laboratory approach to estimate the damage on formations caused by the working fluids as well. The principle is that, if the formation is damaged, the capillary pressure curve will show high entrance pressure and high connate-water saturation, leading the curve to move toward the top right. By comparing the capillary pressure curves of the core before and after its contact with the working fluids, therefore, it is feasible to judge whether the formations are damaged and to evaluate the effect of the additives involved in the working fluids.

The capillary pressure curves are also able to yield the relative permeability data of rock. Related contents will be given in Chap. 10.

Chapter 10
Multiphase Flow Through Porous Medium and Relative Permeability Curve

In real oil reservoirs, the rocks are usually saturated with two or more fluids, such as oil and water, or oil and water and gas. Taking a water-drive reservoir with a local pressure below the saturation pressure as an example, the three phases in it, oil, gas, and water, flow simultaneously and interfere with each other, generating more than one type of capillary effects that resist the flow.

10.1 Characteristics of Multiphase Flow Through Porous Medium

During water drive, the injected water encounters a variety of resisting forces before it displaces the oil out of the pores and channels involved in the porous medium, and the process also results in changes of the quantities and distributions of the phases. For example, the originally continuously distributed oil may be broken into discontinued oil columns and then gradually into individual oil drops as the water injection goes on. In a word, the redistribution of the fluids in a formation results from the mutual constraints between the oil displacement energies and the various resisting forces encountered by the flow. At the same time, the change in fluid distribution conditions also influences in turn the subsequent process of water-displacing oil.

In petroleum reservoir development, researches on these inner changes are required to explain some phenomena and to offer information for the measures to be taken.

© Petroleum Industry Press and Springer-Verlag GmbH Germany 2017
S. Yang, *Fundamentals of Petrophysics*, Springer Geophysics,
DOI 10.1007/978-3-662-55029-8_10

10.1.1   Non-piston-Like Displacement

The ideal water-displacing-oil model is a piston-like oil displacement one: People once imagined that the injected water, which forms a distinct interface with the oil, drives the latter phase in the way like a piston advances in cylinder, and all of the oil contained in the pore system of a formation be driven out at once as shown in Fig. 10.1.

However, the displacement really occurring in the oil-bearing formations conforms to a non-piston-like model. Field data show that the oil wells encounter water breakthrough earlier than expected, and after water breakthrough the wells produce oil and water simultaneously, which means that there are both oil and water flowing in the formation. It is proven by further studies and experiments that the water-displacing oil in the porous medium is a non-piston-like process, and during this process, there are three flowing sections as shown in Fig. 10.2 come into being: the water interval, the two-phase interval, and the oil interval. The micromechanisms for this macrophenomenon are given below.

Fig. 10.1 Ideal model of pistion-like oil displacement

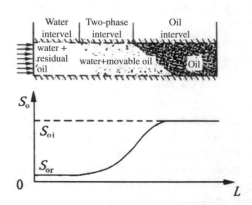

Fig. 10.2 Non-piston-like oil displacement

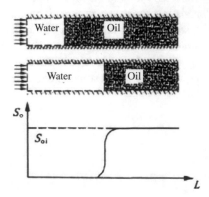

Essentially speaking, the non-piston-like displacement in formation refers to the differences in flowing velocity between the differently sized conducting channels involved in the porous medium. It can be manifested in several respects.

(1) The complexity of the pore structure: the pores and channels involved in a formation vary within a wide range of sizes, wettability characteristics, surface roughness degrees, and tortuosity degrees, so the resisting forces encountered by the oil and water flowing through such a seriously anisotropic medium are enormously different in magnitude, whereupon the flowing velocity therefore differs from one to another.

(2) The existence of capillary pressure: for a water-wet channel, the capillary pressure serves as a driving force, while for an oil-wet channel the capillary pressure presents as a resisting force. No matter whether the capillary pressure is a driving or resisting force, it acts on the oil or water flows to different extends due to its variety in magnitude which is dependent upon the size of the pore openings where the capillary pressure occur. That is to say, even if holding the external driving pressure differential a constant, differences in pressure distribution among the conducing channels will be caused by the varying additional resistances and consequently give rise to difference in the velocity in which they conduct the flowing fluids.

(3) The viscosity difference between oil and water: The two fluids suffer not equivalent viscous forces due to their viscosity difference, and this aggravates the flowing velocity difference between them.

(4) The resisting forces caused by the two-phase flowing of oil and water in capillary tube.

As the large and small channels contained in a porous medium allow the flowing fluids through with different velocities, macroscopically the frontal oil–water interfaces in these channels advance with different velocities, too—in other words, the frontal interfaces do not advance in a manner of keeping abreast of each other but in a rather irregular way, whereupon a mixed zone of coexisting oil and water is formed.

In the flowing sections, the seepage flowing characteristics in several typical pore-channel patterns will be discussed, and on the basis of these discussions, from a microscopic perspective of the processes, the lack of uniformity exhibited by the advancing oil–water interfaces in a porous medium can be illustrated and explained. It deserves to be noted that here the "microscopic" particularly refers to the pore-scale study while the "macroscopic" particularly refers to the reservoir-scale study. Besides, the discussion will be started with the simplest case wherein only a single capillary tube and a single phase are involved and the viscous force is taken into consideration, and then, cases of multi capillary tubes, two phases, and a combined consideration of the viscous force and the capillary force will be analyzed.

10.1.2   Unconnected Capillary Pore-Channels, Single-Phase Fluid Flowing

The flowing of a single-phase in an individual unconnected capillary pore-channel can be represented by that in a single capillary tube. And the condition of single phase takes place in a reservoir in its early development life, the period when the oil phase has not been waterflooded or degasified. As the oil phase exists in the pore space alone, in the system void of capillary force it is only necessary to consider the viscous force.

According to Eq. (6.56) $q = \frac{\pi r^4 \Delta P}{8\mu L}$, the capillary flowing equation, the flowing velocity V of a single-phase fluid through a capillary tube is:

$$V = \frac{r^4 \Delta P}{8\mu L} \qquad (10.1)$$

Obviously, the flowing velocity V is mainly dependent upon the tube radius r and the viscosity of the fluid.

We can assume two channels with the radii, respectively, represented by r_1 and r_2, the ratio of the flowing velocity (V_1 and V_2) of the fluids conducted through the two channels is:

$$\frac{V_1}{V_2} = \left(\frac{r_1}{r_2}\right)^2 \qquad (10.2)$$

or

$$V_1 = V_2 \left(\frac{r_1}{r_2}\right)^2 \qquad (10.3)$$

According to the equation, if holding the pressure differential ΔP, the fluid viscosity μ, and the tube length L as constant, the flowing velocity V in the capillary tube is directly proportional to the square of the tube radius r.

For example, if $r_1 = 10$ μm and $r_2 = 1$ μm, the magnitude of V_1 is 100 times of that of V_2. Thus, this example proves that the seepage flowing through a porous medium subjected to pressure differential dominantly takes place in the large pores and channels. Considering the significant disparity in size between the small pore openings and the large ones, it is reasonable to think that in a porous medium under a given pressure difference, there are actually a portion of small pore openings fail to contribute conducting channels to the seepage flowing. That is to say, in essence the permeability measured in laboratory is an averaged value reflecting a rock's capacity of conducting fluids, and the large pores and channels contained in the rock always contribute more to the fluid-conducting behavior and thus the permeability of it.

Fig. 10.3 Two-phase flow in single capillary tube

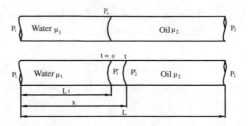

10.1.3   Unconnected Capillary Pore-Channels, Two-Phase Fluid Flowing

Figure 10.3 shows a single capillary tube. The radius of the two capillary tubes is equivalently r, and there is a two-phase (oil and water) fluid flowing within each of them. The water is the wetting phase and has a viscosity of μ_1, while the oil is the non-wetting phase and has a viscosity of μ_2. The capillary at the meniscus between the phases is P_c. Taking both the viscous force and the capillary force into consideration, the equations with respect to the movement of the two-phase interface will be derived below.

(1) Firstly, respectively, consider the flow of each phase in a capillary tube. At any time t, according to the capillary flowing equation we know that:
The flowing velocity of the water phase

$$V_1 = \frac{dx}{dt} = \frac{r^2 (P_1 - P_1')}{8\mu_1 x} \tag{10.4}$$

The flowing velocity of the oil phase

$$V_2 = \frac{dx}{dt} = \frac{r^2 (P_2' - P_2)}{8\mu_2 (L - x)} \tag{10.5}$$

(2) Since the liquids are continuously flowing and the tube radius r is a constant, V_1 is equal to V_2 and equal to the traveling speed of the interface. That is:

$$\frac{dx}{dt} = \frac{r^2 (P_1 - P_1')}{8\mu_1 x} = \frac{r^2 (P_2' - P_2)}{8\mu_2 (L - x)} = \frac{r^2 (P_1 - P_1' + P_2' - P_2)}{8[\mu_1 x + \mu_2 (L - x)]} \tag{10.6}$$

Furthermore, as $P_2' - P_1' = P_c$, the differential equation depicting the movement of the two-phase interface is:

$$\frac{dx}{dt} = \frac{r^2 (P_1 - P_2 + P_c)}{8[\mu_1 x + \mu_2 (L - x)]} \tag{10.7}$$

(3) Through separation of variables and then integration, an expression can be obtained as follows:

$$V = \frac{r^2(P_1 - P_2 + P_c)}{8\sqrt{(\mu_2 L)^2 - (\mu_2 - \mu_1)\left[\frac{r^2 t}{4}(P_1 - P_2 + P_c) + 2\mu_2 L l_t - l_t^2(\mu_2 - \mu_1)\right]}}$$

(10.8)

where L_t—the position of the oil–water interface at the time t.

From the equation above, we can see that in the presence of viscosity difference and capillary forces, the flow of a two-phase fluid within a capillary tube is complicated:

(1) Given a capillary tube with a length L and a pressure differential $(P_1 - P_2)$ over it, the traveling speed V of the interface (i.e., the flowing velocity) varies with the time of displacement t and depends on the viscosity difference between the two phases, the tube radius r, the distance L_t the interface travels, and the total tube length L.
(2) The values of the flowing velocity in tubes of different radius are different from each other.
(3) Even in the same tube, the flowing velocity may be not a constant, depending on the viscosity difference between the two phases. If $\mu_1 < \mu_2$, the flowing liquids increase in velocity. As shown in Fig. 10.4, when several differently sized channels are available for the flowing liquids, the effects discussed above will result in a microscopic fingering in them. In light of this, we can know that a macroscopic fingering will take place in a formation that involve several differently permeable zones, and, generally speaking, the fingering phenomenon tends to be more serious in the case of more serious anisotropy of the formation, larger pore-size differences, or larger viscosity difference between the reservoir oil and the injected liquid (e.g., water). It is to be noted that μ_1 is much smaller than μ_2. The illustration of the macroscopic fingering phenomenon due to the areal anisotropy of a reservoir is given in Fig. 10.5. The more serious the fingering is, the earlier will the oil wells undergo water breakthrough, which is indicative of a longer period of the oil and water

Fig. 10.4 Microscopic fingering in capillary tube

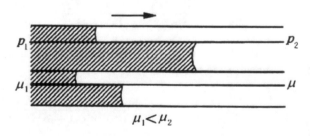

$\mu_1 < \mu_2$

Fig. 10.5 Macroscopic
fingering in reservoir plane

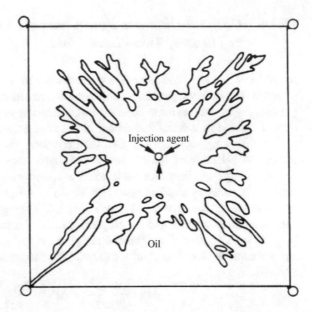

coproduction for the oil wells. It is conceivable that the oil–water relationship
in a waterflooded stratified deposit with serious vertical anisotropy should
exhibit much more complex.

(4) The water-displacing-oil interfaces may advance in a comparatively even and
uniform fashion only when the formation is subjected to a specified suitable
pressure differential. The suitable differential is usually determined by theoretical calculations or in-laboratory simulation experiments.

(5) Consider the simplified case in which the studied process is water displacing
gas and the water is imbibed into the capillary channel only by the capillary
force formed at the meniscus. Obviously, as $P_1 - P_2 = 0$ and $\mu_2 \rightarrow 0$, the
traveling speed of the oil–gas interface is:

$$V = \frac{r\sigma\,\cos\theta}{\sqrt{\mu_1\left(\frac{r\sigma\,\cos\theta\cdot t}{2}\right) + \mu_1 l_t^2}} \tag{10.9}$$

At the moment of $L_t = 0$ and $t \rightarrow 0$, V has its maximum value. This indicates
that the intake velocity of the water into the tube gradually become slower as
time goes on.

(6) In the case of $\mu_1 = \mu_2$ and $Pc = 0$, Eq. (10.8) becomes the velocity calculation
equation for the single-phase liquid flowing through a capillary tube.

10.1.4   Unequal-Diameter Pore-Channels Connected in Parallel, Two-Phase Fluid

As we know, within a rock there is a pore network formed by a number of pores and channels of widely varying sizes, which are interlaced and connected in series or in parallel. Figure 10.6 is a fundamental unit commonly encountered in reservoirs and usually studied as a simplified model of the pore network involved in a rock.

The following sections will give the derivation for the related equations governing the behavior of the oil and water simultaneously flowing within a rock. Assume a large capillary channel of a radius r_1 and with a flow rate q_1 in it; a small capillary channel of a radius r_2 and with a flow rate q_2 in it; and besides, both the two channels are water-wet and have a length L. Taking q as the total flow rate and supposing that the water and the oil present in the channels have the same viscosity ($\mu_1 = \mu_2 = \mu$), in each channel the water phase sustains both the viscous resistance and the capillary force as it drives the oil phase to advance:

$$\Delta P_1 = \frac{8q_1\mu L}{\pi r_1^4}; \quad \Delta P_2 = \frac{8q_2\mu L}{\pi r_2^4} \tag{10.10}$$

$$P_{c1} = \frac{2\sigma \cos \theta}{r_1}; \quad P_{c2} = \frac{2\sigma \cos \theta}{r_2}, \tag{10.11}$$

where

ΔP_1, ΔP_2 The pressure drop caused by the viscous resistance, respectively, in the large and small channels;

P_{c1}, P_{c2} The capillary force, respectively, present in the large and small channels.

Since the two capillary channels are connected in parallel, they must have equivalent pressure at the point A, and so does the pressure at the point B. Therefore, an equilibrium relationship can be built up among the pressures:

Fig. 10.6 Two-phase flow in unequal-diameter pore-channels connected in parallel

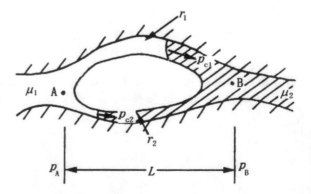

$$P_A - P_B = -P_{c1} + \Delta P_1 = -P_{c2} + \Delta P_2 \tag{10.12}$$

Taking the relationship $q = q_1 + q_2$ into consideration, we can obtain the following two equations by substituting Eqs. (10.10) and (10.11) into Eq. (10.12):

$$q_1 = \left[\frac{4\mu Lq}{\pi r_2^4} - \left(\frac{1}{r_2} - \frac{1}{r_1} \right) \sigma \cos \theta \right] \bigg/ \frac{4\mu L}{\pi} \left(\frac{1}{r_1^4} + \frac{1}{r_2^4} \right) \tag{10.13}$$

$$q_2 = \left[\frac{4\mu Lq}{\pi r_1^4} + \left(\frac{1}{r_2} - \frac{1}{r_1} \right) \sigma \cos \theta \right] \bigg/ \frac{4\mu L}{\pi} \left(\frac{1}{r_1^4} + \frac{1}{r_2^4} \right) \tag{10.14}$$

Then, substitute $q_1 = V_1 \pi r_1^2$ and $q_2 = V_2 \pi r_2^2$ into them and get the ratio of the flowing velocity in the large channel to that in the small one:

$$\frac{V_1}{V_2} = \frac{\frac{4\mu Lq}{\pi r_2^2} - r_2^2 \left(\frac{1}{r_2} - \frac{1}{r_1} \right) \sigma \cos \theta}{\frac{4\mu Lq}{\pi r_1^2} + r_1^2 \left(\frac{1}{r_2} - \frac{1}{r_1} \right) \sigma \cos \theta} \tag{10.15}$$

where

q The total flow rate through the channels connected in parallel;

V_1, V_2 The traveling speed of the oil–water interface, respectively, in the large and small capillary channel;

μ Viscosity of oil or water;

r_1, r_2 The radius, respectively, of the large and small capillary channel.

This equation describes how the traveling speed of the oil–water interface is related to the channel size, the wetting characters of the system, and the fluid viscosity. Several special situations will be discussed here:

(1) In the case that the capillary force is ineffectual, or in other words, $\theta = 90°$ and $\cos \theta = 0$, the oil and water have interface but produce no capillary force in the channel, so the above equation becomes the same as Eq. (10.3). Under such a circumstance the flowing velocity V_1 within the large channel is greater than the flowing velocity V_2 within the small channel, so at the moment when the large channel is submerged by water there are still oil columns left in the small one. As a result, a long period of simultaneous flow of water and oil will be maintained in the small channel because the oil columns are being broken, dispersed, and mixed by the subsequently entering water.

(2) In the case of effective capillary force, or in other words, $\cos \theta \neq 0$ and $P_c \neq 0$, the flowing velocity within the channels, also depending upon the magnitude of ΔP and P_c, does not always obey the relationship $V_1 > V_2$.

(3) From the perspective of improving the displacement efficiency, it is expected that the oil–water interfaces in the parallel large and small channels would simultaneously reach the exit-end (the point B), which leaves no residual oil in

the small. Under such an ideal circumstance, the total flow rate through the conjunction of the parallel channels can be figured out through Eq. (10.15).

Example 10.1 Consider a case wherein the related parameters are as follows: $r_1 = 2 \times 10^{-4}$ cm, $r_2 = 1 \times 10^{-4}$ cm, $\mu = 1$ mPa s, $\sigma = 30$ dyn/cm, and $\theta = 0$. It is required to calculate the total flow rate through the two parallel channels given that $V_1 = V_2$.

Solution: According to Eq. (10.15), when $V_1/V_2 = 1$ the total flow rate is:

$$q = \frac{\pi\sigma\left(r_1^2 + r_2^2\right)r_1 r_2}{4\mu L(r_1 + r_2)} = 1.6 \times 10^{-5} \ \left(\mathrm{cm^3/s}\right)$$

It is shown that the two oil–water interfaces can simultaneously reach the exit-end (the point B) only when $q = 1.6 \times 10^{-5}$ cm^3/s.

When $q < 1.6 \times 10^{-5}$ cm^3/s, as the capillary force is more favorable to the small channel, the flowing velocity in the small one is greater, and thus, the oil–water interface there will reach the exit-end earlier, leaving residual oil in the larger channel.

When $q > 1.6 \times 10^{-5}$ cm^3/s, the viscosity resistance in the small channel, relatively speaking, grows to such a high degree that the flowing velocity in it is surpassed by that in the large one. That is to say, the oil–water interface in the large channel will reach the exit-end firstly and it is in the small channel that there will be residual oil left. Remarkably, the residual-oil drops held up by the narrow constrictions will lead up to the harmful "Jamin's actions/effect," which is a major villain causing stronger flowing resistances, lower energy utilization ratio, and worse water-displacing-oil effect.

Some conclusions can be driven from the given example: water breakthrough will turn up in the oil wells in course of time, followed by gradual increase of the water-cut. That is to say, a considerable portion of crude oil is produced during the water-cut stage. Furthermore, it is impossible for an anisotropic formation to reach 100 % in the water-displacing-oil efficiency or the oil recovery.

10.1.5   Flow of Mixed Liquids Through Capillary Pore-Channels

Consider a capillary channel (length = L, radius = r_o) which is, as shown in Fig. 10.7, filled with evenly and dispersedly distributed oil drops or gas bubbles (radius = r). If we assume that the oil drops (or gas bubbles) advance along with the channel in a beadlike manner, or in other words, the flowing drops (or bubbles) not be distorted or conduct motion relative to the dispersion medium (viscosity = μ), the velocity distribution of the liquids, which perform a laminar flow under the pressure drop ($P_1 - P_2$), displays the shape of a parabola—the liquid flowing closer

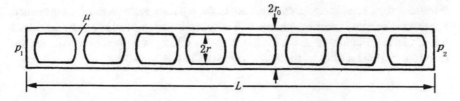

Fig. 10.7 Mixed flow in capillary tube

to the solid wall has lower velocity. Therefore, the flowing velocity of the liquid is related to the ratio of the radius r of the oil drop (or gas bubble) to the radius of the channel r_o.

On the basis of the Eq. (10.1), we already know the flowing velocity when there is only a single phase of liquid is present in the channel: $V_o = \frac{r_o^4(P_1 - P_2)}{8\mu L}$.

Then, in the case that the channel is filled with beadlike oil–water (or oil–gas) mixtures, the flowing velocity becomes:

$$V_L = \frac{(r_o^4 - r_o^4)(P_1 - P_2)}{8\mu L r_o^2}$$

The ratio of the mixture flowing velocity to the single-phase flowing velocity is:

$$\frac{V_L}{V_o} = 1 - \left(\frac{r}{r_o}\right)^4 \tag{10.17}$$

Table 10.1 shows a comparison with respect to the V_L/V_o ratio among three systems with different r/r_o ratio.

It is seen from the table that a larger radius of the oil drop (or gas bubble) in the channel will cause lower flowing velocity of the fluid. Taking account of the extraordinarily high viscosity of the adsorption layer attached to the pore walls, the obstructing effects caused by the dispersed liquid drops and gas bubbles must not be overlooked. However, as it is almost impossible for the real channels to be regularly circle in cross section, the liquid drops or gas bubbles would not enter a real channel as substantially as in the given hypothesis, leading to obstructing effects of lesser severity.

In conclusion, the multiphase flow through a reservoir rock inevitably performs figuring at the microscopic level and non-piston-like displacement at the macroscopic level, and, at the same time, it also gives rise to a variety of resisting effects.

Table 10.1 Relationship between velocity ratio and radius ratio

r/r_o	V_L/V_o
0.9	0.35
0.99	0.04
0.999	0.004

Furthermore, we can imagine that the case of three phases is much more complicated, but what is certain is that the phases will interact, interfere, and interplay with each other, which generates various capillary pressure effects, augment of resistance, and therefore lowered ability of the fluids to move through the rocks.

10.2 Two-Phase Relative Permeability

10.2.1 Effective and Relative Permeability

The terms "phase permeability" and "relative permeability," usually considered to be among the most critical parameters in the calculations of reservoir development, are essential to describing the seepage-flowing characteristics of multiphase fluid in reservoir rocks.

The phase permeability, also called effective permeability, is a relative measurement of the conductance of the porous medium for one fluid phase when the medium is saturated with more one fluid. The phase permeability of any particular in a medium depends upon not only the properties of the rock, but also the fluid saturations of all the phases. Implied by the given definition for phase permeability, the porous medium has a distinct conductance to each phase present in the medium, and, just as K is the accepted symbol for permeability, K_o, K_w, and K_g, are the accepted symbols for the phase permeability, respectively, to oil, water, and gas. The difference between absolute permeability and effective (or phase) permeability below will be specified.

1. Absolute permeability

Example 10.2 Consider a rock sample with a length of 3 cm, a cross-sectional area of 2 cm^2. It is known that if it is 100 % saturated with saline water that has a viscosity of 1 mPa s and subjected to a pressure differential of 0.2 MPa, the flow rate through it is 0.5 cm^3/s. Calculate the absolute permeability of the sample.

Solution: According to the Darcy's law,

$$K = \frac{Q\mu L}{A\Delta P} \times 10^{-1} = \frac{0.5 \times 1 \times 3}{2 \times 0.2} \times 10^{-1} = 0.375 \, (\mu m^2).$$

In another experiment wherein the fluid used to flow through the same sample is an oil with a viscosity of 3 mPa s, the flow rate, under the same pressure differential, is 0.167 cm^3/s. Then, the absolute permeability of this rock sample can also be calculated through the following statement:

$$K = \frac{Q\mu L}{A\Delta P} \times 10^{-1} = \frac{0.167 \times 3 \times 3}{2 \times 0.2} \times 10^{-1} = 0.375 \, (\mu m^2)$$

As seen from the two calculations, the absolute permeability is independent of the fluid flowing through it. Of course, this conclusion is restricted in some limits where extraordinarily low rock permeability or extraordinarily high oil viscosity should be excluded from.

2. Phase permeability

The phase permeability, or effective permeability, refers to the ability of a rock to transmit a particular fluid phase when other immiscible fluids are present in the reservoir.

Although the Darcy's law was originally developed from the experiments performed on the porous medium fully saturated with a homogeneous single-phase, it can also be applied to each phase individually, as long as the equation adopts in calculation the flow parameters corresponding to this individual phase. It follows therefore that the various additional forces caused by the interaction between phases are reflected in the values of the phase permeabilities.

Example 10.3 With the same rock sample as that used in Example 10.2, in another experiment the sample is 70 % saturated with saline water (S_w = 70 %) and 30 % saturated with oil (S_o = 30 %) before it is subjected to a pressure differential also of 0.2 MPa, and the original saturations are held constant throughout the steady seepage-flowing experiment. It is measured that the flow rate of the saline water is 0.30 cm^3/s and the flow rate of the oil is 0.02 cm^3/s. Calculate the effective permeability of the oil phase and the effective permeability of the water phase.

Solution: According to the Darcy's law into which the concept of effective permeability is introduced, the effective permeability of the water phase can be stated as:

$$K_w = \frac{Q_w \mu_w L}{A\Delta P} \times 10^{-1} = \frac{0.3 \times 1 \times 3}{2 \times 0.2} \times 10^{-1} = 0.225 \, (\mu m^2) \qquad (10.18)$$

And the effective permeability of the oil phase can be stated as:

$$K_o = \frac{Q_o \mu_o L}{A\Delta P} \times 10^{-1} = \frac{0.02 \times 3 \times 3}{2 \times 0.2} \times 10^{-1} = 0.045 \, (\mu m^2) \qquad (10.19)$$

The magnitude of an effective permeability depends upon not only of the properties of the rock, but also upon the fluid saturations and distributions within the pore system, which are further intimately associated with the wetting characteristics and the wetting history. Therefore, the phase permeability is a dynamic property reflecting the interaction between the rock and the fluids.

Note that in the above example $K_w + K_o = 0.270 \, \mu m^2 < K = 0.375 \, \mu m^2$. In fact, as a general universal law symbolically expressed as $K_w + K_o < K$, the sum of all the effective permeabilities of a rock is proven to be always smaller than its absolute permeability. The reason for this is that the presence of more than one fluid, causing interference among phases and giving rise to a variety of resisting forces, generally inhibits flow.

3. Relative permeability

Relative permeability is the ratio of effective permeability of a particular fluid at a particular saturation to the base permeability. In multiphase flow in porous media, relative permeability is a dimensionless measure of the effective permeability of each phase, and the calculation of relative permeability allows comparison of the different abilities of fluids to flow in the presence of each other, since the presence of more than one fluid generally inhibits flow.

Anyone among the following three permeabilities can be adopted as the specified datum permeability that is used as the denominator in given concept of relative permeability:

(1) The air permeability, K_a;
(2) The water permeability measured at the 100 % water saturation, K;
(3) The oil permeability measured at the connate-water saturation, K_{swc}.

Taking the first way as an example, the relative permeabilities, respectively, of the oil phase and water phase are:

$$K_{ro} = K_o/K_a$$
$$K_{rw} = K_w/K \tag{10.20}$$

Example 10.4 With the conditions given in the Example 10.3, calculate the relative permeabilities.

The relative permeabilities of water is: $K_{rw} = \frac{K_w}{K} = \frac{0.225}{0.375} = 0.60$

The relative permeabilities of oil is: $K_{ro} = \frac{K_o}{K} = \frac{0.015}{0.375} = 0.12$

Although $S_w + S_o = 100$ %, it is noted that $K_{ro} + K_{rw} = 72$ % < 100 %. As the flow of each phase is inhibited by the presence of the other phases, the sum of relative permeabilities over all phases is always <1.

Besides, it is also noted that when $S_w = 70$ % and $S_o = 30$ %, the water-to-oil saturation ratio is 70/30 = 2.33, but the oil-to-water relative permeability ratio is 0.60/0.12 = 5.

Proven by both theories and experiments, the relative permeability is a function of the fluid saturation, and the curve correlating them is called the "relative permeability curve," which is usually obtained through experiments.

10.2.2   Characterizing Features of Relative Permeability Curve

A typical set of phase permeability curves for a water–oil system with the water being considered the wetting phase is converted into the corresponding relative permeability curves in Fig. 10.8. The figure shows the following distinct and significant features: two curves, three regions, and four characteristic points.

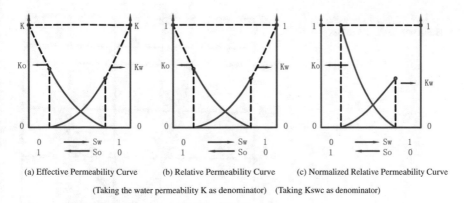

(a) Effective Permeability Curve   (b) Relative Permeability Curve   (c) Normalized Relative Permeability Curve

(Taking the water permeability K as denominator)   (Taking Kswc as denominator)

Fig. 10.8 Effective permeability curve and relative permeability curve

1. Two curves

They are, respectively, the wetting-phase relative permeability curve and the non-wetting-phase relative permeability curve. Obviously, the two curves cross each other to form an X-type chiasma. The vertical axis, denoted by K_{ri}, represents the relative permeabilities of the two phases, and the horizontal axis represents the wetting-phase saturation that ranges from 0 to 1 in magnitude, or the non-wetting-phase saturation that ranges from 1 to 0 in magnitude.

In practical applications what demands special notice is that the relative permeability curve is slightly different in name when the denominator in the ratio expression of relative permeability is assigned in different ways. With the water permeability K that is measured with a rock 100 % saturated by formation water, it is called the ordinary relative permeability curve (Fig. 10.8b); and with the oil-phase relative permeability K_{swc} measured at the connate-water saturation is called the normalized relative permeability curve (Fig. 10.8c).

2. Three regions

Figure 10.9 shows a typical set of oil–water relative curves measured with a water-wetting rock.

A is the oil-flow region. In this region, on the one hand, the water saturation S_w is so small that $K_{rw} = 0$, and, on the other hand, the magnitude of K_{ro} is only a little less than 1 at the high S_o. This characteristic is decided by the water–oil distribution and flow in the pore system of the rock. The reason is that at the low saturations shown in Fig. 10.9, $S_w < S_{wi} = 20$ % the wetting-phase fluid, water, occupies only the corner, small, and narrow pore spaces which do not materially contribute to flow, and thus, changing the saturation in these small pores has a relatively small effect on the flow of the non-wetting phase taking place in the large pores and channels, and the relative permeability to oil is lowered quite slightly. It is also

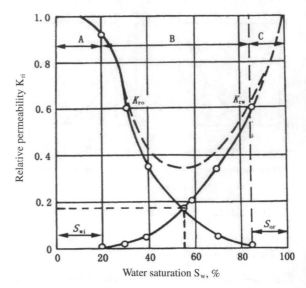

Fig. 10.9 Typical oil–water relative permeability curves

noted that before the saturation S_{wi} where the water will cease to flow, the water distributed at the pore edges and corners and over the rock grain surfaces exists as the discontinuous phase in the rock, not allowed to flow or, in other words, maintained a relative permeability of zero. Therefore, the saturation S_{wi} is referred to as the connate-water saturation, irreducible saturation, or coexisting-water saturation—all the terms are used interchangeably.

B is the mixture-flow region where the oil and the water are flowing together. The principal characterizing features of the curve within this region are as follows: as the water saturation S_w increases, the relative permeability to water K_{rw} increases in magnitude while the relative permeability to oil K_{ro} decreases. From a microscopic perspective, the wetting phase will cease to be continuously distributed in the pore system as long as its saturation exceeds a certain value (S_{wi}), and once subjected to an impressed pressure differential it can flow. However, although the K_{rw} increases while the K_{ro} decreases, in the early stage of the rise of the wetting-phase saturation the latter is still higher than the former one because the wetting phase flowing through the small pore openings and across the solid walls of the large openings encounters greater resisting forces and travels longer distance than the non-wetting phase that occupies the mid-space of the large pore openings.

As the wetting-phase saturation gets higher and higher, the wetting phase begins to occupy the main channels in the rock and its relative permeability curve is increased dramatically while the relative permeability to the non-wetting phase is rapidly reduced. The magnitude of K_{rw} increases because the water, flowing under the pressure differential at the saturations above S_w, gradually forms into more and more interconnected flowing channels in the pore space of the rock. Simultaneously, the non-wetting phase (oil), sharing smaller and smaller saturation and gradually shoved to loss more and more flowing channels, obviously has lower and lower relative permeability. When the non-wetting phase is reduced to a certain degree, it will lose its continuity and break into oil drops, which may cause the harmful Jamin effect.

Besides, in this region where both the oil and the water are flowing in the presence of each other, since each phase suffers interaction and interference from the other, the sum of the oil and the water relative permeabilities, $K_{ro} + K_{rw}$, is considerably reduced as a result of the resisting forces caused by the capillary forces. As illustrated by the dashed curve in Fig. 10.9, the sum $K_{ro} + K_{rw}$ reaches its minimum value at the intersection point of the two curves.

C is the water-flow region. As shown in the figure, the relative permeability to non-wetting begins to be zero (or, symbolically expressed, $K_{ro} = 0$) when the oil saturation is lowered to the residual-oil saturation, where the non-wetting phase ceases to lose its macroscopic fluidity. At the same time, the wetting phase has already occupied almost all the main channels in the rock. As the oil phase is broken into drops and dispersed in the water phase, Jamin effect may be caused by the drops when narrow constrictions are encountered, bringing out great resistances against the water flow. Therefore, a higher residual-oil saturation, being indicative of a larger number of oil drops present in the pore system, will cause greater resistances against the water flow and the point of the corresponding point on the water relative permeability curve will be farther away from the 100 % line, and vice versa.

In addition, because the irreducible wetting phase can be sequestered in the dead pores, the extremely slim spaces, and across the rock grain surfaces, in amount it exceeds the residual non-wetting phase that is broken and dispersed in the mid-pore spaces. That is to say, the lowest wetting-phase saturation is higher than the lowest non-wetting-phase saturation, symbolically expressed as $S_{wi} > S_{or}$.

3. Four characteristic points

They are the connate-water saturation S_{wi}, the residual-oil saturation S_{or}, the water relative permeability at the residual-oil saturation K_{rw}, and the intersection point of the two curves (the isoperm point).

With the characteristic points, more information, for example, the wettability to be discussed in the following section can be obtained from the curves. Besides, by virtue of the initial oil saturation and the residual-oil saturation, it is also feasible to calculate the water-displacing-oil efficiency for a core or reservoir.

$$\text{Oil displacement efficiency} = \frac{\text{initial oil saturation} - \text{residual oil saturation}}{\text{initial oil saturation}}$$
$$= \frac{S_{oi} - S_{or}}{S_{oi}} = \frac{1 - S_{wi} - S_{or}}{1 - S_{wi}}$$

$$(10.21)$$

Taking Fig. 10.9 as an example, its water-displacing-oil efficiency is:

$$\text{Oil displacement efficiency} = \frac{0.80 - 0.15}{0.80} = 0.81$$

Generally speaking, it is hard to achieve a 100 % water-displacing-oil efficiency and even in the best of circumstances, the efficiency can only reach 80 % or so.

4. Common features of the relative permeability curve for systems with wetting and non-wetting phase

Proven by large amounts of experiments, ordinary systems with a wetting phase and a non-wetting phase have common features revealed by the curves in Fig. 10.9. The previous discussion for the oil–gas system is also applied to the relative permeability curves for an oil–gas system, where the oil is the wetting phase while the gas is the non-wetting phase. Figure 10.10 shows that the general trends of the curves are correspondingly similar.

Hence, conclusions about these common features can be drawn as follows:

(1) For each phase, there always, respectively, exists a lowest saturation at which the phase starts to flow, that is to say, each phase is not allowed to flow until the saturation of it gets to be higher than its lowest saturation. The lowest saturation for the wetting phase is higher than the lowest saturation for the non-wetting phase.

(2) When two phases of fluid are simultaneously present in a rock, ability of the two fluid phases to pass through the rock is lowered as a result of the Jamin effect which is caused by the capillary pressure. Therefore, the sum of the relative permeability to the two phases is smaller than 1, and, obviously shown in the figure, the two relative permeabilities are equal to each other at the point where the sum $K_{rw} + K_{ro}$ reaches its minimum.

Fig. 10.10 Oil–gas relative permeability curves

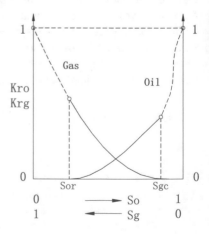

(3) It is found that the permeability of a phase increases with the increase of the saturation of it, whether it is a wetting phase or a non-wetting phase. But the increase of the non-wetting phase with its increasing saturation is faster than that of the wetting phase.

10.2.3 Factors Affecting Relative Permeability

Relative permeability, although being a function of the wetting-phase saturation, is also affected by some other factors such as rock physical property, fluid physical property, rock wettability, saturation history, and experimental conditions. The effect of these factors on relative permeability of a multiphase fluid system will be discussed in the following Sections.

1. Effect of rock pore structure

The size, geometric shape, and other composite characteristics of the pore structure involved in a rock directly affect the relative permeability curves for the rock. Figure 10.11 shows sets of curves for different types of rocks.

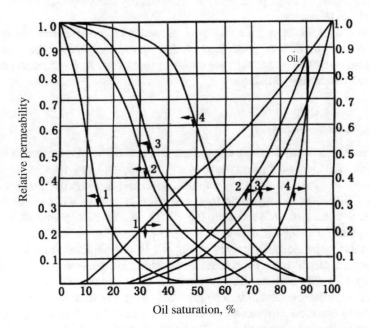

Fig. 10.11 Relative permeability curves of different types of media (by Amyx 1960). *1* capillary tube; *2* dolomite; *3* uncemented sandstone; and *4* cemented sandstone

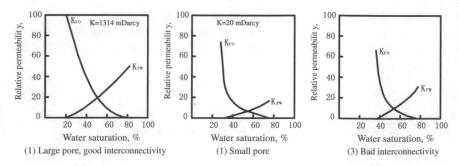

Fig. 10.12 Impact of pore size and interconnectivity on sandstone relative permeability

Based on experiments with sand rocks of different pore structures and permeabilities, Morgan in 1970 proposed a comparison of oil–water relative permeabilities as shown in Fig. 10.12, where we can make several conclusions as follows:

(1) Sandstones of high permeability and large pores have a wide range of two-phase flow region and low connate-water saturation.
(2) Sandstones of small pores and good connectivity have high connate-water saturation and a narrow range of two-phase flow region,
(3) Sandstones of small pores and bad connectivity have low K_{ro} and K_{rw} at the ends of the curves.
(4) Sandstones of large pores but bad connectivity have relative permeability curves from that of good-connectivity rocks; the permeability curves of bad-connectivity sandstones are closer to that of the rocks of small pores but good connectivity in shape.

2. Effect of rock wettability

The petroleum industry has long recognized that wettability of reservoir rock has an important effect on multiphase flow of oil, water, and gas through reservoirs. Generally speaking, the relative permeability to the non-wetting phase (the oil) gradually decreases in magnitude from the strongly wet ($\theta = 0°$) to the strongly non-wet. By contrast, as shown in Fig. 10.13, the relative permeability to the wetting phase gradually increases.

The relative permeability curves in Fig. 10.14 are obtained from experiments conducted on natural cores the wettability of which is alternated by adding surface active agent of different concentrations into the oil–water system. As seen from the curves, the relative permeability to oil gradually increases from the strongly oil-wet (curve 5) to the strongly water-wet (curve 1), whereas the relative permeability to water gradually decreases, and thus, the point of intersection of the corresponding curves gradually moves to right.

Fig. 10.13 Relative
permeability curve under
different conditions of
wettability (Imbibition
Method, by Owen and
Archer, 1971) *1* θ = 180°; *2*
θ = 138°; *3* θ = 90°; *4*
θ = 47°; *5* θ = 0°

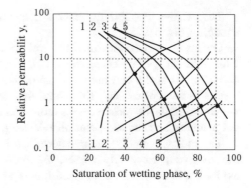

Fig. 10.14 Effect of rock
wettability on relative
permeability in the case of
water displacing oil (by Owen
and Archer, 1971)

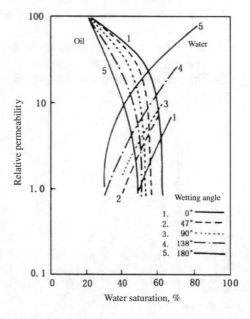

Subjected to a system of semilogarithmic coordinates, the curves show more distinctive characteristics at the far ends of the curves close to the terminal point of saturation.

Because of its effect on the oil/water distribution, wettability influences the relative permeabilities of the flowing fluids. In water-wet rocks, the water, distributed in the small pore openings and at the edges and corners of the pores, exerts quite small effect on the permeability to oil; in oil-wet rocks with the same saturation conditions, the water is distributed in the mid-space of the pores and channels in the form of oil drops or continuous water flow, having great impact on the flow of the oil phase. Besides, as a portion of oil is attached to the solid surface as films, the relative permeability to oil at the same oil saturation tends to be lower.

The relative permeability curves measured for a strongly water-wet rock are shown in Fig. 10.15.

Owens et al. once measured the end-point relative permeability to oil for rocks with different wetting characters $K_{ro}(S_{wi})$, which was actually the relative permeability to oil at the connate-water saturation in the beginning of experiment. The results as shown in Table 10.2 show that with the increase of the wetting angle, which indicates the rock becomes more and more strongly oil-wet, the oil effective permeability and the end-point oil relative permeability K_{ro} gradually decrease.

The rock wettability can be determined by virtue of the characteristic points on the relative permeability curves. According to the study of Craig, the rocks exhibiting the characteristics listed below are water-wet:

(1) The connate-water saturation $S_{wi} > 20$–$25\ \%$
(2) The water saturation at the point of intersection of two relative permeability curves $S_w > 50\ \%$
(3) The water relative permeability at the largest water saturation $K_{rw} < 30\ \%$, which is a common occurrence.
(4) The water relative permeability at the connate-water saturation $K_{rw} = 0$.

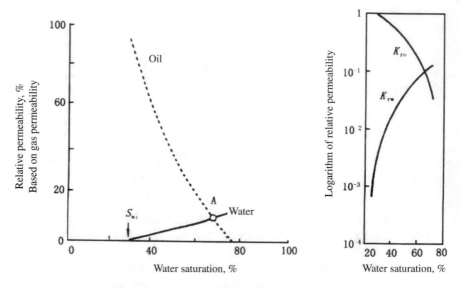

Fig. 10.15 Oil–water relative permeability curves of strongly water-wet rock

Table 10.2 Relationship between wetting angle and oil relative permeability

Contact angle	0°	47°	90°	138°	180°
Hydrophilicity	Strong ← weak				
The end-point relative permeability to oil K_{ro}	0.98	0.83	0.80	0.67	0.63

And the rocks exhibiting the characteristics listed below are oil-wet:

(1) The connate-water saturation $S_{wi} < 15$ %
(2) The water saturation at the point of intersection of two relative permeability curves $S_w < 50$ %.
(3) The oil relative permeability at the connate-water saturation >50 %, or even approaches 100 %.
(4) The water relative permeability at the largest water saturation $K_{ro} = 0$.

In consideration of the great effect of wettability on the relative permeability curves, the initial wettability of the experimented rock in the measurements should be maintained constant, so as to obtain indeed representative relative permeability curves.

3. Physical properties of fluids

(1) The effect of fluid viscosity

Before the 1950s, it is considered that the relative permeabilities had nothing to do with the fluid viscosity, but later it turned out that the relative permeability of the non-wetting phase increases with the increase of the viscosity ratio (non-wetting/wetting) in the case that the viscosity of the non-wetting phase is comparatively high, and it even can reach 100 %; and, by contrast, the relative permeability to the wetting phase is independent of the fluid viscosity.

This phenomenon can be explained by the theory of water film proposed by Coton: As the portion of wetting-phase liquid adsorbed on the solid surface exists as a wetting film, the flow of the adjacent high-viscosity non-wetting phase passing by it, to some extent, can be considered as a sliding motion in which the wetting phase provides lubrication.

The effect of the viscosity ratio decreases as the pore size gets larger. When the rock permeability is higher than 1 μm^2, this effect due to the viscosity ratio is negligibly small.

Figure 10.16 shows that the relative permeability curves are measured with different viscosity ratios. It indicates that the viscosity ratio does not exert considerable effect until the oil saturation becomes high enough. This is because only at high saturation can the oil occupy and flow through sufficient pore-channels where the viscosity ratio may perform a more remarkable role in affecting the relative permeability; when the water saturation is high, the number of pore-channels occupied by oil is accordingly reduced, which gets the oil phase flowing through the large pore-channels and sustaining smaller effect from the viscosity ratio.

(2) Effect of the surface active substances in the fluids

According to the relevant studies, the oil and the water phases simultaneously flowing in a porous medium can be in one of the following three states: (a) the oil is the dispersed phase and the water is the dispersion medium; (b) the oil is the

Fig. 10.16 Effect of the
oil-to-water viscosity ratio on
relative permeability. *1*—
82.7; *2*—74.5; *3*—42.0; *4*—
5.2; *5*—0.5

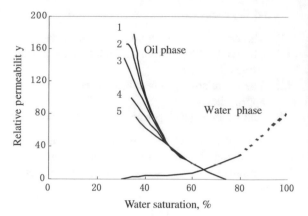

dispersion medium and the water is the dispersed phase; (c) the oil and the water are
in a state of emulsification. During the seepage-flowing process through a rock
these three states are capable of transforming themselves into each other under
given conditions. What a disperse system depends on what and how much surface
active substances as well as polar compounds are contained in the oil and water—
these substances affect the oil–water interfacial tension and the adsorption of the
fluids on the rock surface. Consider two oil–water systems, respectively, in the state
(a) and (b) and observe their relative permeability curves shown in Fig. 10.17 after
adding surfactant to the systems.

As the dispersion medium, in comparison with the dispersed phase, is more
capable to permeate through the rocks, it is indicated by the two sets of curves that
$K_{roa} < K_{rob}$, $K_{rwa} > K_{rwb}$.

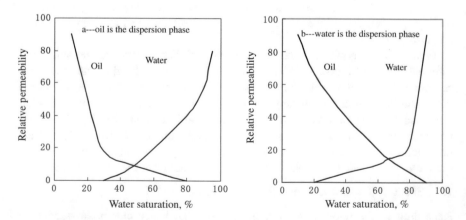

Fig. 10.17 Relative permeability curves to the dispersed phase and the dispersion medium (1980)

4. The effect of saturation history

As was discussed for the capillary pressure data, there is also a saturation history effect for relative permeability. The saturation history is a significant factor in the control of spatial distribution of the fluids and the wetting hysteresis and capillary pressure hysteresis effects in the pore system. Therefore, there is a difference in permeability when changing the saturation history, which can be seen in the relative permeability curves measured, respectively, in the drainage and imbibition processes.

If the rock core is initially saturated with the wetting phase and relative permeability data are obtained by decreasing the wetting-phase saturation while non-wetting fluid flowing into the core, the process is classified as "drainage." And if the data are obtained by increasing the saturation of the wetting phase, the process is termed "imbibition." In laboratory, the relative permeability curves can be measured through either the imbibition process or the drainage process.

There are two viewpoints with respect to the effect of the saturation history on the relative permeability data:

(1) The first viewpoint shown in Fig. 10.18 considers that the relative permeability to the wetting phase is a function only of its own saturation and has no association with the saturation history. The strongly water-wet rocks or strongly oil-wet rocks are more prone to show this characteristic. However, for the non-wetting phase its relative permeability measured in the imbibition process is always lower than that measured in the drainage process.
(2) The second viewpoint (Osoba et al. 1951) considers that, as shown in Fig. 10.19, both the relative permeability of the wetting phase and that of the non-wetting phase are affected by the saturation history.

Fig. 10.18 Relative permeability curves of drainage process and imbibition process

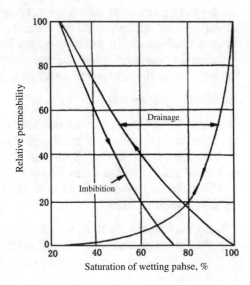

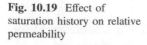

Fig. 10.19 Effect of saturation history on relative permeability

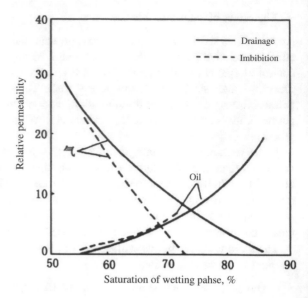

Something can be drawn as analogies of two viewpoints: the relative permeability of the non-wetting phase is far more significantly affected by the saturation history than that of the wetting phase; and, besides, the drainage curve and the imbibition curve of the wetting phase are always more closer to each other.

The difference in the permeability curves obtained from the imbibition process and the drainage process is called "hysteresis." This difference in relative permeability curves is due to the capillary pressure hysteresis. As previously mentioned, the capillary pressure hysteresis is caused by more than one aspect of factors, such as the wetting hysteresis and the variation of the capillary radius. All of these hysteresis effects can be exhibited by the imbibition and drainage curves.

Since the permeability measurements are subjected to hysteresis, it is important to duplicate, in the laboratory, the saturation history of the reservoir.

5. Effect of temperature

Domestic and overseas scholars still have different viewpoints about the effect of temperature on the oil–water relative permeability data:

Viewpoint 1: The study of Miller and Ramey, based on experiments performed on consolidated core and Berea sandstone, proves that temperature has no effects.

Viewpoint 2: As the temperature is raised, the K_{ro} will increase and the K_{rw} will decrease as shown in Fig. 10.20.

(1) As the temperature rises, the connate-water saturation is increased.
(2) As the temperature rises, the oil relative permeability at the same water saturation point increases to some extent, whereas the water relative permeability decreases.

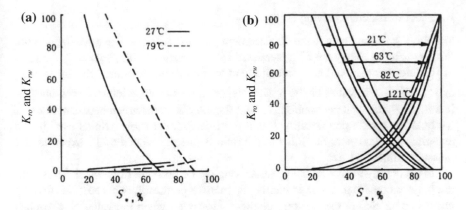

Fig. 10.20 Effect of temperature on relative permeability curves **a** experimental results (Weibrandt 1975); **b** calculated results (Honarpour 1986)

(3) As the temperature rises, the rock becomes more water-wet.

The mechanism of the increase of the relative permeability to oil: The increase in temperature, directly causing fiercer thermal motion of the oil molecules and thus lower oil viscosity, causes the adsorption film on the rock surface to be thinner and the channel space for flow to be larger. As a result, the oil flow, against which the resistance in the channels is lowered, exhibits improved relative permeability through the rock.

The reason for the increase of connate-water saturation is that the polar substances in oil are desorbed off the rock surface at high temperature, leaving the solid surface to be more hydrophilic and more water-wet. Consequently, the original oil-containing channels which are coated with water films changes to water-containing channels. Besides, increase in temperature also leads to expansion of rock, which alters the pore structures to some extent and thereby affects the relative permcability.

Viewpoint 3: As the temperature increases, both K_{ro} and K_{rw} increase. The possible reason supporting this viewpoint is that the layer constructed by the active substances from oil becomes thinner on the rock surface when the temperature is raised. At lower temperature, the oriented adsorption of the active substances on the rock surface takes the form of a colloid layer, reducing the cross-sectional area of the channels for water flow and thus increasing the resistances against flowing. But when the temperature is increased, the thermal motion of the molecules will go fiercer and the adsorbance will be reduced. As a result, the adsorbed layer becomes thinner and the relative permeabilities to both oil and water become higher.

The generally accepted view is the Viewpoint 2 given above. The effect of temperature on relative permeability is an important issue in the thermal recovery.

6. Effect of different driving factors

The driving factors include the displacement pressure, the pressure gradient, and the flowing velocity. They are all generalized into the term "π ratio," which is the ratio of the microscopic capillary-force gradient to the driving pressure gradient ($\pi = \frac{\sigma}{\nu\mu}$ or $\pi = \frac{\sigma L}{K \cdot \Delta P}$). The value of the π ratio depends on the experimental pressure differential ΔP, the rock permeability K, and the interfacial tension σ between fluids. If we hold the rock's permeability K and the interfacial tension σ fluids used in the experiment as constant, the value of the ratio is directly related to the experimental pressure differential ΔP.

It is generally acknowledged that, as long as the pressure differential ΔP is kept not large enough to give rise to inertia, the relative permeability curve to the driving phase is unrelated to the pressure gradient. However, when the value of π reaches from 2×10^2 to 10^7, correlations between the two will be built up as shown in Fig. 10.21.

According to Fig. 10.21, we can get that, both the relative permeabilities to the two phase increase with the decrease of the value of the π ratio, leading to wider mixture section where the two phases flow together with each other.

Obviously, this phenomenon has something to do with the flow of the discontinuous phase.

Once the interfacial tension drops or the driving pressure gradient increases to such a degree that the discontinuous phase is able to overcome the Jamin effects, the discontinuous phase starts to flow and increases in amount with the further increase of the driving force gradient, which makes the two-phase flow section becomes wider and the average saturation decreases.

The discontinuous phase requires great pressure gradient to flow. If holding the other conditions constant, the driving force gradient must be increased hundredfold before driving the discontinuous phase to move. However, in field practice it is almost impossible to raise the injection pressure to such a degree. Therefore, the alternative approach that is practical is to lower the interfacial tension. If the oil–water interfacial tension is lowered to 0.01 mN/m minus, the residual-oil saturation, responding to every slight change of the pressure gradient, can be obviously decreased. This is the theory of low interfacial tension displacement. Proven by

Fig. 10.21 Relative permeability curves to different π ratios (1980)

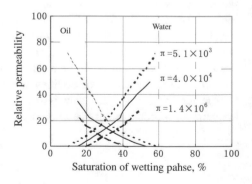

experiments, a π ratio smaller than 10^5 is required to result in decrease of the residual-oil saturation. Of course, for different rock pore structures, the critical values are different.

Besides, some specified similarity conditions should be satisfied so as to make sure that the relative permeability curves measured in laboratory do reflect the underground situations. For example, the in-laboratory experiments should have the same π ratio with the real formation. Note that the π ratio of real formations approximately ranges from 10^6 to 10^7. It is also proposed by Aifeluos that the driving force gradient exerts no effect on the relative permeabilities in the case of $\pi = \frac{\sigma L}{K \cdot \Delta P} \geq 0.5 \times 10^6$, which agrees with our discussion above.

In simulating experiments of the seepage-flowing process taking place in real formations, the similar π ratios are commonly used to study the influence of the ratio of microscopic capillary-force gradient to the driving force gradient.

In a word, there is many a factor affecting the relative permeability, and in analysis and usage of the curves engineers must take care of the accordance between the experimental conditions and the real formation situations.

10.3 Three-Phase Relative Permeability

10.3.1 Relative Permeabilities for Pseudo-Three-Phase Flow System

In dealing with real rocks with three coexisting three phases one of which shares too low a saturation to flow, the situation can be simplified and treated as a two-phase system. Such a simplified flow is referred the pseudo-three-phase flow, so as to handle the problems with respect to the relative permeability.

Taking a water-wetting rock with three phases (oil, gas, and water) as an example, we can attribute the gas phase into the oil saturation if the gas saturation is so low that the gas phase is unable to flow, whereupon the system can be regarded as a two-phase one consisting only oil and water. On the other hand, if it is the water that shares a low saturation and remains in the connate state, the system, with the wetting-phase oil and the non-wetting-phase gas, can be reduced to a two-phase one where the water is considered as a part of the solid and, equivalently, the porosity is decreased to some degree.

Example 10.5 In a system consisting of oil and water, the initial oil saturation is 80 %. After water drive, the residual non-wetting-phase (oil) saturation is 15 % while the wetting-phase (water) saturation is 85 %. When the system is developed under a pressure slightly lower than the saturation pressure, a situation of coexisting three phases of oil, water, and gas will occur. It is known that if the waterflooded area has a residual gas saturation of 10 % its residual-oil saturation will be 5 %. Calculate the oil displacement efficiency of the waterflooded area.

Solution: According Eq. (10.21), the oil displacement efficiency is:

$$\frac{0.80 - 0.05}{0.80} = 94\,\%$$

As seen from the perspective of saturation variation only, the existence of a small amount of gas contributes positively to improving oil displacement efficiency.

The simplification from the three-phase system to the two-phase one comprised only of the wetting phase and the non-wetting phase is for convenience's sake in practical engineering applications. In 1970, Stone et al. had already proposed a method in which the three-phase relative permeabilities could be represented by corresponding two-phase relative permeability and saturation data:

(1) In the case of low gas saturation, the system is a water–hydrocarbon (oil and water) two-phase one where the water is the wetting phase and the hydrocarbon is the non-wetting phase.
(2) In the case of low water saturation, the system is a gas–liquid (oil and water) two-phase one where the liquid is the wetting phase and the gas is the non-wetting phase.

10.3.2 Relative Permeabilities for Real Three-Phase Flow System

If each of the three phase, the oil, the gas, and the water, respectively, shares a considerable saturation, a set of three-phase relative permeability curves as shown in Fig. 10.22 are needed to determine which one, or ones, of the three can flow. In Fig. 10.22, the curves marked with the symbols (a), (b), and (c), respectively, are representative of the relative permeability curves for the oil, gas, and water phase.

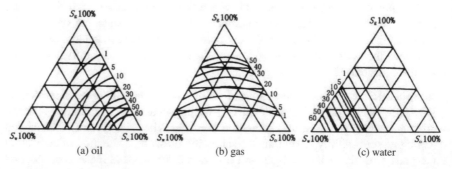

Fig. 10.22 Three-phase (oil, gas, and water) relative permeability as function of saturation (by Leverett and Lewis 1941)

As seen from the figure, the relative permeability to the wetting phase (water) is only associated with the saturation of itself and independent of the variation of the saturations of the other two phase. The reason for this is that the wetting phase occupies the both the main pore space and the small pores and channels.

However, the relative permeability to the other two non-wetting phases is a function of all the three saturations. Although for a water-wetting medium both the oil and the gas are the non-wetting phases, but as the oil, in comparison with the gas, is preferable to wet the solid surface and the oil–water interfacial tension is smaller than the water–gas interfacial tension, the oil phase can occupy the similar pore space as the water phase does. So, when the water saturation is low, the oil phase will go to inhabit most of the small pores. If the oil saturation is maintained a constant while the water saturation varies, the occupancy of the oil in the pore system will change and thereby the relative permeability to the oil phase will consequently vary. As shown in Fig. 10.22a, for an oil saturation of 60 % and a water saturation of 40 %, the relative permeability to the oil is approximately 34 %; for the same oil saturation and a lowered water saturation of 20 %, the relative permeability to the oil is about 38 %; and for an oil saturation of 60 % and a water saturation of zero, the relative permeability to the oil is 18 % or so. Therefore, it is seen that by changing the water and gas saturation, the flow characteristics of the oil are changed so that the oil assumes more tortuous paths.

Leverett in 1941 proposed these three-phase relative permeability curves measured with water-wetting media. For different rocks and fluids, the curves are different in shape.

Taking the 1 % relative permeability to each phase as the starting point at which the phase begins to flow, we can get a triangular diagram as shown in Fig. 10.23 by plotting the relative permeability data as lines of constant-percentage (1 %) relative permeability. This diagram indicates that three situations may occur in a three-phase

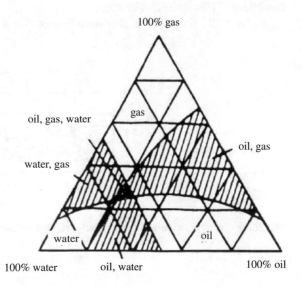

Fig. 10.23 Correlation between the three-phase saturations and the flow regions (by Leverett and Lewis)

system in light of the saturation: the single-phase flow region, the two-phase flow region, and the three-phase flow region. It is noted from the figure that the saturation region in which simultaneous flow of all three phases occurs is quite small. Therefore, it is evident that in most cases two-phase relative permeability curves are quite satisfactory, whereas the drawing of the three-phase ones can be avoided.

The three-phase systems are much more complex than the two-phase systems in behavior and much more difficult to being studied. The normally mentioned relative permeabilities refer to the two-phase relative permeabilities.

10.4   Measurements and Calculations of Relative Permeability Curves

There are two direct ways to measure relative permeability curves from the perspective of experiment principle, including the steady-state method and the unsteady-state method, and there are also three indirect ways to calculate them—calculation from capillary pressure data, calculations from field-performance data, and calculation based on empirical formulas.

10.4.1   Steady-State Method

1. Principle and procedure

The steady-state method used in the laboratory to measure relative permeability is based on the Darcy's law as well as the definition formulas of the relative permeabilities—Eqs. (10.18)–(10.20).

Figure 10.24 shows the experimental procedure of the steady-state method.

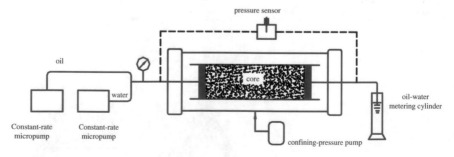

Fig. 10.24 Steady-state method for measuring relative permeability

(1) After extraction and clean, the core is extracted, cleaned and then evacuated using a vacuum pump and saturated with water (or oil).
(2) The core is put into the core-holder where the permeability of the single-phase water (or oil) can be measured.
(3) By using a micropump with constant discharge, the two fluids are introduced at a predetermined fluid ratio into the core;
(4) The fluids are flowed through the core until the produced ratio at the outlet end is equal to the injected ratio at the inlet end. At this time, the core system is considered to be in a steady-state flow condition and the existing saturations are considered to be stable. Measure the pressure differential between the two core-ends with a pressure sensor; calculate the flow rates at the outlet according to the varying volume of the oil and the water in the graduated cylinder; calculate the water saturation according to the accumulative volumes.
(5) On the basis of the data obtained above, figure out the relative permeabilities at the experimented water saturation.
(6) More relative permeability data at other saturation conditions can be obtained by changing the discharge rates of the micropumps, respectively, for the two phases and repeating the measuring procedure (3)–(5).
(7) Finally, the relative permeability curves can be made from the experimental results.

2. Determination of saturation

The saturations of the fluids are determined in one of the three methods mentioned below:

(1) Material balance method (also called volumetric method)

According to the principle of material balance, the volumetric balance for the experimented core is maintained in such a way: the accumulative volume injected into the core—the accumulative volume produced form the core = the volume left in the core.

For the phase j:

$$V_{\mathrm{T}}f_j - V_{j\mathrm{T}} = V_{\mathrm{P}}(S_j - S_{j1}) \qquad (10.22)$$

so

$$S_j = \frac{V_{\mathrm{T}}f_i - V_{j\mathrm{T}}}{V_{\mathrm{P}}} + S_{j1} \qquad (10.23)$$

where

V_T The total volume of the water and the oil injected into the core before the steady state is reached, cm^3;

V_{jT} The total volume of the phase j produced from the core, cm^3;

f_j Volume percentage of the phase j in the total flow rate;

S_j The saturation of the phase j at the test point;

V_p Pore volume of the core, cm^3;

S_{jI} The initial saturation of the phase j at the test point.

(2) Weighting method

Due to the density difference between oil and water (or oil and gas), the core is of different weight when saturated with the two phases in different proportions. Therefore, at each test point, we can obtain the mass of the fluids contained in the core by subtracting the known dried-core mass W_d from the weighted total mass W. Then, calculate the water (or oil) saturation through the equations given below.

$$W - W_d = V_p S_w \rho_w + V_p (1 - S_w) \rho_o$$

$$S = \frac{(W - W_d) - V_p \rho_o}{V_p (\rho_w - \rho_o)} \tag{10.23}$$

where

S_w Water saturation in the core, fraction;

W Total mass of the core containing fluids, g;

W_d Mass of the dried core, g;

V_p Pore volume of the core, cm^3;

ρ_o, ρ_w Oil density and water density, g/cm^3.

In the case that the density difference between the two phases is comparatively large, the weighing method is a quite good way to determine the fluid saturations. However, the principal deficit of it is that the core is required to be removed from the holder and weighted for each saturation point.

3. Electric-resistivity method

Electrodes have been inserted in the test section, and the saturations are determined by measurement of the core resistivity (see Chap. 7 for more information). This method is based on the fact that oil and water are different in electric resistivity (or electric conductivity).

10.4.2   Unsteady-State Method

The unsteady-state method provides relative permeability data of a core by measuring the dynamic behavior of the external water injection, and so, it is also called the external water drive method.

1. Principle of unsteady-state method

The unsteady-state method for relative permeability measurement is based on Buckley–Leverett's theory about the displacement of oil by water in the core sample, which considers that the distribution of the oil and water saturations within the rock is a function of the time and distance. As the oil and water saturations as well as distributions in rock vary with the time and the distance, the displacement of oil by water is unsteady process, and therefore, this way for relative permeability measurement is named "unsteady-state method."

According to the aforementioned theory and the conclusion that the relative permeability to oil and water in porous medium is a function of the saturation, the flow rates of the fluids at a certain cross section of the studied core also vary with time. So, the relative permeability curves can be obtained by accurately measuring the oil and water flow rates at a constant pressure (the constant-pressure method) or the pressure variations during the displacement at a constant flow rate (the constant-flow method).

2. Computing formulas

The relative permeability data are computed through the following formulas when measured by means of the unsteady-state method:

$$K_{ro}(S_{we}) = f_o(S_{we}) \left[d\left(\frac{1}{\overline{V}(t)} \right) \Big/ d\left(\frac{1}{I \cdot \overline{V}(t)} \right) \right] \tag{10.24}$$

$$K_{rw}(S_{we}) = K_{ro}(S_{we}) \frac{\mu_w f_w(S_{we})}{\mu_o f_o(S_{we})} \tag{10.25}$$

$$S_{we} = S_{wi} + \overline{V_o}(t) - f_o(S_{we}) \cdot \overline{V}(t) \tag{10.26}$$

$$I = \frac{\mu_o u L}{K \cdot \Delta P(t)} = \frac{\mu_o Q(t) L}{K A \Delta P(t)} \tag{10.27}$$

where

$K_{ro}(S_{we})$	Relative permeability to oil at the outflow-end saturation;
$K_{rw}(S_{we})$	Relative permeability to water at the outflow-end saturation;
$\overline{V}(t)$	Dimensionless cumulative water injected;
V_t	The cumulative water injected;
V_p	The pore volume of the core;
$\overline{V_o}(t)$	Dimensionless cumulative oil produced;

V_o	The cumulative oil produced;
$f_o(S_{we})$	Oil-cut at the outflow end (the ratio of oil produced compared to the volume of total liquids produced);
$f_w(S_{we})$	Water-cut at the outflow end (the ratio of water produced compared to the volume of total liquids produced);
S_{we}	Water saturation at the outflow end, fraction;
S_{wi}	Connate-water saturation of the core;
μ_o, μ_w	Respectively, the viscosity of oil and water, mPa s;
I	The ratio of the flowing ability at a certain time to that at the initial time;
K	Absolute permeability of the core, μm^2;
A	Cross-sectional area of the core, cm^2;
L	Length of the core, cm;
$Q(t)$	Fluid production rate at the outflow end at the time t, cm^3/s;
$\Delta P(t)$	Pressure difference between the two core-ends at the time t, 10^{-1} MPa;
u	Velocity of seepage flowing, cm/s

Details of the deviation of the formulas are presented in the "Appendix A." More information concerning Buckley–Leverett's theory about the displacement of oil by water will be given in the course of mechanics of flow through porous media.

3. Experiment procedure

Taking the constant-flow method (or constant-velocity method) as an example, the experiment procedure, as shown in Fig. 10.25, involves the steps listed below:

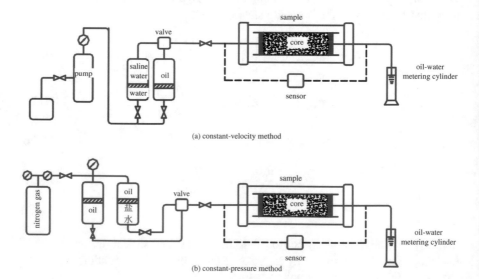

(a) constant-velocity method

(b) constant-pressure method

Fig. 10.25 Unsteady-state method for measuring relative permeability

(1) Preparation and cleaning of core sample;
(2) Measure the absolute permeability of the core: insert the core saturated with water into the core-holder, apply confining pressure to it and heat it to the reservoir temperature; then, with water as the experimental fluid, measure the absolute permeability K of the core 100 % saturated by water.
(3) Saturate the core with oil and create a state of connate water: conduct oil-displacing-water experiment at a constant rate and get the relative permeability curves during the drainage process; the cumulative oil volume injected to the core is 10 times of the pore volume (PV) and the connate-water saturation locates on the terminal point, where the relative permeability to oil at the connate-water saturation can also be read.
(4) Displace oil by water: conduct water-displacing-oil experiment at a constant rate and get the relative permeability curves during the imbibition process; the cumulative water volume injected to the core is 10 times of PV and the residual-oil saturation locates on the terminal point, where the relative permeability to water at the residual-oil saturation can also be read.
(5) Data processing: resort to Eqs. (10.24)–(10.27).

Note that the experiment procedure given above is particularly for water-wet rock cores.

The unsteady method can be carried out in a relatively short time, but the data interpretation of it is more complicated.

The pore space progressively becomes more and more oil-wet due to film rupture as the capillary pressure is raised.

10.4.3 Relevant Problems in the Measurement of Relative Permeability

1. End effect

The end effect, also called capillary end effect, is an important issue in coreflood experiments. When measuring the relative permeability curves by means of the steady-state method, we must confront the question of how to eliminate the errors due to the "end effect," so as to ensure the experimental accuracy.

Essentially, the end effect arises from capillarity at the outlet end of a porous medium through which a two-phase fluid flows, and it is characterized by: (1) the saturation of the wetting phase increases within a certain distance from the outlet end of the core; (2) there is a transient hysteresis before the outlet end witnesses a breakthrough of the water.

Before the wetting phase (water) arrives at the outlet end, as shown in Fig. 10.26a, the distribution of the water saturation keeps being normal, the oil–water meniscus is concave toward the outlet, and the capillary pressure Pc exists as the driving force for water displacing oil; however, as the water is going to flow out

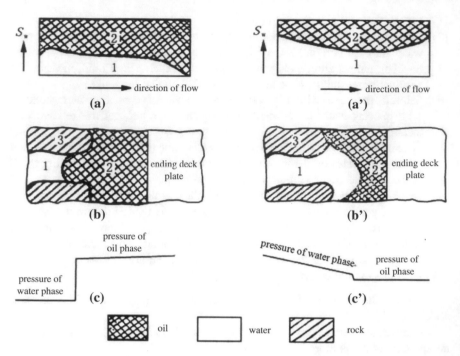

Fig. 10.26 Fluid distribution at the outflow face of a water-wet core **a, a'** saturation distribution of fluid; **b, b'** water invasion at the outflow face; **c, c'** pressure distribution

of the core, as illustrated in Fig. 10.26b, the capillary pressure tries to block the outflow as a result of the deformation of the meniscus and the wettability reversal, leading to higher water saturation in the vicinity of the core outlet and hysteresis of the outflow.

Capillary end effects arise from the discontinuity of capillarity in the wetting phase at the outlet end of the core sample.

Proven by both experiments and theoretical calculations, the end effect is limited to a reach within 2 cm away from the outlet and this distance is dominantly decided by the flow rate (or pressure difference) used in the experiment. It is found that the higher the flow rate, the smaller the influenced length. Therefore, in order to make the errors caused by end effect negligible, it is one of the feasible means to increase the pressure difference.

An alternate approach, as shown in Fig. 10.27, is to, respectively, add a porous medium of 2 cm long in front of and behind the core before test. This approach is called the "three-section" method.

Oil of appropriate viscosity should be used in the tests. This is because the displacement process may follow a piston-like pattern if low-viscosity oil is used for water-wet cores. As a result, the state of simultaneous flow of the two phases cannot be reached, so only the connate-water saturation or the residual-oil saturation can be measured. Besides, as high flow rate is usually used in displacement to

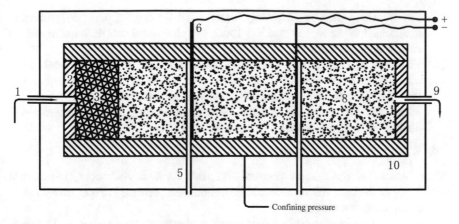

Fig. 10.27 Three-section method to eliminate end effect *1* inlet for fluid flow; *2* end cap; *3* high-permeability orifice meter; *4* artificial core; *5* orifice for pressure measurement; *6* electrode; *7* core; *8* artificial core; *9* outlet for water flow; and *10* rubber sleeve

eliminate the end effect, the water, preferentially flowing through the large pore openings, broke through the core so soon that the obtained relative permeability curves are similar to that of those oil-wet cores. Furthermore, the film formed by the wetting phase, water, on the rock surface, which wet up the oil to some extent, may render relative permeability to oil exceeds 100 %. In conclusion, the unsteady-state method for relative permeability measurement is not competent for strongly water-wet cores.

As stated above, although an increment of flow rate brings in decline of the end effect at the outlet, it also causes enlargement of the end effect at the inlet, where the wetting phase is inhaled into the core of strong wettability and consequently the non-wetting phase is driven toward the outlet in an instantaneous period time. When water and oil are simultaneously admitted into the strongly water-wet core, a portion of the injected water will be spontaneously imbibed into core from the inlet and transiently results in oppositely directed flow of the oil out of the inlet of the core. As the wetting phase (water) continuously enters the core, a region of unsteady-state nonlinear flow is formed. Gradually, as the displacement continues, the water is redistributed along with the core under the action capillary force and the end effect at the inlet becomes more and more obvious.

Note that the end effect impacts only the saturation and recovery data yielded from the in-laboratory coreflood experiments, whereas its impaction on real formations is negligible.

2. About the unsteady-state method

As two fundamental assumptions are built up in the deviation of the equations, the following two conditions must be satisfied when the unsteady-state Eqs. (10.24)–(10.27) are used for calculating the relative permeability curves:

(1) The flow rate must be adequately high so that the driving pressure gradient is maintained far above the capillary force, which ensures that the capillary effect causes negligible errors.
(2) The flow rate at all the rock cross sections during the linear coreflood is constant, or in other words, both of the two phases of fluids are incompressible. In the case that one of the phases is gas, the pressure is maintained adequately high to ensure that the gas expansion caused by the pressure difference is negligibly small.

3. Wettability of rock sample

When measuring the relative permeability data for reservoir rocks, the natural wettability of the reservoir rock should be maintained before it is experimented. It is verified by experiments that:

Prolonged contact with oil-based mud or the filtrate of oil emulsion mud tends to alternate the rock surface to be neutral-wet or oil-wet; the water-based mud also causes more or less change of the rock wettability, but the neutral and acidic mud at pH value has little effect. Therefore, in any circumstance where coring is performed it is better to use water-based mud that contains comparatively less active agents and lower water loss.

The core sample that had been exposed to air for too long tends to yield higher residual-oil saturation than "fresh core" in measurement. Proven by experiments, the original water-wet reservoir rock sample will change into water-wet after being exposed to air for a week. It is generally considered that the oxidation processes that occur to the rocks experiencing exposure to air greatly change the wetting characteristics of them. Therefore, packing methods should be taken to protect the cores from the pollutions due to air. With regard to the in situ treatment of the fresh pores on well site, the common and effective way is to pack the cores with polyethylene films, then wrap aluminum foil around it, and finally seal it off with paraffin wax or plastics.

If cores maintained in initial wetting state of nature are not available, it is suggested to use manually restored cores. Manual restoration refers to a process in which the core is cleaned, dried, and then processed to restore its initial conditions of wettability and saturation. Obviously, the method of manual restoration requires known wettability of the studied reservoir rocks. So, only natural cores in their initial state are feasible for the tests before the rock wettability is determined. Even if the rock wettability is already known, the manually restored can be used only in the case that the rock is strongly water-wet or strongly oil-wet. This is because it is rather hard to restore the core back to an intermediate state of wettability.

Given that the rock has been determined to be water-wet by contact-angle measurement or imbibition test on well site, it is suggested that the steps listed below be adopted to conduct a relative permeability measurement.

(1) Saturate the core with water;
(2) Displace the water with refined oil (e.g., white oil) until a state of coexisting-water saturation required for simulation is reached;

(3) Measure the relative permeability data.

For oil-wet rocks, it is almost impossible to restore its initial coexisting-water saturation and distribution. Therefore, it is better to completely saturate the oil-wet core with oil before measuring the relative permeability data.

It is common to let the strongly water-wet cores obtained from manual restoration undergo a high-temperature treatment, in which, for example, the cores are put in an oven and exposed to a high temperature of 400 °C for 5 h. But the strongly oil-wet cores obtained from manual restoration are usually treated with silicon oil or other oil-wet surfactants.

4. Fluids used in experiment

It is suggested to use refined white oil or kerosene as the oil phase and to use nitrogen as the gas phase. With regard to the water phase, what kind of liquid should be adopted depends on the method in which the saturation is measured. The usual choices for the water phase are distilled water and brine.

5. The other

For the oil-wet cores, the coexisting-water saturation exerts negligible effect on the measured relative permeability data as long as it is lower than 20 %. However, for the water-wet cores, the coexisting-water saturation exercises a determining influence on the measured relative permeability data. Therefore, in dealing with water-wet cores the coexisting-water saturation created in it must be near the natural conditions as closely as possible. The overburden pressure has no effect on the relative permeability measured in laboratory.

To sum up, factors of all aspects, involving those related to cores, fluids, experiment procedure, experiment conditions, should be taken into account when measuring the relative permeability curves, so that data really capable of representing reservoir conditions can be obtained.

10.4.4   Calculation of Relative Permeability Curves from Capillary Pressure Data

Known from the preceding discussions: (1) The rock permeability can be calculated from the capillary pressure curves according to the pore-throat distribution characteristics presented by them; (2) since the relative permeability is a function of the saturation of the fluids and the magnitude of capillary pressure also directly depends on saturations, equations can be built up for the calculation of relative permeability from the capillary pressure data.

1. Calculation of Absolute Permeability of Rock—the Model of Bundles of Capillary Tubes

Purcell in 1949 presented the computing formula for permeability on the basis of the model of bundles of capillary tubes. According to the Poiseuille's law, the flow rate through a single capillary tube is:

$$q = \frac{\pi r^4 \Delta P}{8 \mu L} \tag{10.28}$$

The pore volume provides by a single capillary tube is: $V = \pi r^2 L$, so $\pi r^2 = V/L$.

By transforming the formula $P_c = \frac{2\sigma \cos \theta}{r}$ into $r^2 = 4(\sigma \cos \theta)^2 / P_c^2$, we can obtain another form of the flow rate through a single capillary tube:

$$q = \frac{\pi r^4 \Delta P}{8 \mu L} = \frac{(\sigma \cos \theta)^2 \Delta P V}{2 \mu L^2 P_c^2} \tag{10.29}$$

Supposing that the capillary number involved in a rock is n, the total flow rate through the rock is:

$$Q = \frac{(\sigma \cos \theta)^2 \Delta P}{2 \mu L^2} \sum_{i=1}^{n} \frac{V_i}{P_{ci}^2} \tag{10.30}$$

Known form the Darcy's law, the rate of the seepage flowing through a real rock is:

$$Q = \frac{K A \Delta P}{\mu L} \tag{10.31}$$

By equaling Eqs. (10.30) and (10.31), we get:

$$K = \frac{(\sigma \cos \theta)^2}{2} \sum_{i=1}^{n} \frac{V_i}{P_{ci}^2} \tag{10.32}$$

It is assumed that the ratio of the volume V_i of an individual capillary channel to the total volume V_p of all the capillary channels in the rock equals the saturation S_i of this capillary channel in the whole pore system. That is:

$$S_i = V_i/V_p, \quad V_p = V_i/S_i \tag{10.33}$$

As the rock volume is AL, the porosity is $\varphi = V_p/AL = V_i/ALS_i$. So:

$$V_i = \phi A L S_i \tag{10.34}$$

By substituting the above equation into Eq. (10.32), we get:

$$K = \frac{(\sigma \cos \theta)^2}{2} \phi \sum_{i=1}^{n} \frac{S_i}{P_{ci}^2} \tag{10.35}$$

Supposing the studied capillary tube varies continuously in radius and with the difference between the imaginary rock and the real rock in consideration, a modification coefficient λ should be introduced into the equation:

$$K_w = 0.5(\sigma \cos \theta)^2 \phi \lambda \int\limits_{S=0}^{S=1} \frac{dS}{P_c^2} \tag{10.36}$$

For a given system composed of oil, water, and rock, all of the coefficients present before the sign of integration are constant. So, the permeability depends only on the integration of the inverse square of the capillary pressure.

Draw the relation curve shown in Fig. 10.28 between the inverse square of the capillary pressure $(1/P_c^2)$ with the saturation. The integration $\int_0^1 \frac{dS}{P_c^2}$ in the Eq. (10.36) corresponds to the enveloped area below this relation curve. Then, the absolute permeability K of rock can be obtained from Eq. (10.36).

2. Calculation of phase permeability and relative permeability to oil/water—the model of bundles of capillary tubes

Consider the case in which the capillary pressure curves are measured with water-wet rocks by oil displacing water. When the external pressure difference applied on the core is large enough to overcome a certain capillary force $(P_c)_i$, there is only oil present and flowing in the pore-channels of radius larger than $r_i \frac{2\sigma \cos \theta}{(P_c)_i}$; and, conversely, there is only water present and flowing in the pore-channels of radius smaller than r_i. If the corresponding water saturation in such a case is S_i, the water saturation in the pore-channels of radius smaller than r_i must be lower than S_i. Illustrated in Fig. 10.28, the water-containing pore-channels correspond to the part on the left of S_i, while the oil-containing pore-channels correspond to the part on the right of S_i.

Thence, we can obtain the effective permeability, respectively, to oil and water at the water saturation S_i:

$$K_w = 0.5(\sigma \cos \theta)^2 \phi \lambda \int\limits_0^{S_i} \frac{dS}{P_c^2} \tag{10.37}$$

$$K_o = 0.5(\sigma \cos \theta)^2 \phi \lambda \int\limits_{S_i}^1 \frac{dS}{P_c^2} \tag{10.38}$$

And the relative permeability, respectively, to oil and water:

$$K_{rw} = \frac{K_w}{K} = \frac{\int_0^{S_i} \frac{ds}{P_c^2}}{\int_0^1 \frac{ds}{P_c^2}} \tag{10.39}$$

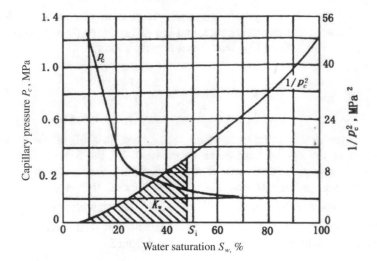

Fig. 10.28 Calculation of absolute permeability and relative permeability curve from capillary pressure curves

$$K_{ro} = \frac{K_o}{K} = \frac{\int_{Si}^{1} \frac{ds}{P_c^2}}{\int_{0}^{1} \frac{ds}{P_c^2}} \tag{10.40}$$

By calculating the relative permeability data at different water saturations, the oil–water relative permeability curves.

The equations given above are deduced on the basis of the model of bundles of capillary tubes, the structure of which still has a gap in comparison with that of real porous medium. The errors due to the imperfection of the model entail some modifications of the calculating equations. Thanks to an explosion of work by scholars, more than one correction formulas have been presented for use. Besides, another point needed to be mentioned is that the method of calculating absolute permeability and relative permeability data from capillary pressure curves is of particular importance to the determination of the permeability of rock debris. This is because it is hard to measure the permeability of the rock debris by using those conventional approaches.

10.4.5   Calculation of Relative Permeability from Empirical Equations

Since the 1950s, researchers, analyzing and generalizing large amounts of field data about relative permeability, have proposed several empirical equations for the relative permeability calculation by means of numerical simulation and mathematical statistics. These equations involve both two-phase and three-phase cases

Table 10.3 Empirical formulas for oil–gas relative permeability

Type of rock	K_{ro}	K_{rg}
Unconsolidated sand, good sorting	$(S*)^3$	$(1 - S*)^3$
Unconsolidated sand, bad sorting	$(S*)^{3.5}$	$(1 - S*)^2(1 - S*^{1.5})$
Cemented sand, carbonate rock	$(S*)^4$	$(1 - S*)^2(1 - S*^2)$

$S* = \frac{S_o}{1-S_{wi}}$, S_{wi} is the connate-water saturation

Table 10.4 Empirical formulas for oil–water relative permeability

Type of rock	K_{ro}	K_{rw}
Unconsolidated sand, good sorting	$(1 - S_w*)^3$	$(S_w*)^3$
Unconsolidated sand, bad sorting	$(1 - S_w*)^2(1 - S_w*^{1.5})$	$(S_w*)^{3.5}$
Cemented sand, carbonate rock	$(1 - S_w*)^2(1 - S_w*^2)$	$(S_w*)^4$

In the Table: $S_w^* = \frac{S_w-S_{wi}}{1-S_{wi}}$, S_{wi} is the connate-water saturation

such as oil–gas, oil–water, oil–gas–water, and microemulsion-oil–water. Two of these empirical equations will be introduced here.

1. The Rose method—relative permeability to oil (wetting phase) and gas (non-wetting phase)

Related relative permeability empirical formulas proposed by Rose are shown in Table 10.3.

The other symbols have the same significances as before.

2. The Rose method—relative permeability to oil and water

Related relative permeability empirical formulas proposed by Rose are shown in Table 10.4.

The major defect of the Rose method is that it requires the residual saturations of the two phases to be accurately known.

10.4.6   Calculation of Relative Permeability from Field Data

This method is a calculating procedure utilizing field data (e.g., the productions of oil, gas, and water, the pressure, the gas-to-oil ratio, the water-to-oil ratio) to calculate the average relative permeabilities of the formation.

1. Calculation of oil–gas relative permeability of dissolved-gas-drive reservoirs

When the reservoir pressure of an oil reservoir drops to be lower than the saturation pressure and has not been waterflooded, the originally dissolved gas will be liberated from the liquid. Reservoirs in such a state are called dissolved-gas-drive reservoirs. If it is assumed that (1) the hydrocarbons are distributed evenly in the pore system, (2) the pressure drop in the gas is the same as the pressure drop in the oil, (3) the force of gravity and the bottom-hole pressure drawdown be take no account of, the productions of oil and gas can be calculated from the radial fluid flow formula:

For the oil phase:

$$Q_o = \frac{2\pi h K_{ro} K}{B_o \mu_o In \frac{r_e}{r_w}} (P_e - P_w) \tag{10.41}$$

For the gas phase:

$$Q_g = \frac{2\pi h K_{rg} K}{B_g \mu_g In \frac{r_e}{r_w}} (P_e - P_w) \tag{10.42}$$

where

Q_o, Q_g Respectively, oil and gas flow rates expressed in terms of surface conditions of temperature and pressure, m^3/d;

B_o, B_g Volume factors, respectively, of oil and gas;

P_e, P_w Pressure at the supply edge, bottom-hole pressure, $\times 10^{-1}$ MPa;

r_e, r_w Drainage radius, oil-well radius, m;

h Effective thickness of the formation, m.

The production gas-to-oil ratio R is:

$$R_t = \frac{Q_g}{Q_o} = \frac{B_o}{B_g} \cdot \frac{\mu_o}{\mu_g} \cdot \frac{K_{rg}}{K_{ro}} \tag{10.43}$$

If taking in consideration the solubility R_s of gas in oil under reservoir conditions, the total gas-to-oil ratio becomes:

$$R_t = R_s + \frac{B_o}{B_g} \cdot \frac{\mu_o}{\mu_g} \cdot \frac{K_{rg}}{K_{ro}} = R_s + F \frac{K_{rg}}{K_{ro}} \tag{10.44}$$

$$F = \frac{\mu_o}{\mu_w} \cdot \frac{B_o}{B_w}$$

where

R_s Solubility R_s of gas in oil under reservoir conditions, m^3/m^3;

R_t Total production gas-to-oil ratio, m^3/m^3;

F Parameter, obtained from high-pressure physical-property data.

Based on the equation above, it is known that:

$$\frac{K_{rg}}{K_{ro}} = \frac{R_t - R_S}{F} \tag{10.45}$$

The saturations of oil and gas can be computed through using the material balance method:

$$S_L = S_o + S_w = (1 - S_{wi})\frac{N - N_p}{N} \cdot \frac{B_o}{B_{oi}} + S_{wi} \tag{10.46}$$

$$S_g = 1 - S_L \tag{10.47}$$

where

S_L, S_g	Respectively, saturations of the liquid and gas phases;
N	Initial geological reserve, m^3;
N_p	Cumulative produced oil, m^3;
B_{oi}, B_o	Volume factors of oil, respectively, at the initial and current reservoir pressure.

By utilizing Eqs. (10.44)–(10.46) given above and substituting the high-pressure physical-property parameters and the production data, the K_{rg}/K_{ro}—S_L relation curve can be obtained.

It is hard for real reservoirs to satisfy the postulated conditions of this method. As there is always pressure drop existing nearby the well bore, for example, the phenomenon of gas channeling readily occurs to the flowing gas. So, the real gas-to-oil ratio is always higher than the value from calculation. In order to overcome the drawbacks due to the inaccuracy of the production data needed by the material balance method, the data are selected by means of statistical method which helps avoid great errors.

2. Calculation of oil–water relative permeability of waterflooded reservoirs

In a similar way, the computing formulas for the oil–water relative permeability of waterflooded reservoirs can be deduced. After the oil well has witnessed the breakthrough of water, the relation curve between the oil-to-water relative permeability ratio and the saturations is usually calculated as follows:

$$\frac{K_{rw}}{K_{ro}} = \frac{R_w}{\frac{\mu_o}{\mu_w} \cdot \frac{B_o}{B_w}} = \frac{R_w}{F} \tag{10.48}$$

where

R_w Production water-to-oil ratio.

By making use of the oil and water productions at a certain stage, the water-to-oil ratio can be obtained:

$$R_{\mathrm{w}} = Q_{\mathrm{w}}/Q_{\mathrm{o}} \qquad\qquad (10.49)$$

The average saturations can be obtained by using the material balance method:

$$S_{\mathrm{o}} = \frac{1}{1 + \frac{B_{\mathrm{o}}}{B_{\mathrm{w}}} \cdot R_{\mathrm{w}}} \qquad\qquad (10.50)$$

$$S_{\mathrm{w}} = 1 - S_{\mathrm{o}} \qquad\qquad (10.50)$$

10.5   Use of Relative Permeability Curves

The relative permeability data are essential to reservoir studies involving multiphase flow of fluids in porous medium. In reservoir development, they are used in the parameter calculation, the dynamic analysis, the numerical simulation, etc.

10.5.1   Calculation of Production, Water–Oil Ratio, and Fluidity

When oil and water are simultaneously produced, the productions can be calculated according to the Darcy's law:

$$Q_{\mathrm{o}} = \frac{K_{\mathrm{ro}}KA\,\Delta P}{\mu_{\mathrm{o}}L} = \frac{K_{\mathrm{o}}A\Delta P}{\mu_{\mathrm{o}}L} \qquad\qquad (10.51)$$

$$Q_{\mathrm{w}} = \frac{K_{\mathrm{rw}}KA\,\Delta P}{\mu_{\mathrm{w}}L} = \frac{K_{\mathrm{rw}}A\Delta P}{\mu_{\mathrm{w}}L} \qquad\qquad (10.52)$$

The water-to-oil ratio is:

$$\frac{Q_{\mathrm{w}}}{Q_{\mathrm{o}}} = \frac{\frac{K_{\mathrm{w}}A\Delta P}{\mu_{\mathrm{w}}L}}{\frac{K_{0}A\Delta P}{\mu_{\mathrm{o}}L}} = \frac{\frac{K_{\mathrm{w}}}{\mu_{\mathrm{w}}}}{\frac{K_{0}}{\mu_{\mathrm{o}}}} = \frac{\lambda_{\mathrm{w}}}{\lambda_{\mathrm{o}}} = M \qquad\qquad (10.53)$$

If the viscosity of the fluids is held constant, the ratio depends only on the ratio between the effective permeabilities of the water and the oil.

Fluidity λ is the ratio of the effective permeability to the viscosity of a fluid. It reflects the level of difficulty encountered by the fluid when it flows—the greater its value is, the more easily the phase can flow.

The fluidity of the water phase: $\lambda_{\mathrm{w}} = K_{\mathrm{w}}/\mu_{\mathrm{w}}$

The fluidity of the oil phase: $\lambda_{\mathrm{o}} = K_{\mathrm{o}}/\mu_{\mathrm{o}}$

In the case of water-displacing oil, the water-to-oil fluidity ratio is denoted as M:

$$M = \frac{\lambda_w}{\lambda_o} \tag{10.54}$$

The fluidity ratio is an important parameter deciding the sweep coefficient of the displacing medium and the recovery.

10.5.2   Analysis of the Water Production Regularity of Oil Wells

1. Relationship between relative permeability ratio and fluid saturations

In order to bring more convenience in practice, as shown in Fig. 10.29, the relative permeability ratio K_{ro}/K_{rw} data is usually expressed as a function of the water saturation S_w. Subjected to a system of semilogarithmic coordinates, the relation curve shows a linear middle segment with two curved end segments. The linear segment indicates the range where the two phases flow simultaneously, which is commonly used in practice.

Fig. 10.29 Relationship between relative permeability ratio and water saturation

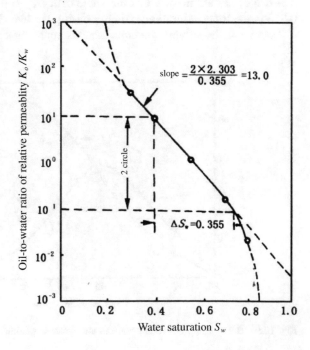

Proven by numerous measurements with respect to the relative permeability ratio curve, this characteristic in shape is possessed by most of rocks. This straight line can be expressed as:

$$\frac{K_{ro}}{K_{rw}} = \frac{K_o}{K_w} = ae^{-bS_w}$$

(10.55)

where

a The intercept of the line;
b The slope of the line.

The coefficients, a and b, are decided by the characteristics of the relative permeability curves, and their values vary with the parameters such as rock permeability, pore-size distribution, fluid viscosity, interfacial tension, and wetting characteristics. They can be obtained by the diagrammatical method.

Example 10.6 Calculate the values of a and b according to the curves given in Fig. 10.30.

Solution 1 (the diagrammatical method):

Read the curves directly and get: the intercept $a = 1222$, the slope $b = \frac{2 \times 2.303}{0.355} = 13$. In the calculating equation of b, 2.303 is the conversion coefficient used between the common logarithm and the natural logarithm, 2 indicates that two cycles are involved as the value of K_o/K_w varies from 10^{-1} and 10^{1}, and 0.355 is the water saturation variation during the two circles, ΔS_w.

Solution 2 (by solving the simultaneous equations):

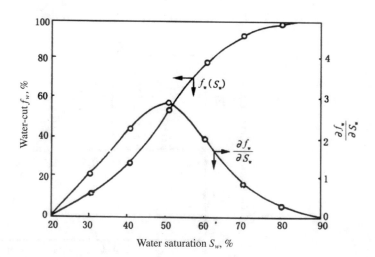

Fig. 10.30 Relationship between water-cut and water saturation

For example, when $S_w = 0.3$, $K_o/K_w = 25$; and when $S_w = 0.7$, $K_o/K_w = 0.14$. Substitute two groups of data into Eq. (10.55).

$$\begin{cases} 25 = ae^{-0.3b} \\ 0.14 = ae^{-0.7b} \end{cases}$$

By solving this set of simultaneous equations we can get that $a = 1222$ and $b = 13$, which are the same as the results yielded form the diagrammatical method.

2. Aids in Analyzing Water Production Regularity

The water production regularity refers to the variation of the water-cut with the increase of water saturation in formation. The water-cut, defined as the ratio of the water production to the total liquid production from the well, is an important index in the dynamical analysis of reservoir development. The following formula is called the fractional flow equation:

$$f_w = \frac{Q_w}{Q_w + Q_o} = \frac{K_w/\mu_w}{K_w/\mu_w + K_o/\mu_o} = \frac{1}{1 + \left(\frac{K_o}{K_w}\right)\left(\frac{\mu_w}{\mu_o}\right)} \tag{10.56}$$

For an oil reservoir where the viscosity ratio μ_w/μ_o is a constant, the water-cut depends only on the oil-to-water relative permeability ratio. Furthermore, as the relative permeability is a function of the water saturation, the water-cut is a function of the water saturation, too. The functional relationship between them is shown in Fig. 10.30. As shown by the figure, even if well produce 100 % water, it is not be 100 % water-saturated in the necessary that a formation, which indicates that the formation still holds a certain amount of residual oil.

By substituting Eq. (10.55) into Eq. (10.56), we can get the expression correlating f_w and S_w.

$$f_w = \frac{1}{1 + (\mu_w/\mu_o)ae^{-bS_w}} \tag{10.57}$$

From Eqs. (10.56) and (10.57), we can recognize that:

(1) The water-cut f_w increases with the increase of the water-to-oil fluidity ratio $M = \frac{(K_w/K_o)}{(\mu_w/\mu_o)}$.
(2) The more viscous the oil (i.e., $\mu_o \gg \mu_w$), the higher the water-cut f_w. This is rightly the reason why the viscous oil formation suffers quite high water-cut once water breakthrough occurs.
(3) With the increase of the water saturation S_w in the formation, the water-cut f_w rises. Therefore, oil wells at different locations of the oil–water transition zone have different water-cuts. Likewise, the water-cut will gradually increase after the formation where the oil well is located is waterflooded. However, as shown in Fig. 10.30, S_w and f_w do not exhibit functional relationship of direct proportion, but obey a relationship of power series.

(4) The relationship between the ascending rate $(\partial f_w / \partial S_w)$ of water-cut and the water saturation can be obtained by calculating the partial derivative of the water-cut f_w in Eq. (10.57) with respect to S_w.

$$\frac{\partial_w}{\partial S_w} = \frac{(\mu_w / \mu_o) bae^{-bS_w}}{[1 + (\mu_w / \mu_o) bae^{-bS_w}]^2} \tag{10.58}$$

The physical significance of this equation is the variation of the water-cut corresponding to a unit (e.g., 1 %) increase of the water saturation. It is equal to the slope of the $f_w(S_w)$ curve given in Fig. 10.30. Indicated by the curve, the ascending rate of water-cut is small at the low water saturations; as the saturation becomes higher, the water-cut increases rapidly later on; however, within the interval of comparatively high water saturation, the ascending rate of water-cut is lowered again. As a result, the curves exhibit a shape of "slow at the two ends, fast at the middle." This theory can also be used to explain the characteristic curve of water drive (i.e., the $\sum Q_w - \sum Q_o$ relation curve).

Knowledge about the ascending regularity of the water production of wells helps a lot in predicting production behaviors, based on which measures can be taken in advance to prevent the oil wells from being waterflooded too soon.

10.5.3 Determination of Vertical Distribution of Oil and Water in Reservoir

Based on the capillary pressure curves, the height of each saturation above the free water surface is obtained. For the homogeneous formation, the distribution of the water and oil in the formation, in other words, the oil saturation distribution with height in the formation, can be determined by using the set of relative permeability curves combined with the set of capillary pressure curves. For example, oil providing region, water providing region, oil and water providing region can be determined.

Figure 10.31 shows the process of determining division of the pay zone (pure-oil pay zone, mixture pay zone, and pure-water zone). The oil–water contact by the use of the two sets of curves. The region above the plane where the point A is located represents the pure-oil pay zone of the formation; the region between the points A and B represents the mixture-flow zone from which oil and water are simultaneously produced; the region between B and C represents the pure-water flow zone; and the region below B is the zone 100 % saturated with water. B is the oil–water interface, and C is the free water surface.

If we representing the capillary pressure with the liquid column height above the oil–water interface, the height of the point A stands for the lowest shut height within which the formation with such a pore system produces pure oil. The formation is

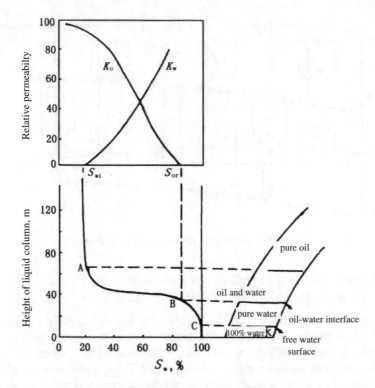

Fig. 10.31 Determination of the oil–water contact in formation

capable of paying pure oil as long as the shut height is greater than this mentioned lowest shut height, and the larger the gap, the larger the thickness of the oil-bearing zone.

It can be seen that the use of the combination of the oil–water relative permeability curves and the capillary pressure curves makes us able to accurately determine the height of the points A and B, the height of the oil–water interface, as well as the thickness of the zone where oil and water are simultaneously produced.

10.5.4   Determination of Free Water Surface

As was mentioned in the discussion of capillary pressure curves, it is necessary to determine the free surface level in order to properly calculate fluid distribution within the oil–water transition zone.

From the relative permeability curves which have been presented, it should have become apparent that the point at which 100 % water flow occurs is not necessarily the point of 100 % water saturation, as shown in Fig. 10.32. It is recognized that two water tables exist. These two water tables are:

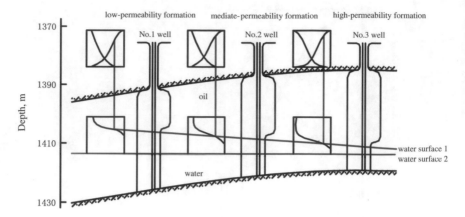

Fig. 10.32 The two water tables and the free water surface in oil reservoir

(1) The free water or zero capillary pressure level;
(2) The level below which fluid production is 100 % water.

These two definitions for the water tables in an oil reservoir are shown in Fig. 10.32. Note that the water by definition 2 rises as the permeability in the formation decreases while the water table by definition 1 is a horizontal surface, providing static conditions prevail in the reservoir. From a production engineering standpoint, an oil–water contact defined as the highest point of water production is useful. From a reservoir engineering standpoint, an oil–water contact defined by the zero capillary pressure is more appreciate.

The actual location of water Tables 10.1 and 10.2 can be determined by the use of electric logs, drill-stem tests, and relative permeability and capillary pressure data. For example, from electric logs and drill-stem tests it is possible to determine the depth (h_2) at which 100 % water flow occurs or the point of zero oil permeability; from relative permeability the engineers can determine what the fluid saturations must be at the point of zero oil permeability ($K_{ro} = 0$). When a fluid saturation determined from well test data and relative permeability curves is used, the capillary pressure can be determined and the height (d_h) of this saturation above the free water surface or zone of 100 % water saturation can be calculated. With the symbol h_1 as the depth of the free water surface, $h_1 = h_2 + d_h$.

The existence of two distinct water tables materially aids in explaining the occurrence of some tilted water tables. It is noted that the more permeable the formation, the more closely the pore structure approaches a supercapillary system and the smaller the divergence between the two water tables.

10.5.5   Calculation of Displacement Efficiency and Water Drive Recovery

1. Calculation of displacement efficiency and water drive recovery

The oil recovery is defined as the ratio of the volume of oil extracted from the reservoir to the volume of the underground initial reserve. It is also expressed as the product of the volumetric sweep coefficient (also called sweep coefficient) and the displacement efficiency:

$$E_R = \frac{V_{o\ out}}{V_{oi}} = \frac{V_{oi} - (V_{o\ usw} + V_{o\ sw})}{V_{oi}} = E_v \cdot E_D \tag{10.59}$$

where

E_v Sweep coefficient;
E_D Displacement efficiency;
V_{oi} Volume of initial oil reserve, m^3;
$V_{o\ out}$ Volume of oil extracted out, m^3;
$V_{o\ usw}$ Volume of the unswept remaining oil, m^3;
$V_{o\ sw}$ Volume of the residual volume within the swept area, m^3.

The sweep coefficient refers to the extent to which the injected working agents spread or sweep in the oil-bearing formation. The volumetric sweep coefficient is defined as the volumetric percentage taken up by the swept portion in the oil-bearing formation. For example, consider a reservoir with an area of A and an average thickness of h, if the swept area by the working agents is A_s and the swept thickness is h_s, the volumetric sweep coefficient is:

$$E_v = \frac{A_s}{A} \cdot \frac{h_s}{h} \tag{10.60}$$

where

E_v Volumetric sweep coefficient, also directly called sweep coefficient for short;
A_s, h_s Respectively, the swept area and swept thickness by the working agents.

Within the swept region, it is not a piston-like pattern in which the working agents displace the oil. Although the water passes through the pores, it fails to completely drive the oil there away. That is to say, because of the complexity of the microscopic pore structure of the rocks, only a portion of oil that dwells in the comparatively large pore openings can be displaced by the oil, whereas a certain amount of residual oil is left in some pores and channels. So, the sweep coefficient characterizes, in micropresentation, the extent to which the reservoir oil is cleaned out by the injected working agents.

The displacement efficiency E_D could be determined by water-displacing-oil tests in laboratory or by calculations based on the relative permeability curves:

$$E_D = \frac{S_{oi} - S_{or}}{S_{oi}} = 1 - \frac{S_{or}}{S_{oi}} \qquad (10.61)$$

where

S_{oi} Initial oil saturation;
S_{or} Residual-oil saturation.

In dealing with in-laboratory tests, the water-displacing-oil recovery is equal to the displacement efficiency because the sweep coefficient in the small core can be considered as 100 %.

As the oil recovery of a reservoir being developed is the product of the volumetric sweep coefficient and the displacement efficiency, the greater the sweep coefficient E_v, or the greater the displacement efficiency E_D, the higher the oil recovery E_R. Therefore, the way to improve the oil recovery lies in two aspects: raising the sweep coefficient and raising the displacement efficiency. And, what is more, the key to raise the displacement efficiency is reducing the residual-oil saturation.

2. Factors affecting residual-oil saturation

The residual-oil saturation, mainly decided by the non-piston-like characteristics of the process of oil displacement, is affected by many factors. In view of micro-analysis, the factors exerting effects on the residual-oil saturation include rock physical property, pore structure, wetting characteristics, fluid property, interfacial characteristics, and experimental conditions.

For a water-wetting core, whether the residual-oil drops after water-displacing-oil test is able to flow or not depends on the pressure difference between the two core-ends, which is built up by human, and the additional capillary force at the meniscus of oil drop. In other words, it is decided by the combination of the driving force and resisting force applied on the oil drops.

The driving force for the oil drop can be expressed as the pressure gradient $\Delta P/i$, in which i is the length of the oil drop and ΔP is the pressure difference applied on it. In dealing with the additional resisting force, it is feasible to consider a comparatively simple case as shown in Fig. 9.11, where the second type of resisting effect is given. Then, calculate according to Eq. (9.25). As the parameters in this equation except σ are quite difficult to be determined, the quantitative description of the resisting force concerns itself only with σ.

Define the concept of capillary number as the ratio of the driving force to the resisting force applied on the oil drop, which is mathematically expressed as $\frac{\Delta P/l}{\sigma}$. For a rock with specified wetting characteristics and capillary radius, whether the oil drops in it is able to flow depends on the capillary number $\frac{\Delta P/l}{\sigma}$. As long as $\frac{\Delta P/l}{\sigma}$ reaches a specified value, the oil drops begin to flow. By transforming the Darcy's law, the pressure gradient can be expressed as the product of the water viscosity μ and the seepage-flowing velocity v, so the definition formula of the capillary number (N_c) given above could take another form as follows:

$$N_c = \frac{v\mu}{\sigma} \tag{10.62}$$

N_c is dimensionless and represents the ratio of the driving force to the resisting force applied on the oil drops, given that oil and water are simultaneously present in a porous medium with specified wettability and permeability.

Proven by a plurality of experiments, the residual-oil saturation is associated with the capillary number in a good way. Table 10.5 gives some experimental results from three rock cores. It is seen that:

(1) The residual-oil saturations are different for the different cores although under the same displacing conditions.
(2) The residual-oil saturation decreases when the viscous force is increases or the interfacial tension is reduced.

Taking the wettability into consideration, Moore and Slobed defined the capillary number as the following equation:

$$N_c' = \frac{v'\mu}{\sigma \cos \theta} \tag{10.63}$$

where

v' Real velocity

The experimental relation curve between N_c' and the residual-oil saturation S_{or} is shown in Fig. 10.33. It is noted that the residual-oil saturation decreases with the increase of N_c'.

Table 10.5 Correlation between residual-oil saturation and capillary number

	Displacing conditions				Residual-oil saturation (fraction)		
	v (mm/s)	μ_o/μ_w	Σ (mN/m)	N_c	TorPedo core	Elgin core	Berea core
Initial displacing conditions	0.007	1.0	30	2.33E−04	0.416	0.482	0.495
Change the viscous force	0.7	1.0	30	2.33E−02	0.338	0.323	0.395
	0.007	0.055	30	2.33E−04	0.193	0.275	0.315
Change the capillary force	0.007	1.0	1.5	4.67E−03	0.285	0.275	0.315

Fig. 10.33 Relationship
between residual-oil
saturation S_{or} and capillary
number N_c'

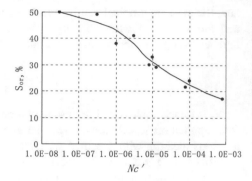

Abrams replaced the seepage-flowing velocity with the effective velocity $v/(S_{oi} - S_{or})$, took account of the effect of the water-to-oil viscosity ratio on the capillary number, and finally proposed the definition of mutant capillary number (N_{cam}):

$$N_{cam} = \frac{v\mu}{(S_{oi} - S_{or})\sigma \cos\theta} \left(\frac{\mu_w}{\mu_0}\right)^{0.4} \tag{10.64}$$

where

S_{oi} Initial oil saturation before water injection;
S_{or} Residual-oil saturation after water injection.

The resultant curves from several experimented cores are given in Fig. 10.34. When $N_{cam} < 10^{-6}$, the curves slope gently and the residual-oil saturation varies little. This is the normal range of the capillary number for ordinary water-displacing-oil processes, where the capillary force dominates in governing the displacement. As N_{cam} increases, the residual-oil saturation decreases. The capillary force and the viscous force rival with each other in affecting the residual-oil saturation within

Fig. 10.34 Relationship
between residual-oil
saturation S_{or} and capillary
number N_{cam}

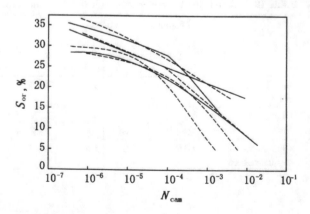

the range of $10^{-5} < N_{cam} < 10^{-4}$. In the range of higher capillary number where the residual-oil saturation becomes quite small, the viscous force, overwhelming the viscous force, plays a dominant role in governing the displacement.

As known from Eq. (9.25), the additional capillary resisting force is independent of the length of the oil drop. However, the longer the oil drop in length, the smaller the additional capillary-force gradient applied over the oil drop, and the more easily the drop can be displaced when the corresponding viscous force is specified. For example, consider a sandstone with a pore-throat radius of $r = 10^{-4}$ cm and oil drops contained in it. It is assumed that the length L of the oil drop is 100 times of the pore-throat radius, that is, $L = 10^{-2}$ cm; and that the oil–water interfacial tension is 36 mN/m and $\theta = 0$. In such a case, the capillary resisting force gradient applied over the oil drop is:

$$\frac{dpc}{dl} = \frac{2\sigma}{rl} = \frac{2 \times 36}{10^{-2} \times 10^{-2}} = 72 \times 10^6 \left(\frac{mN}{m} / cm^2\right) = 7.2\,MPa/cm$$

Generally speaking, the level to which the pressure gradient applied for the oil displacement by water can be achieved is about 2.8×10^2 Pa/cm = 2.8×10^{-4} MPa/cm. This is a value far smaller than the capillary-force gradient mentioned above, leading the oil drops to motionlessly stagnate. If the length of the oil drops is increased to 10^2 cm, the capillary-force gradient will be reduced to 7.2×10^{-4} Mpa/cm, which is much closer to the pressure gradient built up by human and gives the oil drops better chance of flowing. Conversely, it is easy to know that lower residual-oil saturation, indicating smaller sized residual-oil drops, requires larger driving pressure gradient.

10.5.6   Other Usage of Relative Permeability Data

The relative permeability curves have many other uses such as aid in determining the wettability of rock. Further discussion about this point will not be given here.

The three-phase relative permeability curves are commonly used in the reservoir dynamical analysis and calculations during the secondary recovery and the tertiary recovery.

Relative permeability is a critical parameter for evaluation of reservoir performances and widely used in petroleum engineering. More knowledge about it will be presented in related subsequence courses.

Appendix A
Unit Conversation Tables

1. Length

Centimeter (cm)	Meter (m)	Inch (in)	Foot (ft)
1	0.01	0.3937	0.03281
100	1	39.37	3.281
2.54	0.0254	1	0.0833
30.48	0.3048	12	1

2. Mass

Gram (g)	Kilogram (Kg)	Pound (lbm)	Ton (t)
1	10^{-3}	2.205×10^{-3}	10^{-6}
10^3	1	2.205	10^{-3}
453.6	0.4536	1	4.536×10^{-4}
10^6	10^3	2.205×10^3	1

3. Volume

Liter (liter) (L)	Cubic meter (m^3)	Cubic inch (in^3)	Cubic foot (t^3)	Gallon (USA) (Usgal)	Barrel (bbl)
1	10^{-3}	61.03	3.53×10^{-3}	0.264	
10^3	1	6.1×10^4	35.32	2.64×10^2	6.29×10^{-3}
1.64×10^{-2}	1.64×10^{-5}	1	5.79×10^{-4}	4.33×10^{-3}	6.29
28.32	2.83×10^{-3}	1728	1	7.48	1.03×10^{-4}
3.785	3.79×10^{-3}	231	0.134	1	0.178
159	0.159	9.71×10^3	5.615	42	2.38×10^{-2}
					1

© Petroleum Industry Press and Springer-Verlag GmbH Germany 2017
S. Yang, *Fundamentals of Petrophysics*, Springer Geophysics,
DOI 10.1007/978-3-662-55029-8

4. Pressure

Kilopascal (kPa) (kN/m^2)	Kilograms per square centimeter (kgf/cm^2)	Pound per square inch (psi)	Atmospheric pressure(atm)	Mercury column (0 °C)		Water column (15 °C)	
				Millimeter (mm)	Inch (in)	Meter (m)	Inch (in)
1	1.019×10^{-2}	0.145	9.86×10^{-3}	7.498	0.29522	0.102	4.0165
9.814×10	1	14.22	0.9678	735.5	28.96	10.01	3.94×10^{-3}
6.8948	7.03×10^{-2}	1	6.8×10^{-2}	51.71	2.036	0.7037	27.7
1.014×10^2	1.0333	14.70	1	760^4	29.92	10.34	4.072×10^2
0.1334	1.360×10^{-3}	0.01934	1.316×10^{-3}	1	0.0394	0.01361	0.5358
3.3873	3.453×10^{-2}	0.4912	3.34×10^{-2}	25.4	1	0.3456	13.61
9.8039	9.99×10^{-2}	1.421	9.67×10^{-2}	73.49	2.892	1	39.37
0.2490	2.538×10^{-3}	3.61×10^{-2}	2.456×10^{-3}	1.87	7.35×10^{-3}	2.54×10^{-2}	1

5. Coefficient of heat conductivity

Watt/(meter × °C) [W/(m × °C)]	Kilocalorie/ (meter × hour × °C) [kcal/(m × h × °C)]	Calorie/ (centimeter × second×°C) [cal/(cm × s × °C)]	British thermal unit/(inch × hour × F) [Btu/(h × ft × F)]
1	0.8598	2.389×10^{-3}	0.5778
1.163	1	2.778×10^{-3}	0.672
4.186×10^2	360	1	241.9
1.73	1.48819	4.136×10^{-3}	1

6. Specific heat

Joule/(gram × °C) [J/(g × C)]	Calorie/(gram × °C) [kcal/(g × °C)]	Btu/lbm × F
1	0.2391	0.2389
4.186	1	0.9994
4.18676	1.0007	1

7. Thermal capacity

British thermal unit/ (Inch^3 × F) [Btu/ft^3 × F]	Kilojoule/(Meter^3 °C) [KJ/(m^3 °C)]	Kilocalorie/(meter^3 × °C) [kcal/(m^3 × °C)]
1	67.27	16.07
0.01486	1	0.2389
0.06223	4.186	1

8. Kinetic viscosity

Poise[gram/(centimeter × second)] [g/(cm × s)]	Millipascal second (mPa s)	Kilogram/ (meter·second) [kg/ (m s)]	Pound/ (inch × second) [lbm/(ft × s)]
1	1×10^2	0.1	6.72×10^{-2}
1×10^{-2}	1	1×10^{-3}	6.72×10^{-4}
10	1×10^3	1	0.672
14.882	1.4882×10^3	1.4882	1

9. Temperature

$°F = 1.8 °C + 32;$	Celsius temperature, °C
$°C = 5/9(°F–32)$	Fahrenheit temperature, F
$°K = °C + 273$	Rankine temperature, R
$°R = °F + 460$	Absolute temperature, K
$°C = 5/9(R–492)$	

10. Interfacial tension

1dyn/centimeter=10^{-3}newton/meter=1millinewton/meter;
1dyn/cm=10^{-3}N/m=1mN/m; 1Erg/square centimeter=1dyn/cm

11. Density of crude oil: $API = \frac{141.5}{\gamma_o} - 131.5$

Appendix B
Vocabulary of Technical Terms
in Chapters

Terms	Chapter
Petrophysics	
Viscous crude	1
Heavy oil reservoir	1
Wax content	1
Wax-bearing crude	1
Black oil	1
Wax precipitation	1
Borehole bottom	1
Condensate field	1
Gas condensate	1
Solution gas expansion	1
Dead oil	1
Base crude	1
Crude oil	1
Medium crude oil	1
Heavy crude oil	1
Associated gas	2
Principle of corresponding	2
Dry gas	2
Lean gas	2
Gas deviation factor	2
Wet gas	2
Pseudo-correspondence temperature	2
Pseudo-correspondence pressure	2
Natural gas	2
Specific gravity of natural gas	2
Gas formation volume factor	2
Density of nature gas	2

(continued)

© Petroleum Industry Press and Springer-Verlag GmbH Germany 2017
S. Yang, *Fundamentals of Petrophysics*, Springer Geophysics,
DOI 10.1007/978-3-662-55029-8

(continued)

Terms	Chapter
Gas pseudo-critical temperature	2
Gas pseudo-critical pressure	2
Gas compressibility	2
Gas viscosity	2
State equation of natural gas	2
Initial gas formation volume factor	2
Saturated temperature	3
Saturation pressure	3
Differential liberation	3
Multistage separation	3
Multistage separator	3
Retrograde condensate	3
Retrograde condensate phenomenon	3
Retrograde condensate pressure	3
Miscible flood	3
Contact liberation	3
Pseudo-component	3
Gas-condensate reservoir	3
Equilibrium vapor phase	3
Triangular phase diagram	3
Tricomponent model	3
Flash separation	3
Flash liberation process	3
System	3
Gas solubility	3
Phase state of hydrocarbon system	3
Degasification	3
Gas breakout	3
Undersaturated oil reservoir	3
Undersaturated oil	3
Phase	3
Phase boundary curve	3
Phase equilibrium	3
Phase state equation	3
Phase volume	3
Phase viscosity	3
Phase composition	3
P-X diagram	3

(continued)

(continued)

Terms	Chapter
Phase diagram of reservoir hydrocarbon	3
Gas and oil separator	3
Methane gas injection	3
Component	3
PVT cell	4
Apparent velocity	4
Unstable state flow	4
Reservoir fluid compressibility	4
Reservoir oil	4
Two-phase formation volume factor reservoir oil	4
Solution gas–oil ratio of reservoir	4
Reservoir oil analysis	4
Coefficient of kinematical viscosity	4
Dynamic viscosity	4
PVT investigate	4
Bottom hole static pressure	4
Flowing pressure of bottom hole	4
Bottom hole temperature	4
Bottom hole pressure	4
Borehole fluid sampler	4
Salinity	4
Rheological behavior	4
Bubble point pressure	4
Gas-cap	4
Gas-cap drive	4
Gas factor	4
Gas–oil ratio	4
Brine salinity	4
Water salinity	4
Aqueous phase viscosity	4
Water compressibility	4
Water hardness	4
Material balance equation	4
Field water	4
Original formation oil volume factor	4
Natural bottom hole temperature	4
Original bubble-point pressure	4
Initial gas–oil ratio	4

(continued)

(continued)

Terms	Chapter
Initial solution gas–oil ratio	4
Original pressure	4
Initial reservoir pressure	4
Initial oil formation volume factor	4
Oil compressibility	4
Kinematical viscosity	4
Viscometer	4
Viscosity–temperature curve	4
Non-interconnected void	5
Stagnant pore	5
Irreducible oil saturation	5
Residual oil saturation	5
Bedding plane structure	5
Sedimentary structure	5
Initial oil saturation	5
Reserve bed	5
Secondary pore	5
Coarse sandstone	5
Single-porosity structure	5
Multipore media	5
Anisotropic medium	5
Macroirregularity	5
Basal cement	5
Matrix porosity	5
Cement	5
Absolute porosity	5
Pore throat	5
Throat-to-pore ratio	5
Type of pore–throat structure	5
Cumulative distributed curve of pore size	5
Pore-blocking	5
Porous cement	5
Pore structure	5
Pore pressure	5
Particle size distributed curve	5
Connected pore space	5
Fracture porosity	5
Fissured oil field	5

(continued)

(continued)

Terms	Chapter
Fluid saturation	5
Bed of interest	5
Mud stone	5
Dirty sand	5
Shaly clastic reservoir	5
Bentonite	5
Mean pore size	5
Buried hill-type reservoir	5
Triple-porosity system	5
Irreducible water	5
In situ water saturation	5
Irreducible water saturation	5
Double-porosity media	5
Double-porosity reservoir	5
Clastic rock	5
Carbonate rock	5
Carbonate reservoir	5
Natural fractured formation	5
Fine-grained sandstone	5
Compaction	5
Porosity of rock	5
Compressibility of pore space of rock	5
Compressibility of reservoir rock	5
Apparent viscosity	5
Shaly	5
Petroleum reservoir core sample	5
Effective porosity	5
Net pore volume	5
Tortuosity	5
Primary pore	5
Original water saturation	5
Initial gas saturation	5
Virgin porous media	5
Initial fluid saturation	5
Clay	5
Clay swelling	5
Bulk compressibility of reservoir	5
Complex compressibility of reservoir	5

(continued)

(continued)

Terms	Chapter
Poiseuille equation	6
Formation damage	6
Non-Darcy flow	6
Slip effect	6
Near-wellbore formation damage	6
Absolute permeability	6
Density of fracture	6
Critical residual oil saturation	6
Bundle of capillary tubes	6
Compatibility	6
Mean permeability	6
Areal heterogeneity	6
Flow velocity	6
Permeability/heterogeneity	6
Permeability	6
Permeability impairment	6
Laboratory core test	6
Water-sensitive formation	6
Acid stimulation	6
Non-damage formation	6
Linear flow	6
Permeability of rock	6
Acid sensitivity	6
Natural permeability	6
Anti-clay welling agent	6
Specific heat	7
Electric resistivity of reservoir rock	7
Heat diffusivity of reservoir rock	7
Thermal conductivity	7
Heat capacity	7
Heat conduction coefficient of rock	7
Heat capacity of rock	7
Surfactant	8
Non-wetting phase	8
Separate phase	8
Macromolecule	8
Macromolecular surfactant	8
Angle of contact	8

(continued)

(continued)

Terms	Chapter
Contact angle hysteresis	8
Surface viscosity	8
Static adsorption	8
Strongly oil wet	8
Angle of wetting	8
Wetting phase	8
Wettability	8
Wettability reversal	8
Wettability hysteresis	8
Water blocking	8
Adsorption	8
Pendant drop method	8
Core wettability	8
Cationic surfactant	8
Oil-wet formation	8
Original reservoir wettability	8
Porous diaphragm method	9
Enhanced flow resistance	9
Jamin's effect	9
Interfacial tension	9
Capillary force	9
Capillary displacement pressure	9
Capillary pressure	9
Capillary pressure curve	9
Capillary imbibition	9
Capillary retention	9
Drainage capillary pressure curves	9
Imbibition displacement	9
Intrusive mercury curve	9
Water–oil transition zone	9
Primary imbibition curves	9
Spontaneous capillary penetration	9
Spontaneous imbibition	9
Imbibition capillary pressure curve	9
Imbibition	9
Swept region	10
Conformance efficiency	10
Sweep efficiency	10

(continued)

(continued)

Terms	Chapter
Recovery percent of reserves	10
Final residual oil saturation	10
Oil-producing zone	10
Oil-producing capacity	10
Single-phase flow	10
Multiphase flow	10
Multiphase region	10
Multiple-phase flow	10
Heterogeneous formation	10
Macroscopic sweep efficiency	10
Chemical flooding	10
Homogeneous formation	10
Two-phase region	10
Flow ability coefficient	10
Mobility thickness product	10
Mobility	10
Mobility ratio	10
Capillary number	10
Areal coverage factor	10
End effect	10
Displacement efficiency	10
Displacement pressure	10
Oil displacement process	10
Triple-phase region	10
Tongued advance	10
Produced water–oil ratio	10
Distribution of remaining oil	10
Water displacement recovery	10
Displacement of oil by water	10
Water-phase saturation	10
Water relative permeability	10
Water–oil ratio	10
Bypassed pocket of oil	10
Enhanced oil recovery	10
Volumetric sweep efficiency	10
Breakthrough	10
Microscopic displacement efficiency	10
Bypassed areas	10

(continued)

(continued)

Terms	Chapter
Relative permeability	10
Phase permeability	10
Laboratory core flood	10
Core displacement test	10
Primary depletion	10
Primary depletion recovery	10
Oil–gas mobility ratio	10
Oil–water contact	10
Oil–water relative permeability ratio	10
Oil phase	10
Oil phase saturation	10
Efficient permeability	10
Original gas–water contact	10
Original oil–water interface	10
Ultimate residual oil saturation	10
Ultimate waterflood recovery	10
Ultimate oil recovery	10

Appendix C
Charts of Equilibrium Ratio

See Figs. C.1, C.2, C.3, C.4, C.5, C.6, C.7, C.8, C.9, C.10, C.11 and C.12.

© Petroleum Industry Press and Springer-Verlag GmbH Germany 2017
S. Yang, *Fundamentals of Petrophysics*, Springer Geophysics,
DOI 10.1007/978-3-662-55029-8

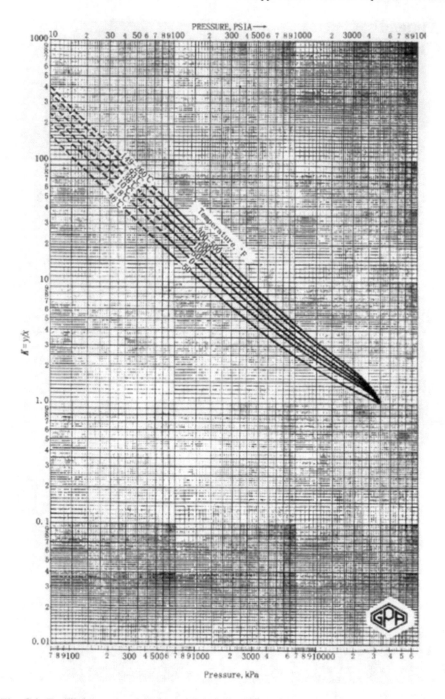

Fig. C.1 Equilibrium ratio of methane, 34.47 MPa (5000 psia) convergence pressure

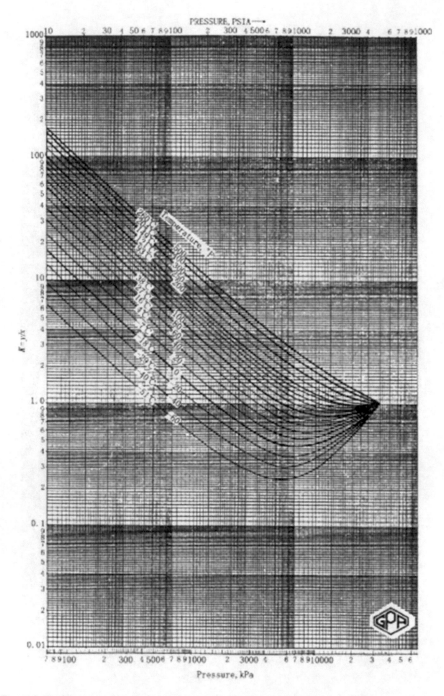

Fig. C.2 Equilibrium ratio of ethane, 34.47 MPa (5000 psia) convergence pressure

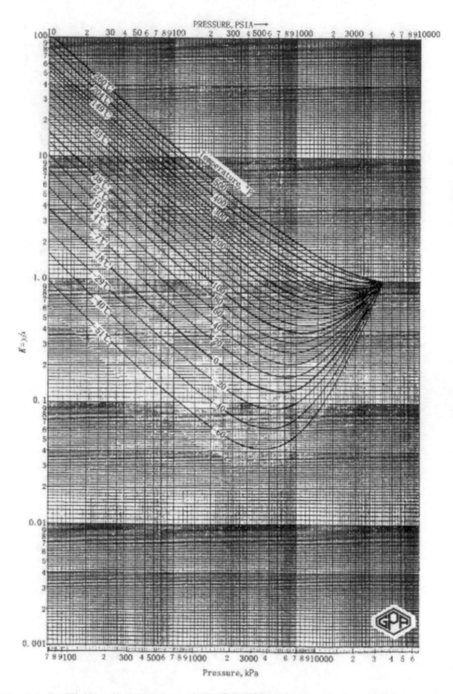

Fig. C.3 Equilibrium ratio of propane, 34.47 MPa (5000 psia) convergence pressure

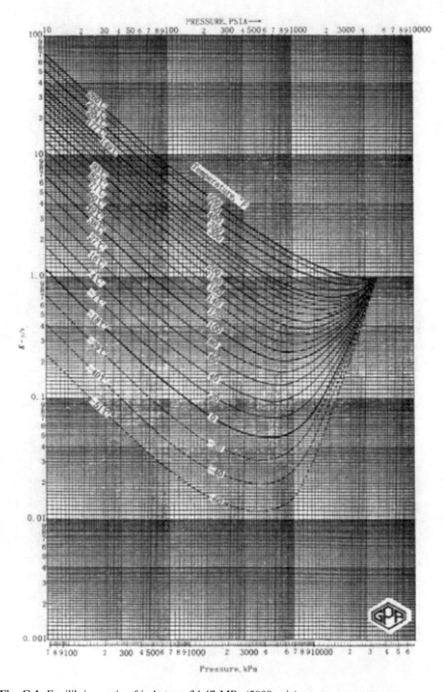

Fig. C.4 Equilibrium ratio of isobutane, 34.47 MPa (5000 psia) convergence pressure

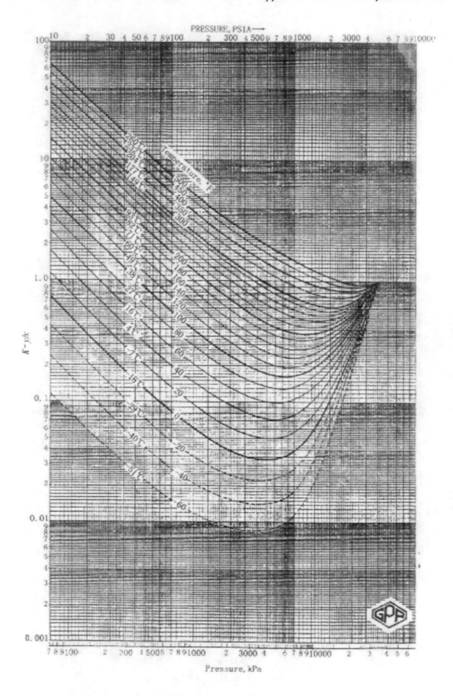

Fig. C.5 Equilibrium ratio of normal butane, 34.47 MPa (5000 psia) convergence pressure

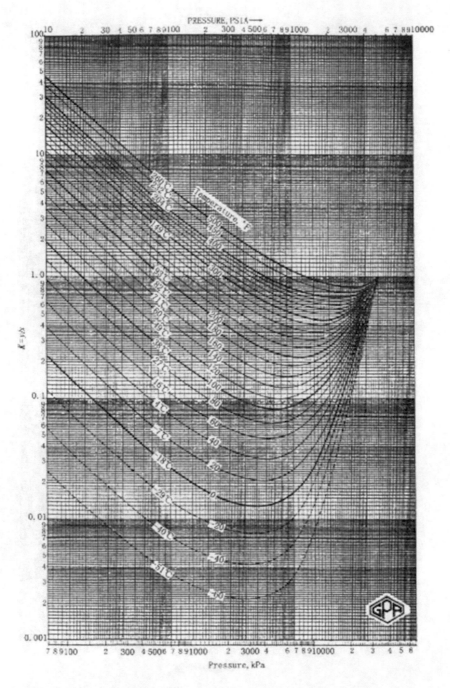

Fig. C.6 Equilibrium ratio of isopentane, 34.47 MPa (5000 psia) convergence pressure

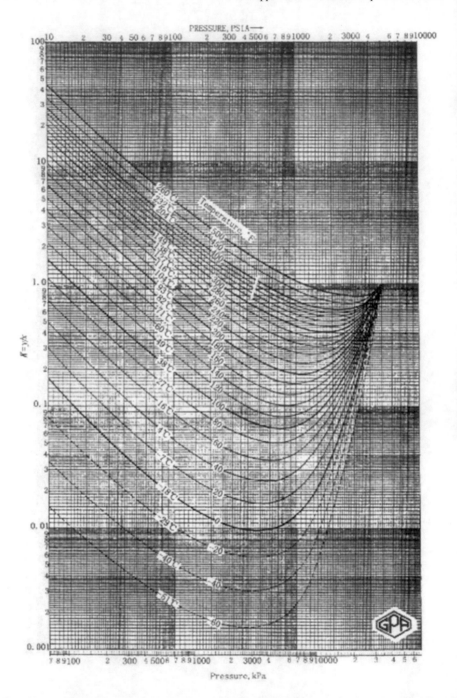

Fig. C.7 Equilibrium ratio of normal pentane, 34.47 MPa (5000 psia) convergence pressure

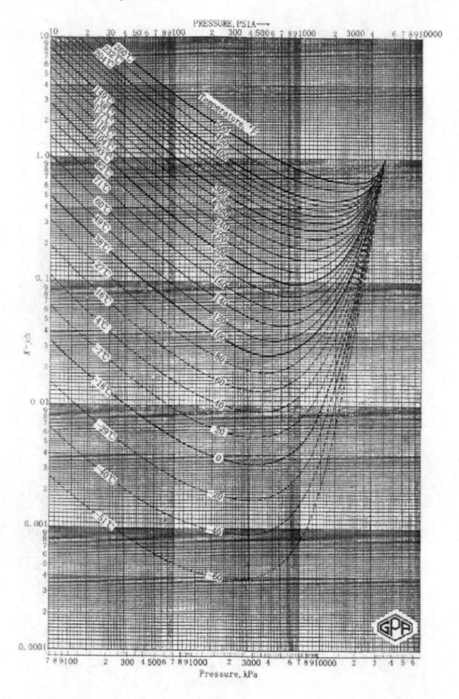

Fig. C.8 Equilibrium ratio of hexane, 34.47 MPa (5000 psia) convergence pressure

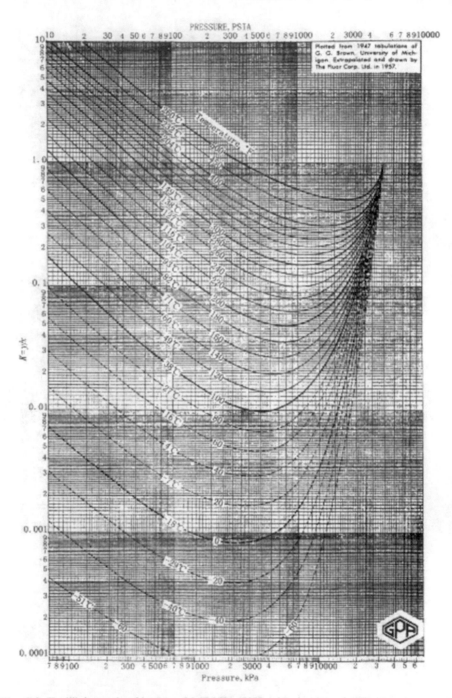

Fig. C.9 Equilibrium ratio of heptane, 34.47 MPa (5000 psia) convergence pressure

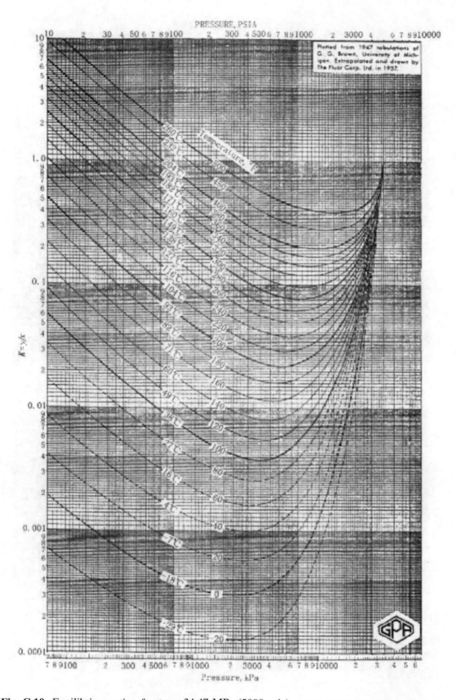

Fig. C.10 Equilibrium ratio of octane, 34.47 MPa (5000 psia) convergence pressure

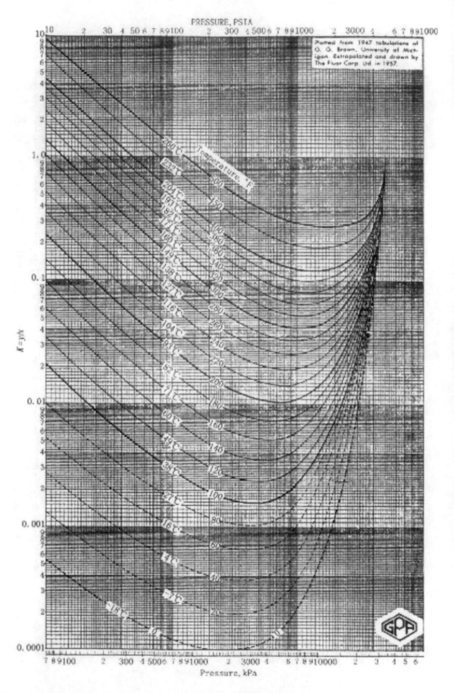

Fig. C.11 Equilibrium ratio of nonane, 34.47 MPa (5000 psia) convergence pressure

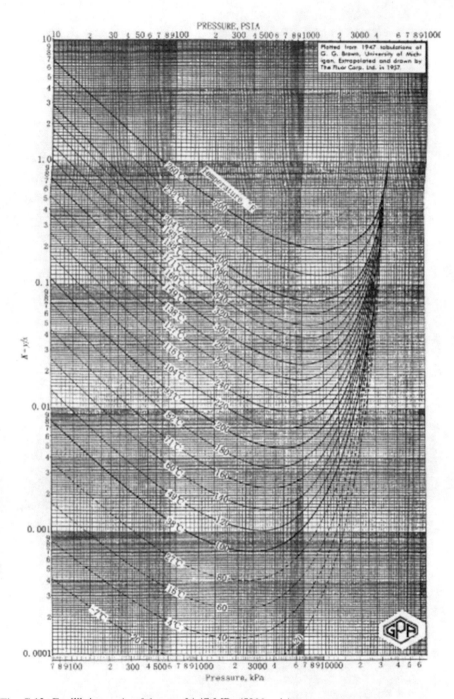

Fig. C.12 Equilibrium ratio of decane, 34.47 MPa (5000 psia) convergence pressure

References

V.N.Kobranova. Petrophysics. 1986

Society of Professional Well Log Analysts. Petrophysics

Thomas, E.C. Petrophysics.the society of Petroleum Engineers, 1994

Tiab, Djebbar. Petrophysics: theory and practice of measuring reservoir rock and fluid transport properties. Butterworth-Heinemann, 1996

Howard B. Bradley...et al. Petroleum engineering handbook.Society of Petroleum Engineers, 1987

Archer, J. S.Petroleum engineering: principles and practice.Graham & Trotman. 1986

Mian, M. A. Petroleum engineering handbook for the practicing engineer. vol.II. PennWell Books. 1992

John Lee and Robert A. Wattenbarger. Gas Reservoir Engineering.society of petroleum engineers. 1996

Erling Fjaer et al.Petroleum related rock mechanics.Elsevier, 1992.

YangShenglai,WeiJunzhi.Fundamentals of PetroPhysics,Beijing:Petroleum Industry Press.2004

Tiab, Djebbar, Donaldson, Erle C.Petrophysics.Petroleum Industry Press. 2007

Sandler P.L.The petroleum programme.Petroleum Industry Press. 1991

Pirson, S.T. Oil Reservoir Engineering, McGranHill New York. 1958

Morris Musket,Physical Principles Of Oil Production,New York, MCGRAW-HILL Book Company INC,1949

Richardson,S.G.Flow Through Porous Media, Handbook of fluid Dynanices Section l6,McGraw. Hill Book Company, New York,1961

API:Technical Data Book-Petroleum Refining,Chapter9 (1980)

Holder G D and Sloan E D. Handbook of Gas Hydrate Properties and Occurrence, DOE/MC, 1983:1239-1546

Amott.E.Observations relating to the wettability of porous rock.Trans.AIME.V216,1959:156

Brown,Harry W. Capillary pressure Investigation, Trans.AIME.1951

Slobod,R.L.,Adele chambers,and W.L.Prehn,Jr. Use of Centrifuge for determining connate water. Residual oil,and Capillary pressure curves of small core sample. Trans AIME l951

Johnson E F, Bossler,D.P. and Nauman,V. O. Calculation of relative permeability from displacement experiments, Trans AIME Vol216.1959: 370

Merliss, F.E., Doane, J.D. and Rzasa,M. J. Influence of rock and fluid properties and immiscible fluid.flow behavior in porous media. AIME Annual-meeting. New Orleans, 1955

Lo H Y and Mungan N. Effect of Temperature on water-oil relative permeabilities in oil-water and water-wet systems. SPE4505, 1973

Edmondson, T.A. Effect of temperature on waterflooding. Can.JPT. Vol10. 1965:236

Miller, M.A. and Ramey.H.J., Effect of temperature on oil / water Relative permeabilities of unconsolidated and consolidated sand. SPE 12116, 1983

White, R., and Brown, G. G., Phase Equilibria of Complex Hydrocarbon Systems at Elevated Temperature and Pressures, Ind. Eng. Chem. 1942(37): 1162

© Petroleum Industry Press and Springer-Verlag GmbH Germany 2017
S. Yang, *Fundamentals of Petrophysics*, Springer Geophysics,
DOI 10.1007/978-3-662-55029-8

502

Richardson, J.G., Calculation of waterflood recovery from steady-state relative permeability data. Trans AIME, Vol210. 1957: 373

Crag, F.F.Jr. Errors. calculation of gas injection performance from laboratory date, JPT. Vol8. 1952. 23

Odeh, A.S. Effect of viscosity ratio on relative permeability. Trans AIME Vol216. 1959:346

Lefebve du Prey E.J., Factors affecting liquid-liquid relative permeabilities of a consolidated porous medium. SPEJ. Vol2. 1973:39

Printed in the United States
By Bookmasters